Materials in Advanced Manufacturing

This book introduces the latest processing technologies for a variety of materials in advanced manufacturing and applications. Design criteria and considerations of processing or devices are theoretically introduced, and numerical simulation and experimental study are included.

FEATURES

- Covers a variety of materials, including hard materials, soft materials, metals, and composites
- Describes nanotechnology approaches, modern piezoelectric techniques, and physical and mechanical studies of the structure-sensitive properties of the materials
- Reviews advanced manufacturing for antenna applications and embroidered RFID tags for wearable applications
- Considers additive manufacturing of cellular solids and metal additive manufacturing
- Discusses advanced materials for sound absorption

Aimed at engineers, researchers, and advanced students in materials processing and advanced manufacturing, this work helps readers to understand which processing technology is suitable for a specific material and the design rules for a particular application.

Advances in Materials Science and Engineering

Series Editor:
Sam Zhang

Hierarchical Micro/Nanostructured Materials: Fabrication, Properties, and Applications
Weiping Cai, Guotao Duan, and Yue Li

Biological and Biomedical Coatings Handbook, Two-Volume Set
Sam Zhang

Nanostructured and Advanced Materials for Fuel Cells
San Ping Jiang and Pei Kang Shen

Hydroxyapatite Coatings for Biomedical Applications
Sam Zhang

Carbon Nanomaterials: Modeling, Design, and Applications
Kun Zhou

Materials for Energy
Sam Zhang

Protective Thin Coatings and Functional Thin Films Technology, Two-Volume Set
Sam Zhang, Jyh-Ming Ting, and Wan-Yu Wu

Fundamentals of Crystallography, Powder X-Ray Diffraction, and Transmission Electron Microscopy for Materials Scientists
Dong ZhiLi

Materials for Devices
Sam Zhang

Materials in Advanced Manufacturing
Yinquan Yu and Sam Zhang

For more information about this series, please visit https://www.routledge.com/Advances-in-Materials-Science-and-Engineering/book-series/CRCADVMATSCIENG

Materials in Advanced Manufacturing

Edited by

Yinquan Yu
Sam Zhang

CRC Press
Taylor & Francis Group
Boca Raton London New York

CRC Press is an imprint of the
Taylor & Francis Group, an **informa** business

First edition published 2023
by CRC Press
6000 Broken Sound Parkway NW, Suite 300, Boca Raton, FL 33487-2742

and by CRC Press
4 Park Square, Milton Park, Abingdon, Oxon, OX14 4RN

CRC Press is an imprint of Taylor & Francis Group, LLC

ISBN: 978-1-032-02157-7 (hbk)
ISBN: 978-1-032-02158-4 (pbk)
ISBN: 978-1-003-18214-6 (ebk)

DOI: 10.1201/9781003182146

Typeset in Times
by codeMantra

Contents

Series Preface

Materials form the foundation of technologies that govern our everyday life, from housing and household appliances to handheld phones, drug delivery systems, airplanes, and satellites. Development of new and increasingly tailored materials is key to further advancing important applications with the potential to dramatically enhance and enrich our experiences.

The *Advances in Materials Science and Engineering* series by CRC Press/Taylor & Francis is designed to help meet new and exciting challenges in Materials Science and Engineering disciplines. The books and monographs in the series are based on cutting-edge research and development, and thus are up to date with new discoveries, new understanding, and new insights in all aspects of materials development, including processing and characterization and applications in metallurgy, bulk or surface engineering, interfaces, thin films, coatings, and composites, just to name a few.

The series aims at delivering an authoritative information source to readers in academia, research institutes, and industry. The Publisher and its Series Editor are fully aware of the importance of Materials Science and Engineering as the foundation for many other disciplines of knowledge. As such, the team is committed to making this series the most comprehensive and accurate literary source to serve the whole materials world and the associated fields.

As Series Editor, I would like to thank all authors and editors of the books in this series for their noble contributions to the advancement of Materials Science and Engineering and to the advancement of humankind.

Sam Zhang

Foreword

Smartphone chips, biomedical devices, and electrical actuators are common concepts and application examples that need creation of certain "materials" to realize. In other words, these need advanced manufacturing to achieve high accuracy, high value-added target, and high performance. For this end, advanced material processing technologies need to be developed. To improve the productivity of advanced manufacturing, wireless sensor network with an innovative design of the antenna is important to conduct machining processing data transmission and real-time monitoring of intelligence manufacturing and machining process optimization.

Putting emphasis on the cutting-edge material processing technologies, advanced high-precision positioning actuators, and industrial wireless sensors in advanced manufacturing, this book introduces various latest processing technologies from hard material to soft material, from metal to composite in advanced manufacturing and applications. Design criteria and considerations of processing or devices are theoretically introduced, and numerical simulation and experimental study are also included. The reader will be able to understand which processing technology is suitable for a special material and understand the design rules in the applications after he/she grasps the knowledge in this book.

There are a total of 8 chapters in the book. Chapter 1 focuses on Advanced Manufacturing of Hard Magnetic Materials, Chapter 2 on Advanced Manufacturing of Flexible Piezoelectric Arrays, Chapter 3 on Application and Research Trends of These Arrays, Chapter 4 on Manufacturing of Carbon Fibers, and Chapter 5 on Laser Microprocessing of sapphire. This book also covers additive manufacturing, which is described in Chapter 6 on Metals and Chapter 7 on Lattice Structures. Finally, Chapter 8 touches on Advanced Manufacturing for Wearable Applications.

This book is written for engineers with senior bachelor's degrees, preferably in materials science or mechanical engineering. No prior knowledge of manufacturing is required. This book is also suitable as textbook at the senior undergraduate or graduate levels. The contents of this book can be taught as a one-semester advanced course in advanced manufacturing area.

The editors would like to take this opportunity to thank all chapter authors for their zest in participation in this book project and their hard working to make this book a reality, and thank Allison and Gabrielle, the editors at CRC Press, for their guidance and assistance in formulation and publication of the book.

Finally, Prof. Yu would like to acknowledge the funding of the National Natural Science Foundation of China (52067006), Foreign Expert Project of the Ministry of Science and Technology (G2021022002L), Jiangxi Science and Technology Innovation High-End Talent Project (jxsq2019101027), and Prof. Sam Zhang of the Fundamental Research Funds for the Central Universities (SWU118105), which brought him to his current university.

Yinquan Yu
School of Mechatronics and Vehicle Engineering
East China Jiao Tong University
Nanchang, China

Sam Zhang
School of Materials and Energy
Southwest University
Chongqing, China

Editors

Yinquan Yu (余银犬) was born and brought up in the famous city, Anqing, China. He earned a Bachelor of Engineering in mechanical engineering in 1991 at Chongqing University (Chongqing, China) and a Master of Engineering in mechatronics and a PhD in Electrical and Computer Engineering (ECE) at the National University of Singapore (Singapore) in 2008 and 2014, respectively. He was a Scientist at the Agency for Science, Technology and Research (A*star), Singapore. In January 2019, he joined the School of Mechatronics and Vehicle Engineering, East China Jiaotong University, Nanchang, China, and assumed duties of the Director of the Institute of Intelligent Manufacturing and Precision Research and Deputy Director of the Key Laboratory of transport tools and equipment of the Ministry of Education.

Prof. Yu was elected as a Fellow of SAS (2020), a member of ASME, Singapore executive committee member ASME (2017–2018), a Senior member of IES (2018) and IEEE (2019), and a mentor of the Lean Launch Pad Singapore (2018). He was also elected into the Double Thousands Program of Jiangxi Province (2019) and the Top Overseas Experts Program by the Ministry of Science and Technology of the PRC (2021). He was appointed as the Chairman of the Organizing Committee of the third International Symposium on Power Electronics and Control Engineering (2020), Chairman of the local committee of the IEEE second International Conference on Big Data Artificial Intelligence and Internet of Things Engineering (2021), Chairman of the fourth International Symposium on Power Electronics and Control Engineering (2021), Chairman of International Conference on Smart Transportation, Energy and Power (STEP 2021), Chairman of the third International Conference on Mechanical Engineering and Intelligent Manufacturing (MEIM 2022), and Guest Editor of the special issue of wireless sensors and artificial intelligence in intelligent manufacturing of the *International Journal of Distributed Sensor Networks*. He is also appointed as reviewer of many international journals such as *Magnetic Journal*, *Industrial Electronic Technology Journal of the American Society of Electrical Engineers*, and *International Society for Electromagnetic Research of the Society of Electrical Engineers*.

His current research interests include smart manufacturing, smart permanent magnet synchronous motor (SPMSM), machine health monitoring and fault diagnosis, mechatronics and robotics, shock absorption and noise reduction of high-speed and high-precision equipment, intelligent manufacturing, and precision machining.

Sam Zhang (Sam Zhang Shanyong, 張善勇), FRSC, FTFS, FIoMMM, academically known as Sam Zhang, was born and brought up in the famous "City of Mountains" Chongqing, China. He earned a Bachelor of Engineering in materials in 1982 at Northeastern University (Shenyang, China), a Master of Engineering in materials in 1984 at Iron & Steel Research Institute (Beijing, China), and a PhD in ceramics in 1991 at the University of Wisconsin–Madison, USA. He was a tenured full professor (since 2006) at the School of Mechanical and Aerospace Engineering, Nanyang Technological University Singapore. In January 2018, he joined the School of Materials and Energy, Southwest University, China. He serves as the founding and current Director of the Centre for Advanced Thin Films and Devices (http://fmae.swu.edu.cn/s/fmaenew/yjzx/).

Prof. Zhang was elected as a Fellow of Royal Society of Chemistry (FRSC), a Fellow of Thin Films Society (FTFS) in 2018, and a Fellow of Institute of Materials, Minerals and Mining (FIoMMM) in 2007.

Prof. Zhang was a Principal Editor for *Journal of Materials Research* (USA) for thin films and coating field (since 2003). Prof. Zhang has been serving the world's first "Thin Films Society" (www.thinfilms.sg) as its founding and current president since 2009. Prof. Zhang has authored/edited 14 books, of which 13 are published with CRC Press/Taylor & Francis. The most recent titles are published in August 2021: *Protective Thin Coatings Technology* (https://www.routledge.com/Protective-Thin-Coatings-Technology/Zhang-Ting-Wu/p/book/9780367542504), *Functional Thin Films Technology* (https://www.routledge.com/Functional-Thin-Films-Technology/Zhang-Ting-Wu/p/book/9780367541774), and *Materials for Devices,* currently in press at CRC. Currently, Prof. Zhang is working on a four-volume *Materials for Energy Handbooks* in Chinese to be published by Chongqing University Press. Of these books, *Materials Characterization Techniques* has been adopted as textbook by more than 30 American and European universities alone as of October 2015. That book had also been translated into Chinese and published by China Science Publishing Co. in October 2010 and distributed nationwide in China (also available online at Amazon.cn). Prof. Zhang is also the Series Editor for *Advances in Materials Science and Engineering* Book Series published by CRC Press/Taylor & Francis. A full list of his books is at Amazon: https://www.amazon.com/-/e/B001JPBX42

Prof. Zhang's current researches center in the areas of films and coatings for solar cells and hard yet tough nanocomposite coatings/high-entropy alloy coatings for tribological applications by physical vapor deposition; over the years, he has published 14 books and authored/co-authored over 360 peer-reviewed international journal papers with 32.7 citations per article and h-index: 57 and total citation of over 11,600 times as of 17 June 2022 (https://publons.com/researcher/2817766/sam-zhang/).

Contributors

Guojie Chen
School of Physics and Optoelectronic Engineering
Guangdong-Hong Kong-Macao Joint Laboratory for Intelligent Micro-Nano Optoelectronic Technology
Foshan University
Foshan, People's Republic of China

Yang Jingwen
Institute of Precision Machining and Intelligent Equipment Manufacturing
East China Jiaotong University
Nanchang, People's Republic of China

Anna Pui Yi Lau
Laxcen Technology, Inc.
Singapore

Roberto Leidhold
Faculty of Electrical Engineering and Information Technology
Otto-von-Guericke University Magdeburg
Magdeburg, Germany

Xinwei Li
Department of Mechanical Engineering
National University of Singapore
Singapore

Yulong Liu
Department of Electrical and Electronic Engineering
Southern University of Science and Technology
Shen Zhen, People's Republic of China

Hai Tao Si
Laxcen Technology Inc.
Singapore

Xinzhu Wang
Chongqing Key Laboratory of Heterogeneous Material Mechanics
College of Aerospace Engineering
Chongqing University
Chongqing, China

Linji Xu
Chongqing Key Laboratory of Heterogeneous Material Mechanics
College of Aerospace Engineering
Chongqing University
Chongqing, China

Terry Tao Ye
Department of Electrical and Electronic Engineering
Southern University of Science and Technology
Shen Zhen, People's Republic of China

Mengxia Yu
Department of Electrical and Electronic Engineering
Southern University of Science and Technology
Shen Zhen, People's Republic of China

Wei Zhai
Department of Mechanical Engineering
National University of Singapore
Singapore

Han Zhang
Key Lab of Noise and Vibration Control
Institute of Acoustics
Chinese Academy of Sciences
Beijing, People's Republic of China

Hao Zhao
3D Printing Technology
Beijing, People's Republic of China

Wang Zhongke
Singapore Institute of Manufacturing Technology
Singapore

Garrison Zong
3D Printing Technology
Beijing, People's Republic of China

1 Advanced Manufacturing of Hard Magnetic Materials

Yang Jingwen and Yinquan Yu
East China JiaoTong University

Roberto Leidhold
Otto-von-Guericke University

CONTENTS

DOI: 10.1201/9781003182146-1

1.1 INTRODUCTION

Advanced manufacturing (AM) is inseparable from high-end manufacturing equipment that can achieve micro- even nano-level machining accuracy, and one of the key devices that determine the machining accuracy of equipment is the driving device of equipment – actuator. With the rapid development of permanent magnet synchronous motor (PMSM) drive technology and the significant reduction of cost, the high-efficiency and energy-saving PMSM with closed-loop control almost completely replaces other types of motors, such as induction motor, in AM equipment.

Hard magnetic materials are also called "permanent magnetic (PM) materials". The magnetic energy product of PM materials directly affects the power conversion efficiency of PMSM. The PM is usually installed on the rotor of the inner rotor motor and located inside the motor, so the heat resistance of the PM material is also a big consideration. If the internal temperature of the motor exceeds the Curie temperature of the PM, the PM will begin to demagnetize, which will not only reduce the efficiency of the motor but also affect the stability of the motor to a certain extent. Moreover, serious demagnetization will even damage the Insulated Gate Bipolar Transistor (IGBT) of the motor driver. This may cause the entire AM production line to shut down. This situation is absolutely intolerable in AM. Therefore, the research on the properties of hard magnetic material is of great significance to improve AM efficiency, machining accuracy and reliability.

1.2 HARD MAGNETIC MATERIAL

Magnetism, the same as force, heat, sound, electricity and light, is one of the basic properties of matter. PM materials can maintain constant magnetic properties once they are magnetized, also known as hard magnetic materials. It has high anisotropy field, high coercivity, large hysteresis loop area and the largest magnetization field required for magnetization to saturation. Magnetoplumbite is the earliest type of PM material discovered and used by human. The development of modern PM materials has successively experienced five stages of carbon steel, alnico system requirements, ferrite, rare earth PM and nanocrystalline composite magnets.

1.2.1 Types of Hard Magnetic Material

Magnetic material is an important functional material, which occupies an extremely important position in modern industry and technology. Magnetic materials are divided into soft magnetic materials, hard magnetic materials, semi-hard magnetic materials and special hard magnetic materials. Among them, PM materials and soft magnetic materials are the most widely used. In addition, PM materials can be divided into cast PM materials, ferrite PM materials, rare earth PM materials and other PM materials. Among them, ferrite PM and Nd-Fe-B rare earth PM are the most commonly used.

Casting PM materials refer to PM produced by casting process, mainly referring to Al-Ni-Co PM. The casting process includes ingredients, smelting, casting, heat treatment, performance testing, mechanical processing, inspection and packaging. Iron, nickel and aluminum are the main components of Al-Ni-Co PM alloys and also contain copper, cobalt, titanium and other elements. The cast Al-Ni-Co PM material has a high Curie temperature T_c, high temperature stability and low temperature coefficient of magnetic induction, but it contains more strategic metals Co and Ni. It was widely used from the 1930s to the 1960s, and it is now mostly used in the instrument industry to manufacture magnetoelectric meters, flow meters, micro motors, relays, etc.

Ferrite PM materials mainly include barium ferrite and strontium ferrite, which have high resistivity and high coercivity, and can be effectively used in large gap magnetic circuits. They are especially suitable for PM in small generators and motors. PM ferrite does not contain precious metals, such as nickel and cobalt. The raw materials used to manufacture it are abundant, the process is simple and the cost is low. It can replace Al-Ni-Co PM to manufacture magnetic separators, magnetic thrust bearings, speakers, microwave devices, etc. However, its maximum magnetic energy product is low, temperature stability is poor and the texture is brittle, fragile and not resistant to shock and vibration. It is not suitable for measuring instruments and magnetic devices with precision requirements. According to different production processes, it can be divided into sintered ferrite, bonded ferrite and injection-molded ferrite. These three production processes can be further divided into isotropic and anisotropic magnets according to the orientation of the magnetic crystal.

Rare earth PM materials mainly include rare earth Co PM materials and Nd-Fe-B PM materials. The former is an intermetallic compound formed by the rare earth elements Ce, Pr, La, Nd, Co, etc., and cobalt. Its magnetic energy product can reach 150 times that of carbon steel, 3–5 times that of alnico PM materials and 8–10 times that of permanent ferrite. It has a high Curie point, stable magnetism and a coercivity of up to 800 kA/m. It is mainly used in low-speed torque motors, starting motors, sensors, magnetic thrust bearings and other magnetic systems. Nd-Fe-B PM material is the third generation of rare earth PM material. Its remanence, coerced force and maximum magnetic energy product are higher than the former, and its mechanical properties are good, the alloy density is low and it is not easy to be broken, which is beneficial to realize the lightness, thinness, miniaturization and super-miniaturization of magnetic elements. However, its Curie point is low, which limits its application. However, its magnetic Curie point is low, which limits its application. According to different production processes, it is divided into sintered Nd-Fe-B, bonded Nd-Fe-B and injection-molded Nd-Fe-B.

The sintered Nd-Fe-B PM is smelted into powder after jet milling. It has a high coercivity value and extremely high magnetic properties. Its maximum magnetic energy product is more than 10 times higher than that of ferrite. Its mechanical properties are also quite good, and it can be cut to process different shapes and drill holes. The maximum working temperature of high-performance products can reach 200°C. Because it is easily corroded, different coating treatments must be applied to the surface according to different requirements. It is very hard and brittle, has high resistance to demagnetization and is cost-effective, but it is not suitable for working in temperatures higher than 200°C environment.

Bonded Nd-Fe-B is a kind of composite PM which mixes Nd-Fe-B powder with resin, plastic or low melting point metal binder, and then makes it by compression, extrusion or injection molding. The product is formed once, without secondary processing, and can be directly made into various complicated shapes. Bonded Nd-Fe-B has magnetism in all directions and can be processed into Nd-Fe-B compression molds and injection molds. It has high precision, excellent magnetic properties, good corrosion resistance and good temperature stability.

Injection-molded Nd-Fe-B can easily be made into thin-walled rings or thin magnets with complex anisotropic shapes.

1.2.2 Structure of Hard Magnetic Material

1.2.2.1 The Crystal Structure of Ferrite PM Materials

Strontium ferrite and barium ferrite have similar crystal structures. Figure 1.1 shows the strontium ferrite unit cell structure, called the M-type structure. Each unit cell contains two $SrFe_{12}O_{19}$ molecules, and there are a total of ten dense layers of oxygen ions, of which two dense layers each contain one Sr ion occupying an oxygen ion site. Iron ions are in five different sublattice positions, which are represented by symbols $2a$, $4f_2$, $12k$, $4f_1$ and $2b$; each Fe^{3+} has a magnetic moment of 5 μ_B (μ_B is called a Bohr magneton). The first three of these five sublattices have an octahedral structure, the $4f_1$ sublattice has a tetrahedral structure and the $2b$ sublattice has a hexahedral structure. Corresponding to this, M-type ferrite is divided into five magnetic sublattices. The ionic magnetic moments of the three sublattices $2a$, $2b$ and $12k$ are arranged in parallel, while the ionic magnetic moments of the $4f_1$ and $4f_2$ sublattices are arranged anti-parallel to the above three sublattices. This results in a strontium ferrite PM single cell with a total magnetic moment of only 40 μ_B.[1] Figure 1.1 also shows a schematic diagram of exchange coupling in a strontium ferrite PM unit cell, and each arrow in the figure represents a Fe^{3+} magnetic moment.[2]

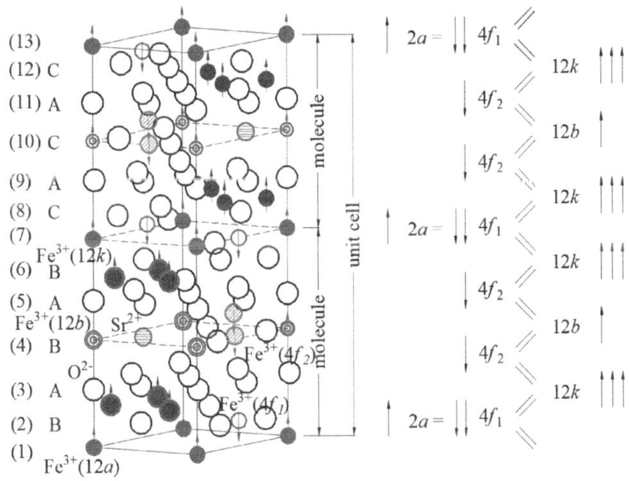

FIGURE 1.1 Schematic diagram of the lattice structure of M-type strontium ferrite ($SrFe_{12}O_{19}$).[1]

1.2.2.2 Crystal Structure of Rare Earth PM Materials

Most of the rare earth chemical elements (represented by RE for rare earth chemical elements) form $RE_2Fe_{14}B$ compounds, which are the matrix phase of the S-REFeB series PM material and are sometimes expressed by the ratio of the amount of its substances, which is called 2:14:1 phase. This phase occupies 96–98% of the S-REFeB PM, so it is also called the main phase. In the compound $RE_2Fe_{14}B$, $Pr_2Fe_{14}B$, (Pr, Nd)$_2Fe_{14}B$, (Ce, Pr, Nd)$_2Fe_{14}B$, (MM, Nd)$_2Fe_{14}B$, (Nd, HRE)$_2Fe_{14}B$, etc. can all be made into practical PM materials. Among them, MM stands for rare earth metals, and HRE stands for heavy rare earth metals. All compounds $RE_2Fe_{14}B$ have the same crystal structure, but their magnetic properties are very different.

Figure 1.2 is a space diagram of the crystal unit cell structure of the compound $Nd_2Fe_{14}B$. It consists of four $Nd_2Fe_{14}B$, with 68 atoms in a unit cell (8 Nd atoms, 56 Fe atoms and 4 B atoms). It belongs to tetragonal crystal (or tetragonal phase for short), the crystal lattice constant is a=0.8792 mm and c=1.2177 nm, and the theoretical density is 7.62 g/cm^3. The whole crystal unit cell (unit) can be regarded as two Nd-rich and B-rich atomic layers and four Fe-rich atomic layers, alternately composed along the c-axis. If the lattice constant c is taken as the z-axis and c=1, it can be seen from Figure 1.3 that Nd atoms and B atoms are mainly distributed in the two atomic layers of z=0 and z=0.5, and Fe atoms are mainly distributed in the second, third, fifth and sixth atomic layers. The density of atoms in the same atomic layer is high, the distance between adjacent atoms is close and the distance between adjacent atoms is farther.

The magnet with high PM characteristics has a special microstructure inside. This special microstructure can not only inhibit the nucleation of reverse magnetic domains or cause domain wall pinning, which not only makes the magnet

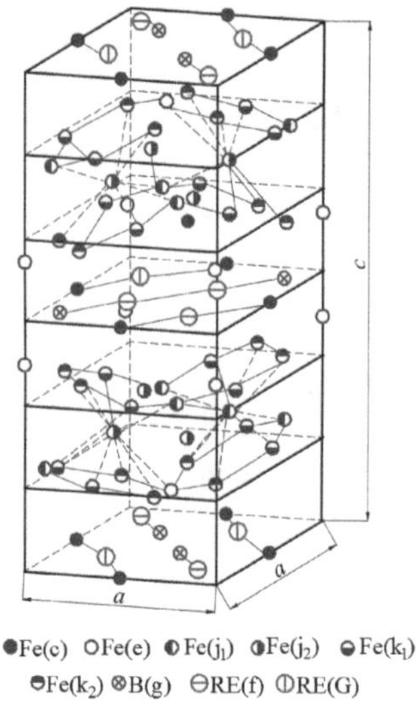

FIGURE 1.2 Space diagram of the crystal unit cell structure of the compound $Nd_2Fe_{14}B$.[3]

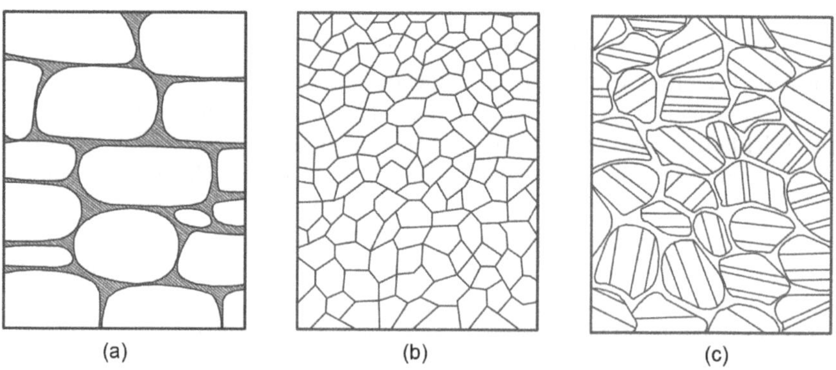

FIGURE 1.3 Ideal microstructures of rare earth PM.[5] (a) the dispersed main phase is completely surrounded by the intergranular phase; (b) the grains form clusters; (c) the platelike phase is dispersed inside the main phase grains.

have high intrinsic coercivity, but also ensure that the magnet obtains high remanence and excellent squareness of the decoupling curve. Figure 1.3 schematically depicts typical examples of ideal microstructures of rare earth PM materials prepared by different processes.[4]

As shown in Figure 1.3a, the hard magnetic main-phase particles are completely surrounded and dispersed by extremely thin intergranular phases with large differences in magnetic properties from the main phase. This is the typical microstructure of sintered Nd-Fe-B magnets. The grain boundary phase is an alloy or intermetallic compound with a relatively low melting point and good wettability to the main phase. During the sintering process, the grain boundary phase is the sintered liquid phase, which promotes the densification of the magnet, and infiltrates and divides the main phase. The sintered magnet undergoes heat treatment at a specific temperature to adjust the interface between the liquid phase and the main-phase crystal grains, eliminate grain boundary defects and improve the coercivity of the magnet. Liquid-phase sintering can reduce the interfacial energy, produce high-density materials and contribute to eliminate sharp corners of the crystal grains.

Figure 1.3b shows a cluster composed of fine crystal grains with a size close to or smaller than the critical diameter d_c of a single domain. There are no boundary phases or only a small amount of boundary phases between the crystal grains. Since the crystal grain size is equivalent to the critical size of a single domain, not only a single crystal grain is in a single-domain state, but also multiple adjacent crystal grains form a domain. The domain wall is pinned by the grain boundary phase to increase the coercivity of the material. The microstructure can be obtained directly by rapid quenching, or by annealing the over-quenched amorphous state. Hydrogenation–disproportionation–dehydrogenation–recombination (HDDR) treatment of as-cast alloys under fire hydrogen atmosphere can also transform the large grains of the main phase into such fine grain clusters. The d_c of rare earth PM materials requires the size of the fine grains to be on the order of 10–100 nm.

Figure 1.3c shows that the thin plate-like phase is scattered inside the main-phase crystal grains and is significantly different from the main phase, but it is compatible with the main phase in terms of crystal structure. The secondary phases are dispersed between the hard magnetic main phases with fine precipitates. The precipitate can be compared at least one dimension with the domain wall thickness δ_W, which exerts a strong pinning effect on the movement of the domain wall. The Sm_2Co_{17} type magnet is a typical representative of the hard magnetization precipitated in this way.

1.2.3 Physics of Hard Magnetic Material

PM materials are mainly used to provide a magnetic field or magnetic force. In addition to the basic magnetic performance parameters obtained by the hysteresis loop, the important indicators for measuring PM materials include the stability of the material and other properties.

1.2.3.1 Basic Magnetic Properties

The magnetism of matter comes from the magnetism of atoms, that is, the atomic magnetic moment of each atom. Atomic magnetic moment can be divided into two parts: nuclear magnetic moment and electronic magnetic moment. Nuclear magnetic moment is the carrier of nuclear magnetic resonance detection signal, but its value is much smaller than electronic magnetic moment, which is often neglected in the study of magnetic materials. The electronic magnetic moment is composed of its orbital magnetic moment and spin magnetic moment. Therefore, the atomic magnetic moment generally refers to the vector sum of the electron's orbital magnetic moment and spin magnetic moment. The unit of atomic magnetic moment is Bohr magneton μ_B, that is, 1 μ_B=9.274×10^{-24} J/T.

According to the different magnetic properties of the material, it can be divided into weak magnetism and strong magnetism. Weak magnetism can only be established on a microscopic level when there is an external magnetic field and appears on a macroscopic level. Magnetism increases as the magnetic field increases, and its direction may be parallel (paramagnetic) or anti-parallel (diamagnetic) to the magnetic field. Strong magnetization can spontaneously magnetize microscopically without an external magnetic field. However, in general, to obey the principle of the lowest energy and minimize the magnetostatic energy, the spontaneous magnetization distribution of a substance is naturally divided into several small areas. The spontaneous magnetization direction in one area is the same, but the spontaneous magnetization direction of different areas is different. These small areas are called magnetic domains. In the absence of an external magnetic field, the vector sum of the spontaneous magnetization of each magnetic domain is zero, the total magnetic moment of the system cancels each other out and there is no external magnetism. But in the case of an external magnetic field, due to the movement of the domain wall or the change of the spontaneous magnetization direction, the total magnetic moment of the system is no longer zero, and it usually exhibits strong magnetism. According to the different arrangements of atomic magnetic moments on the atomic scale, strong magnetism can be divided into ferromagnetism (the magnetic moments of neighboring atoms are arranged in parallel) and anti-ferromagnetism (the magnetic moments of neighboring atoms are the same but arranged in anti-parallel), ferrimagnetism (the magnetic moments of neighboring atoms are different but arranged in anti-parallel) and spiral magnetism (the magnetic fields of neighboring atoms are arranged in a spiral), etc. In addition to anti-ferromagnetism, these magnetisms are usually called ferromagnetism in a broad sense. As the temperature rises, thermal disturbance will destroy the exchange interaction that drives the atomic magnetic moments to be aligned parallel or anti-parallel, so that the thermodynamic average magnetic moment gradually decreases or even approaches zero, leading to the demagnetization of the material. The critical temperature at which ferromagnetism disappears is called the Curie temperature T_c. When the temperature is higher than the Curie temperature, because the thermal motion is stronger than the exchange interaction, the substance no longer spontaneously magnetizes and the magnetism weakens. The higher the Curie temperature of a ferromagnetic substance, the better its thermal stability. The anti-ferromagnetic substance also has a critical temperature at which the spontaneous magnetization is zero and the anti-ferromagnetism disappears, which is called the Neel temperature T_N.

The magnetic size and direction of ferromagnetic materials are usually expressed by the magnetization vector $\boldsymbol{M}$, which represents the vector sum of the atomic magnetic moments per unit volume, in A/m (ampere/meter), and Gs (Gauss) in the Gauss unit system. In the same magnetic domain, the value of magnetization $\boldsymbol{M}$ is M_s, which is called spontaneous magnetization. For a piece of ferromagnetic material with $\boldsymbol{M}=0$, when the external magnetic field strength is $\boldsymbol{H}$ [the unit is A/m, Oe (Oersted) in the Gauss unit system], to minimize the interaction energy, the internal domain wall of the material moves or the atomic magnetic moment deflects toward the $\boldsymbol{H}$ direction. As $\boldsymbol{H}$ increases, $\boldsymbol{M}$ will gradually tend to the $\boldsymbol{H}$ direction. The vector sum of the external magnetic field intensity $\boldsymbol{H}$ and the magnetization $\boldsymbol{M}$ is called the magnetic induction intensity (or magnetic flux density), denoted by $\boldsymbol{B}$, and the unit is T (Tesla) [Gs (Gauss) in the Gauss unit system]:

$$\boldsymbol{B} = \mu_0(\boldsymbol{H} + \boldsymbol{M}) = \boldsymbol{B}_0 + \boldsymbol{J} \tag{1.1}$$

or

$$\boldsymbol{B} = \boldsymbol{H} + 4\pi\boldsymbol{M} = \boldsymbol{H} + \boldsymbol{J} \quad \text{(in the Gauss unit system)} \tag{1.2}$$

In the formula, $\boldsymbol{B}_0 = \mu_0\boldsymbol{H}$ is the vacuum magnetic induction intensity, $\mu_0 = 4\pi\times10^{-7}$ Tm/A is the vacuum permeability (μ_0=1 Gs/Oe in the Gauss unit system, $|\boldsymbol{B}_0|=|\mathbf{H}|$ in terms of value), $\boldsymbol{J}=\mu_0\boldsymbol{M}$ ($\boldsymbol{J}=4\pi\boldsymbol{M}$ in the Gauss unit system) is the magnetic polarization and the unit is also T or Gs (Gauss unit system).

Materials with ferromagnetism are often referred to as magnetic materials. The hysteresis behavior is the most obvious feature of magnetic materials, that is, the hysteresis behavior of the magnetization $\boldsymbol{M}$ changing with the external magnetic field $\boldsymbol{H}$. When $\boldsymbol{H}$ returns to zero from a certain state, $\boldsymbol{M}$ still maintains a certain size and has residual magnetization. Only when a magnetic field is applied in the opposite direction to a certain magnetic field intensity, $\boldsymbol{M}$ will be zero. Therefore, the reciprocal change of $\boldsymbol{M}$ with $\boldsymbol{H}$ is a loop, rather than a single curve that crosses the origin. Figure 1.4 shows a schematic diagram of a typical hysteresis loop, where μ_0M-H is the magnetic polarization intensity hysteresis loop, B-H is the magnetic induction intensity hysteresis loop and the relationship between B and μ_0M is given by Equation (1.1). The main parameters describing the characteristics of the hysteresis loop are as follows:

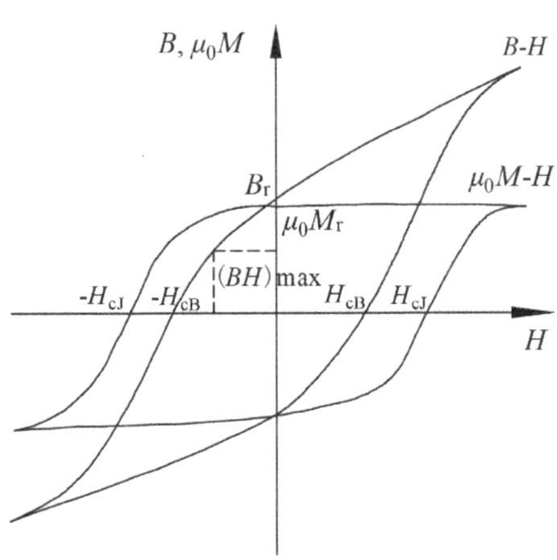

FIGURE 1.4 A schematic diagram of hysteresis loop.[5]

1. When the external magnetic field $H = 0$, the residual magnetic induction intensity B_r, the residual magnetic polarization intensity $J_r = \mu_0 M$ or the residual magnetization intensity M_r are all referred to as residual magnetism; their relationship is $B_r = J_r = \mu_0 M$.
2. The coercivity H_{cB} is the value of the reversal magnetization field when the magnetic flux density $B = 0$.
3. The intrinsic coercivity H_{cJ} is the value of the reversal magnetization field when the magnetic polarization intensity $J = 0$.
4. The maximum magnetic energy product $(BH)_{max}$ is the product of the magnetic induction intensity B and the reversal magnetization field value H on the saturation demagnetization curve (second quadrant part of B-H hysteresis loop), that is, the maximum value of the magnetic product energy. Since the magnetic energy product of other points on the demagnetization curve is rarely mentioned, the maximum magnetic energy product is often directly referred to as the magnetic energy product. The unit of $(BH)_{max}$ is J/m^3, which is the energy density [GOe (Gauss Otes, or Gao Ou) in the Gauss unit system]. The maximum magnetic product energy and the remanence B_r or M_r have the following relationship as $(BH)_{max} \leq B_r^2/(4\mu_0) = \mu_0 M_r^2/4 \leq \mu_0 M_S^2/4$ (theoretical limit value).

Magnetic materials with different hysteresis loops are used in different application fields. Magnetic materials are usually divided into two categories based on the intrinsic coercivity H_{cJ}. One is soft magnetic materials with smaller H_{cJ} ($H_{cJ} < 100$ A/m), and the other is PM materials with larger H_{cJ} (also called hard magnetic material, $H_{cJ} > 1000$ A/m).

The characteristic of a PM is that after being magnetized by an external magnetic field, the magnetic field can still be maintained stably for a long time after the external field is removed. The most comprehensive criterion for the performance of PM materials is the demagnetization curve under various external conditions (external magnetic field includes current detection, temperature, time, radiation, vibration, etc.). But usually only the three parameters (1), (3) and (4) mentioned above and their stability under external conditions are used to determine.

Remanence B_r, coercivity H_c and maximum magnetic energy product $(BH)_{max}$ are the most basic magnetic properties of PM materials. The maximum magnetic energy product $(BH)_{max}$ represents the ability of PM to provide static magnetic field energy. In the case of meeting the same magnetic field value in the same space range, the amount of material with a large $(BH)_{max}$ is less. The greater the remanence B_r, the greater the strength of the magnetic field that the magnet may provide to the space. Materials with large intrinsic coercivity H_{cJ} have strong anti-interference ability. Therefore, the higher the above three parameters, the better the magnetic performance of the PM.

The value of the coercive force H_c is closely related to the intrinsic properties of the material and the fiber structure, and the magnetization reversal process of the material must be considered. In addition to selecting materials with high saturation magnetization M_s, the way to increase J_r can choose materials with high magnetic moment. The coercivity and maximum energy product of different PM materials currently available in the industry are shown in Table 1.1.

TABLE 1.1
Coercivity, Remanence and Maximum Product Energy of Different PM[1] Materials

	H_{cJ}		J_r		$(BH)_{max}$	
Materials	kA/m	Oe	T	Gs	kJ/m³	MGOe
AlNiCo	40–160	500–2000	0.70–1.32	7000–13,200	8–88	1–11
$Ba(Sr)Fe_{12}O_{19}$	159–358	2000–4500	0.30–0.44	3000–4400	8–36	1–4.5
Sm_2Co_{17}	1194–2388	15,000–30,000	1.00–1.30	10,000–13,000	175–255	22–32
$SmCo_5$	1194–2388	15,000–30,000	0.90–1.00	9000–10,000	127–207	16–26
S-Nd-Fe-B	955–2388	12,000–30,000	1.10–1.40	11,000–14,000	239–446	30–56

TABLE 1.2
Curie Temperature of Commonly Used PM Materials[1]

Materials	AlNiCo	$Ba(Sr)Fe_{12}O_{19}$	Sm_2Co_{17}	$SmCo_5$	S-Nd-Fe-B
T_c (°C)	800–860	465	800–850	750	312–420

1.2.3.2 Curie Temperature T_c

The Curie temperature is the highest temperature at which PM materials maintain ferromagnetism. The higher the T_c, the slower the magnetic properties of the material change with increasing temperature, so the higher the thermal stability of the material. Table 1.2 shows the T_c of commonly used PM materials. To return the magnetized PM to the magnetic neutral state, the material must be heated above T_c, and this process is called thermal demagnetization.

1.2.3.3 Stability of PM Materials

The stability of PM materials refers to the ability of their magnetic properties to remain unchanged when they are affected by external factors, such as temperature, external magnetic field, shock and vibration during long-term use. It is expressed by the rate of change η:

$$\eta = \frac{\Delta Z}{Z} \times 100\% \tag{1.3}$$

In the formula, Z is the properties of PM materials, such as coercivity and remanence, and ΔZ is the amount of change in properties.

The thermal stability of PM materials is also often expressed by temperature coefficient, including the temperature coefficient of remanence and the temperature coefficient of coercivity, and the unit is %/°C. The relative change of the remanence caused by the unit temperature change is called the temperature coefficient of remanence α, and the relative change of the coercive force caused by the unit temperature change is called the temperature coefficient of coercivity β.

The calculation formula of the temperature coefficient α and β in the temperature range of material $T_0 - T$ is as follows:

$$\alpha = \frac{J_r(T) - J_r(T_0)}{J_r(T_0)(T - T_0)} \times 100\% \tag{1.4}$$

$$\beta = \frac{H_{cJ}(T) - H_{cJ}(T_0)}{H_{cJ}(T_0)(T - T_0)} \times 100\% \tag{1.5}$$

As the temperature increases, the magnetic properties of most PM decrease, so α and β generally show negative values. The smaller the absolute value, the better the thermal stability of the magnet. The temperature coefficient of remanence of different PM materials is shown in Table 1.3.

1.2.3.4 Physical Properties of PM Materials

PM materials are key components of electromagnetic devices, and different applications have different requirements for their physical properties. Table 1.4 shows the resistivity, density, thermal expansion coefficient, thermal conductivity and specific heat of several PM materials.

TABLE 1.3
Temperature Coefficient of Different PM Materials[1]

Materials	AlNiCo	Ba(Sr)$Fe_{12}O_{19}$	Sm_2Co_{17}	$SmCo_5$	S-Nd-Fe-B
α(%/°C)	−0.03 to −0.02	−0.18 to −0.09	−0.03	−0.05 to 0.04	0.01

TABLE 1.4
Common Physical Properties of Different PM Materials[1]

Materials	AlNiCo	Ba(Sr)$Fe_{12}O_{19}$	Sm_2Co_{17}	$SmCo_5$	S-Nd-Fe-B
Resistivity (Ω·cm)	45×10^{-6}	$>10^4$	$\sim8.6\times10^{-5}$	8.6×10^{-5}	14.4×10^{-5}
Density (g/cm³)	7.0–7.3	4.8–5.0	8.3–8.5	8.1–8.3	7.5–7.6
Thermal expansion coefficient (1/°C)	11	13(//), 8(⊥)	−8(//), 11(⊥)	6(//), 12(⊥)	6.5–7.4(//), −0.5(⊥)
Thermal conductivity (W/m²·K)	—	—	~12	—	—
Specific heat (kcal/kg·°C)	—	0.2	—	0.13	0.12

TABLE 1.5
Hardness and Strength of PM Materials[1]

Materials	AlNiCo	Ba(Sr)$Fe_{12}O_{19}$	Sm_2Co_{17}	$SmCo_5$	S-Nd-Fe-B
Webster hardness	650	530	500–600	8.6×10^{-5}	600
Bending strength (N/m^2)	—	1.27×10^8	1.5×10^8	1.2×10^8	2.457×10^8
Compressive strength (N/m^2)	—	—	8×10^8	$>2\times10^8$	—
Tensile strength (N/m^2)	—	3.97×10^7	3.5×10^7	—	7.87×10^8

1.2.3.5 Mechanical Properties of PM Materials

Many applications have certain requirements on the mechanical properties of PM materials. Table 1.5 shows the hardness and strength of different PM materials.

1.2.4 Computer-Aided Engineering Tools

Ansoft maxwell (Ansoft maxwell EM) is a low-frequency electromagnetic field computer analysis software, which is widely used in various engineering electromagnetic fields. It is based on Maxwell's differential equations and uses a discrete form of finite elements to transform the calculation of electromagnetic fields in engineering into a huge matrix solution. Ansoft can not only numerically calculate individual electromagnetic mechanisms but also jointly simulate the whole system. As an early introduction of electromagnetic field finite element software in China, it is widely used in many industries, such as electrical, mechanical, petrochemical, automotive, metallurgy, water conservancy and hydropower, aerospace, shipping, electronics, nuclear industry and weaponry.

The demand for advanced and easy-to-use digital field simulation technology has grown significantly due to the demand for performance and volume design packages. Maxwell's products are far ahead of other leading companies in terms of practicality and digital capabilities. Maxwell 2D includes AC/DC magnetic, electrostatic and transient electromagnetic fields, temperature field analysis, parametric polarization and optimization capabilities. In addition, Maxwell 2D simulates highly accurate equivalent circuit models for use with SIMPLORER module and other circuit analysis tools of Ansoft.

The wizard-style user interface, accuracy-driven adaptive profiling technology and powerful post-processors make Maxwell 3D the industry's best high-performance 3D electromagnetic design software. Eddy currents, displacement currents, skin effects and proximity effects can be analyzed to obtain the overall characteristics of electromagnetic components such as motors, busbars, transformers and coils. Parameters such as power loss, coil loss, impedance at a certain frequency (R and L), force, torque, inductance and energy storage can be calculated automatically. Graphical results such as magnetic lines of force, B and H distribution diagrams, energy density and temperature distribution for the whole phase can also be given, which can be used to analyze the magnetic field generated by PM, etc.

1.3 MANUFACTURING PROCESS

Since the late 19th century, PM materials have developed very rapidly. Through new technologies and processes such as grain boundary diffusion, grain boundary adjustment, rapid quenching, sintering and bonding, the technology of PM material preparation has been continuously improved and the performance has been enhanced. On the other hand, the requirements of high-performance products such as composition control and microstructure characteristics of PM are becoming more and more critical, which require continuous composition and process improvement.

1.3.1 Grain Boundary Regulation

The application of Nd-Fe-B PM in motors and other fields requires magnets with not only high enough coercivity and magnetic energy product but also high thermal stability. Reducing the material cost while improving the overall magnetic properties of Nd-Fe-B magnets is also an issue to be considered. The sintering and heat treatment process can be adjusted and optimized to smooth and clean the grain boundaries effectively. The grain boundary structure can also be regulated by adding trace elements, which can effectively improve the coercivity and overall hard magnetic properties of the material while reducing the rare earth content of Nd-Fe-B PM.

The microstructure of sintered Nd-Fe-B PM has the following characteristics. First, the grains of the matrix phase $Nd_2Fe_{14}B$ are polygonal. Second, the number of B-rich phase is very small, and it exists in isolated or granular state. Third, the Nd-rich phase is distributed along the grain boundaries or at the corners of the grain boundaries. The Nd-rich phases distributed along the grain boundaries are thinly layered and surround the matrix phase grains, and there may also be a few granular Nd-rich phases distributed inside the grains. Fourth, Nd_2O_3, an oxide of Nd, α-Fe phase and foreign dopants (e.g., chloride) and cavities may also be present. For sintered Nd-Fe-B PM, the mechanism of anti-magnetization is mainly the nucleation and growth of inverse domains. According to the nucleation model, the anti-magnetization process occurs through the formation and growth of anti-magnetization domains at the defects on the surface of $Nd_2Fe_{14}B$ grains, and a smooth and defect-free grain boundary is not conducive to the nucleation of anti-magnetization domains. Therefore, controlling the interface structure between $Nd_2Fe_{14}B$ phase and Nd-rich phase, reducing the grain boundary defects by grain boundary control and obtaining clean and smooth grain boundaries can effectively improve the coercivity, as shown in Figure 1.5.

The ways to improve the coercivity and high temperature stability of Nd-Fe-B include composition optimization and tissue optimization. From the perspective of composition optimization, the use of low-cost rare earths to replace Dy (Tb) and the addition of elements that are beneficial to improve coercivity are the main approaches used so far. From the point of view of tissue optimization, grain

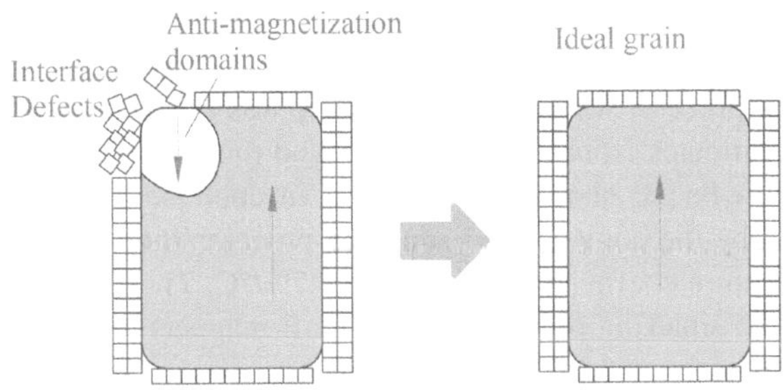

FIGURE 1.5 Diagram of grain boundary control to improve the coercivity of magnets.[1]

refinement and optimization of grain boundaries are the main ways currently considered. In recent years, the grain boundary control technology for Nd-Fe-B materials has been developing rapidly due to the market demand. A large number of studies have been conducted by domestic and foreign researchers, and a series of research results have been obtained from the composition regulation of grain boundaries to the structure regulation.

1.3.1.1 Sintering and Heat Treatment Process Optimization

The preparation process of sintered Nd-Fe-B magnets has a significant influence on the composition, particle size and distribution, orientation density and microstructure, which directly affects the overall performance of the magnets. Among them, the heat treatment process has a significant effect on the microstructure of the magnet, especially the grain boundary phase, so the coercivity of the magnet is particularly sensitive to the heat treatment process. Earlier studies have shown that the Nd-rich phase at the grain boundary of sintered Nd-Fe-B magnets is mainly distributed at the corner of the main-phase $Nd_2Fe_{14}B$ grains, and the Nd-rich phase at the interface of the main-phase $Nd_2Fe_{14}B$ grains is very faint and discontinuous, while the grain boundary of the main phase of the magnet after heat treatment becomes clear and smooth, and the distribution of the Nd-rich phase at the grain boundary becomes more continuous and straight, and the thickness increases. The effect of these changes in the coercivity is mainly reflected in two aspects. First, the smooth and clear boundary of the main-phase $Nd_2Fe_{14}B$ grains can reduce the defects at the grain boundaries, improve the defect density at the main-phase grain interfaces and increase the anisotropic field at the grain boundaries, thus increasing the coercivity of the magnet. Second, the straight and continuous Nd-rich phase at the grain boundary can hinder the coupling effect between the main-phase $Nd_2Fe_{14}B$ grains, thus improving the coercivity of the magnet. This heat treatment mechanism to improve the coercivity of Nd-Fe-B PM has been widely recognized by researchers. In recent years, the relationship between the crystal structure and coercivity of Nd-rich phase at grain boundaries has been further understood with further research work, but there are also some controversies.

The conventional sintering process of Nd-Fe-B mainly adopts a stepwise temperature rise and then holds at about 1080°C for 3–4 hours. Kim et al.[6] used a cyclic sintering process, where the temperature was raised and lowered twice at a rate of 10°C/minutes from 950 to 1050°C, and the prepared sintered Nd-Fe-B magnets $Nd_{13}Dy_2Fe_{79}B_6$ showed an increase in coercive force compared with those prepared by the conventional sintering process from 1955 to 2253 kA/m, and the temperature coefficient β reaches −0.47%/°C. The analysis revealed that during the cyclic sintering process, the Nd-Fe-B magnets formed h-Nd_20_3 phase and amorphous rare earth-rich phase at the grain boundary corner and at the grain boundary, respectively. An amorphous thin band with a width of about 5 nm was formed between the grains of the main-phase $Nd_2Fe_{14}B$. This clear, continuous and smooth amorphous thin band layer not only suppressed the nucleation of reverse domains but also reduced the exchange coupling between the hard magnetic phases and significantly improved the coercivity of the magnet.[7] Sepehri-Amin et al.[8] prepared $Nd_{29.9}Fe_{65.8}B_{3.1}Cu_{1.2}$ amorphous thin ribbons of the same ratio by magnetron sputtering and Lorentz transmission electron microscopy analysis and found that the amorphous thin ribbon phase may be ferromagnetic phase and therefore cannot isolate the main-phase grains effectively. In addition, they found that the aggregation of Cu in the grain boundary phase is beneficial for the improvement of coercivity. Liu et al.[9] suggested that the heat treatment effectively eliminated the thermal stress in the sintered Nd-Fe-B magnets during the sintering process, thus improving the coercivity of the magnets. In addition, Mo et al.[10] found a correlation between the crystal structure of the grain boundary Nd-rich phase and its O content, and they suggested that the grain boundary Nd-rich phase would transform from the hcp structure to the dhcp structure with the increase of O content in the grain boundary phase.

The heat treatment of sintered Nd-Fe-B magnets generally goes through primary tempering treatment and secondary tempering treatment, and the heat treatment temperature is generally determined according to the phase change temperature of the sample. The first-level tempering treatment is near the temperature below the alloy binary eutectic ($L' \rightarrow T_1+T_2$), which is 850–900°C. The secondary tempering treatment is lower than the alloy ternary eutectic ($L' \rightarrow T_1+T_2$+Nd-rich) temperature, which is 480–650°C. Since the actual composition of Nd-Fe-B magnets varies a lot in industry, its heat treatment process should be adjusted accordingly.

Li et al.[11] investigated the changes in the distribution of Nd-rich phase in sintered Nd-Fe-B magnets before and after heat treatment. It was shown that during the heat treatment, the Nd-rich phase at the corner of the grain boundary would diffuse to the surrounding $Nd_2Fe_{14}B$ main-phase thin band layer boundary due to capillary tension, thus increasing the coercivity of the magnet. They also found that the coercivity of the sintered magnets $Nd_{11.7}Pr_{2.8}Fe_{76.8}B_{6.0}Al_{0.5}Cu_{0.1}O_{2.1}$ was increased from 0.9 to 1.18 T after heat treatment at 600°C for 1 hour, and the remanence did not decrease. Three-dimensional atom probe microscopy clearly shows the formation of a thin Cu-rich layer between the rare earth-rich phase and the main-phase $Nd_2Fe_{14}B$. This work also provides a new idea for the subsequent preparation of low or no heavy rare earth, high coercivity sintered Nd-Fe-B magnets by grain boundary diffusion. Vial et al.[12] investigated the mechanism of the effect of tempering on the coercivity of magnets and also found that the boundary between the Nd-rich phase and the main phase of the tempered magnets became clear and smooth, and the Nd-rich phase was continuously distributed along the main-phase boundary and its thickness increased, thus increasing the coercivity of the magnets. Park et al.[13] further investigated the interface between the main phase and the Nd-rich phase after tempering and found that the crystal structure of the Nd-rich phase is transformed from a double hexagonal structure (dhcp-Nd) to a face-centered cubic junction (fcc-NdO*x*) after the magnet is tempered, and the fcc-NdO*x* has a suitable crystallographic relationship with the main phase; thus, the coercivity of the magnet is substantially improved after tempering. Fukagawa et al.[14] suggested that this face-centered cubic structure of the fcc-NdO*x* phase is able to suppress the nucleation of anti-magnetization domains on the surface of the main-phase grains. The structural transformation of the Nd-rich phase during tempering must involve oxygen, and the generation of fcc-NdO*x* has an important relationship with the O content in the Nd-rich phase. The composition and distribution of the RE-rich phase are also very complex, and for magnets with different compositions, the tempering treatment has different effects on the RE-rich phase, and therefore, on the coercivity.

1.3.1.2 Grain Refinement

The effect of grain size on the coercivity of sintered Nd-Fe-B magnets is very obvious. In a certain size range (larger than the single-domain particle size), the coercivity of the magnet increases as the grain size decreases. Assuming that the number of nuclei causing anti-magnetization domains per unit surface area at the grain boundary is certain, there is a certain correlation between the grain size and the coercivity, which is expressed as

$$H_{cJ} = a - b\ln D \tag{1.6}$$

where a and b are constants and D is the main-phase grain size.[15]

For ultra-fine grain Nd-Fe-B magnets, the size of the main-phase $Nd_2Fe_{14}B$ grains is effectively reduced, and the discrete field around the grains is reduced, so the nucleation field of the magnetic field is improved. Unlike the initial magnetization curve of conventional sintered Nd-Fe-B magnets, the initial magnetization curve of ultra-fine grain sintered Nd-Fe-B magnets shows a clear "step" phenomenon, similar to the initial magnetization curve of fast-quenched and thermally deformed magnets. This is due to the fact that the size of the ultra-fine grains is close to the size of single-domain particles, and the domain walls are pinned by the grain boundaries after they disappear inside the grains, resulting in a "step" phenomenon where the initial permeability is very high and the subsequent permeability becomes low. Therefore, the coercivity mechanism of ultra-fine grain sintered Nd-Fe-B magnets is not only a nucleation mechanism but also a combination of nucleation and pinning mechanism. However, when the grain size is smaller than a certain critical value (about 3 μm), the coercivity of the magnet decreases significantly, mainly due to the oxidation of the ultra-fine grains.

1.3.2 Grain Boundary Diffusion

1.3.2.1 Overview of Grain Boundary Diffusion Technology

As shown in Figure 1.5, the reverse magnetization process of sintered Nd-Fe-B magnets is the nucleation and growth of reverse magnetization domains at the defects on the surface of$Nd_2Fe_{14}B$ grains. Therefore, the material coercivity can be improved by increasing the anisotropic field on the grain surface and suppressing the nucleation of the reverse domains at the grain boundaries. Based on this principle, the grain boundary diffusion technique has been proposed. It is mainly based on the principle of elemental diffusion, which allows elements (e.g., Dy and Tb) that are favorable to improve the anisotropy of alloy magnetic crystals to enter the grain boundaries and change the interfacial composition of the grain boundary phases, thus improving the local anisotropy field. Dy or Tb entering the grain boundaries form the $(Nd,Dy)_2Fe_{14}B$ phase at the grain surface, which has a higher magnetic crystal anisotropy than the $Nd_2Fe_{14}B$ phase. This technique is itself an interfacial composition control technique. Unlike the traditional melting addition, the elements of grain boundary diffusion do not need to enter into the crystal, so the heavy rare earth Dy and Tb content can be greatly reduced, while minimizing the reduction of remanent magnetism, which is beneficial to improve the material cost performance. One important reason why Dy or Tb can enter the grain boundary and less into the crystal is that the grain boundary diffusion temperature exceeds the liquid-phase line of the sintered Nd-Fe-B magnets, the grain boundary rare earth-rich phase has become liquid phase and the diffusion rate of Dy (Tb) in the liquid phase is much higher than that in the solid phase.

For the basic principle of grain boundary diffusion, two grain boundary diffusion processes have been proposed in the early days, mainly using monomers or compounds of heavy rare earth elements Dy or Tb as diffusion media. One way is to coat the powder surface with a layer of Dy before pressing, followed by orientation pressing and sintering. This process can be achieved in the airflow mill powder-making process. Another way is to form a Dy or Dy-containing compound layer on the surface of the sintered Nd-Fe-B block by coating and then perform high temperature diffusion. Both of the two approaches have achieved a bit of success so far. The second method is shown in Figure 1.6. Dy or Dy-containing compounds coated or deposited on the magnet surface will diffuse along the grain boundaries at a certain temperature. In 2000, Park et al.[16] sputtered a few microns thick Dy metal film on the surface of sintered Nd-Fe-B magnets and

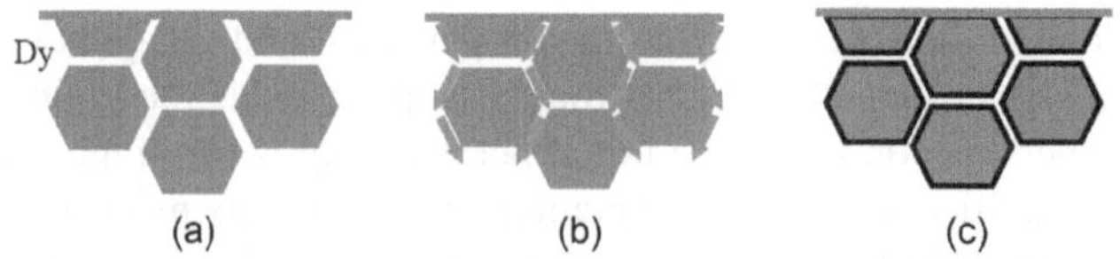

FIGURE 1.6 Schematic diagram of grain boundary diffusion[1]: (a) magnet surface coated with Dy, (b) Dy thermal diffusion and (d) Dy is distributed along the main-phase surface.

enriched Dy on the surface of the main-phase $Nd_2Fe_{14}B$ grains by heat treatment, which increased the magnet coercivity by a factor of two without reducing the remanence. This can be considered the earliest grain boundary diffusion process. Later, Hirota[17] developed a similar but simpler and more practical grain boundary diffusion method by grinding oxides or fluorides of Dy and Tb into a powder with a grain size of <5 μm, mixing it with alcohol 1:1 (mass ratio) to coat the magnet surface and then annealing the diffusion treatment. The researchers found that Dy dissolves in the matrix phase and is mainly enriched near the interface between the matrix phase and the grain boundary phase near the side of the matrix phase, while O and F dissolve in the Nd-rich phase and can improve the grain boundary phase. Researchers also found that in addition to the replacement of the main-phase Nd with Dy can improve the local anisotropic field, N, O and Ni also have some improvement on the overall properties of magnets, such as coercivity, resistance and corrosion resistance.[18,19] Suzuki et al.[20] performed Dy diffusion on sintered Nd-Fe-B magnets using dysprosium fluoride as a raw material to study the effect of Dy diffusion on magnetic properties and microstructure. The coercivity of the sintered Nd-Fe-B magnets after fluoride diffusion by thermal diffusion of Dy was increased from 1.1 to 1.5 T. Microstructural analysis showed the presence of Dy enrichment at the grain boundaries. The results indicated that the substitution reaction on the surface of the main-phase particles to form $(Nd_{0.8}Dy_{0.2})_2Fe_{14}B$ phase was the main reason for the improvement of the coercivity. Komuro et al.[21] investigated the effect of fluoride-coating treatment of Nd, Pr, Dy and Tb on the improvement of the coercivity of the magnets. When the thickness of the magnet was 1 mm, the coercivity of the Dy-F-coated magnet improved by 41% and the remanence decreased by 0.6%. The grain boundary diffusion method is applicable to magnets with a thicknesses of 5 mm and below, and magnets with a thickness of 5 mm can be used in HEV motors. At present, the relevant research has achieved remarkable results and has been gradually industrialized.

The early grain boundary diffusion used compounds of Dy or Tb with the aim of increasing the anisotropic field on the grain surface of the main phase, which can be called the first generation of grain boundary diffusion techniques. To further reduce the heavy rare earth content, in 2010, Sepehri-Amin et al.[22] proposed the use of low melting point eutectic compounds, such as Nd-Cu without heavy rare earths as diffusion media to improve the coercivity of magnets by modulating the grain boundary phase. The study achieved significant results and further reduced the dependence of high coercivity Nd-Fe-B magnets on heavy rare earth Dy and reduced the cost of raw materials. Sepehri-Amin et al.[23] also studied the effect of grain boundary diffusion process of Nd-Cu alloy on the magnetic properties and microstructure of heat-deformed (HID) magnets. $Nd_{70}Cu_{30}$ alloy diffused at 600°C for 2 hours, the H_c of heat-deformed $Nd_{14}Fe_{76}Co_{3.4}B_6Ga_{0.6}$ magnets increased from 1.5 to 2.3 T, M_s decreased from 1.35 to 1 T, (BH)max decreased from 340 to 228 k/m³, and the coercivity thermal stability coefficient improved from −0.55 to −0.45%/°C, while the coercivity thermal temperature coefficient of the sintered Nd-Fe-B magnets prepared by the conventional process is generally −0.69%/°C. Using backscattered electron analysis, it was found that the grain boundary diffusion process of Nd-Cu alloy resulted in a significant increase in the rare earth-rich phase component of the thermally deformed magnets, which is the main reason for the change in magnetic properties of the magnets. In recent years, domestic and foreign research on diffusion media has focused on low melting point rare earth metal eutectic alloys. This can be called the second generation of grain boundary diffusion technology. A heavy rare earth-free sintered Nd-Fe-B magnet with a high coercivity of 2.1 T was prepared at Peking University using $Pr_{68}Cu_{32}$ as the diffusion medium, and it was found that the decrease in the content of ferromagnetic elements Fe and Co in the grain boundary phase was the main reason for the high coercivity.[24] The National Institute for Materials Research, Japan, used Nd-M (M=AI, Cu, Zn, Ga, Mn, Ni, etc.) alloy grain boundary diffusion to significantly improve the coercivity and thermal stability of thermally deformed Nd-Fe-B and found that the improvement of the grain boundary phase composition was the main reason for the improvement of its performance.

1.3.2.2 Grain Boundary Diffusion Technology for Non-Rare Earth Compounds

In the initial grain boundary diffusion process, the diffusion medium is usually chosen from heavy rare earths or heavy rare earth compounds. The nucleation of anti-magnetized magnetic domains was suppressed by increasing the anisotropic field at the particle boundary of the main phase of the magnet, thus increasing the coercivity of the magnet. Later, low melting point eutectic compounds such as Nd-Cu without heavy rare earths were used as diffusion media. The mechanism is mainly to use grain boundaries as diffusion channels, so that these rare earth eutectic alloys can enter the grain boundary phase by diffusion, increase the grain boundary phase content, improve the grain boundary phase distribution and composition and, thus, increase the coercivity. After more than 10 years of research and development, although the grain boundary diffusion technology has made significant progress, but still cannot get rid of the dependence on rare earth elements, the current process invariably requires the use of rare earth compounds or rare earth alloys as the diffusion source.

Earlier studies have shown that doping with low-melting elements such as Al, Cu, Zn and Mg can react with the Nd-rich phase to form a new phase or dissolve it, improve the composition and distribution of the grain boundary phase and increase the coercivity of the magnet.[25] Kim et al.[26] found that the addition of a moderate amount of Cu can effectively modulate the crystal structure of the grain boundary phase, form a low mismatch Cu-rich grain boundary phase c-Nd_2O_3 and improve the coercivity. However, excessive Cu forms grain boundary phase hcp-Nd_2O_3, which increases the lattice mismatch, increases the interfacial roughness between the grain boundary phase and the main phase and decreases the coercivity. As early as 1996, Chen et al.[27] systematically investigated the effect of 20 oxide intergranular additions on $Nd_{22}Fe_{71}B_7$ magnets and found that MgO could increase the coercivity of magnets to some extent and reduce the magnetic loss.

The above work suggests that low melting point metals (alloys) or metal oxides have the potential to be used as diffusion media for grain boundary diffusion to improve coercivity.

1.3.2.3 Deficiency and Development Direction of Grain Boundary Diffusion Technology

The grain boundary diffusion technique has made remarkable progress since it received attention. Numerous practical results have proved that either by surface coating with different types of heavy rare earth compounds or by sputtering or vapor deposition of heavy rare earth metal, monomers can eventually lead to significant improvements in the magnetic properties of sintered magnets.

In terms of technology alone, grain boundary diffusion technology has its distinctive features. Compared with light rare earth elements, such as Pr and Nd, heavy rare earth elements, such as Dy and Tb, are more likely to form RE-Fe-B compounds. In the previous double alloying method or other alloying element addition process, the heavy rare earth elements are added to Nd-Fe-B powder before the sintering process, which results in a large amount of heavy rare earth elements entering the main-phase grains during the subsequent long sintering and tempering process, resulting in a significant decrease of remanent magnetism and magnetic energy product. In contrast, the grain boundary diffusion technology is in the state of sintered Nd-Fe-B blanks, starting from the surface to penetrate into the interior, which can avoid the problem of excessive heavy rare earths in the main phase and solve the problem of decreasing remanent magnetism and excessive addition of heavy rare earths at once. However, on the other hand, the essential process feature that heavy rare earths can only diffuse in the sintered blanks raises another problem, that is, the limited diffusion depth of heavy rare earths inside the magnet and the decreasing gradient distribution of heavy rare earth concentration from the surface to the inside. In other words, the grain boundary diffusion method is limited by the magnet thickness, and the effect of coercivity increase is diminished when the magnet thickness increases. Since the heavy rare earths Dy and Tb are diffused from the surface to the interior of the magnet, Dy and Tb are distributed in a gradient within the magnet.

Therefore, if the grain boundary diffusion technique is to be engineered, it must be able to stably control the depth of diffusion and the level of concentration gradient distribution of heavy rare earths inside the magnet. To find out the main factors that can control the diffusion of heavy rare earths inside the magnet is the common goal of many research works. At the same time, the field of grain boundary diffusion technology also has a wide scope for exploration and improvement if it is combined with the basic theory of Nd-Fe-B magnet microstructure.

1.3.3 Rapid Quenching

Rapid quenching means that the furnace temperature is preheated to a higher temperature than required, and then, the workpiece is loaded and the heat is stopped.

When the furnace temperature drops to the quenching temperature, the heat is applied and the temperature is controlled until the workpiece is burned through and then removed from the quench.

1.3.3.1 Rapid Quenching Nd-Fe-B Magnetic Powder

Fast-quenching process is the most important method to obtain Nd-Fe-B magnetic powder with nanocrystalline structure. During the fast-quenching process, the melt condenses at a very fast-cooling rate, the diffusion of atoms is inhibited, the organization of the solid cannot reach the thermodynamic equilibrium state and the obtained solid is nanocrystalline or amorphous.

1.3.3.1.1 Preparation of Rapid Quenching Nd-Fe-B Magnetic Powder

One of the first research methods used in the discovery of Nd-Fe-B PM materials is the fast-quenching method,[28] which is also the most widely used method for the preparation of Nd-Fe-B magnetic powder.[29]

In the preparation of nanocomposite PM materials by melt quenching method, the material organization and magnetic properties are related to many process parameters, and the quenching speed has the greatest influence on the organization and magnetic properties. The preparation of nanocomposite PM materials by melt quenching method is generally done by two technical ways. One is to prepare nanocomposite magnets with fine and uniform microstructure directly by the optimal fast-quenching speed, and the other is to obtain completely amorphous samples by complete over quenching and then crystallize them under the optimal annealing process. Usually, the magnetic properties of the nanocomposite magnets prepared by the best fast-quenching method are the best, but the window of the best fast-quenching speed is too narrow to be grasped in industrial production, so the research on the fast-quenching process is mainly focused on the latter method mentioned above. The typical production process of fast-quenched Nd-Fe-B is as follows.

The Nd-Fe-B alloy designed according to the composition is placed in a quartz crucible, heated by induction to below the melting temperature, and then, the pressure of the inert gas is applied to eject the melt from the nozzle (usually 0.5–1 mm in diameter) at the bottom of the crucible to a high-speed rotating cold copper or molybdenum roll (the linear speed of the cold series is usually 20–30 m/s), which is rapidly cooled by the cold roll and then thrown out into an amorphous thin strip, the thickness of which is usually 30–80 μm. The thickness of the thin strip is usually 30–80 μm, and the thickness of the thin strip varies according to the speed of the roll, the material of the roll, the pressure of argon, the size of the gap between the roll and the nozzle and the temperature of the melt during injection. The width of the strip depends on the shape of the nozzle.

There are two types of continuous rapid quenching equipment in industrial production: one is overflow type, and the other is injection type. They are shown in Figures 1.7 and 1.8, respectively.

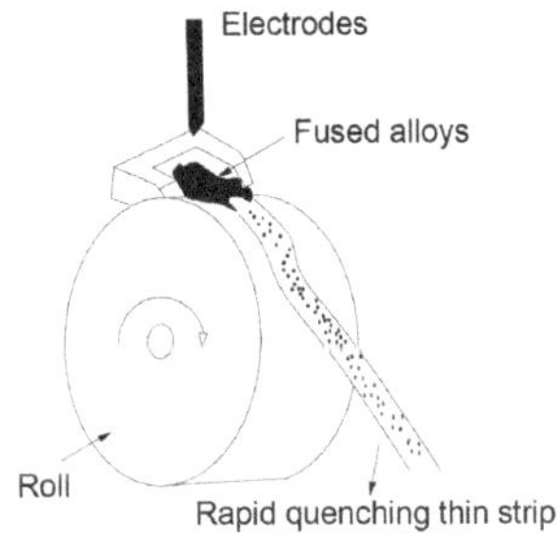

FIGURE 1.7 Schematic diagram of overflow fast-quenching equipment.[30]

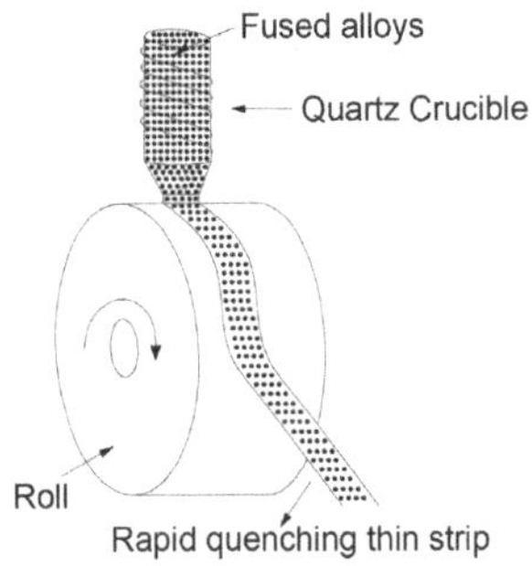

FIGURE 1.8 Schematic diagram of jet fast-quenching equipment.[30]

1.3.3.1.2 Types of Rapid Quenching Nd-Fe-B Magnetic Powder

According to the composition, the rapid quenching Nd-Fe-B alloys can be divided into, (1) high Nd content (16–20 at.%) alloys, (2) near stoichiometric Nd content (11–13 at.%) alloys and (3) low Nd content (<10 at.%) alloys.

1. Rapid quenching Nd-Fe-B magnetic powder with high Nd content

   Rapid quenching Nd-Fe-B powders with high Nd content have some non-magnetic Nd-rich phases in addition to the main-phase $Nd_2Fe_{14}B$. The Nd-rich phase is mainly distributed on the boundary of $Nd_2Fe_{14}B$ grains, which hinders the movement of domain wall and isolates the exchange coupling between $Nd_2Fe_{14}B$ grains, so it has high endogenous coercivity. However, too high Nd content will promote grain growth and reduce the coercivity. At the same time, the remanence magnetization is reduced due to the dilution effect of the non-magnetic Nd-rich phase. In 1983, Croat et al.[31] obtained rapid quenching magnetic powders with magnetic energy product up to 14.1 MGOe in $Nd_{0.13}(Fe_{0.95}\,B_{0.05})_{0.87}$ alloy for the first time by melt-spining and investigated the effect of B content, Nd content and fast-quenching speed on the properties of fast-hardened magnetic powders. The results show that there exists an optimal fast-quenching speed to achieve the best magnetic properties. The magnitude of the optimum fast-quenching speed is related to the specific composition of the alloy.

2. Nearly positive fractional composition rapid quenching Nd-Fe-B magnetic powder

   The fast-hardened Nd-Fe-B magnetic powder with nearly positive fractional composition is basically composed of $Nd_2Fe_{14}B$ single phase, with few non-magnetic and soft magnetic phases, which can obtain high magnetic energy product and coercivity, and has very good comprehensive magnetic properties. At present, most of the high-performance fast-hardened Nd-Fe-B magnetic powders are of this type.

3. Low Nd composition rapid quenching Nd-Fe-B magnetic powder

   In 1988, Coehoorn et al.[32] of the Philips Institute in the Netherlands made amorphous thin strips of $Nd_4Fe_{77}B_{18.5}$ by the fast-quenching method and then annealed and crystallized at 670°C for 30 minutes to obtain isotropic fine-grained powder wood with a composition of 73% Fe_3B phase, 15% $Nd_2Fe_{14}B$ phase and 12% α-Fe phase. The grain size of Fe_3B is 30 nm, the grain size of $Nd_2Fe_{14}B$ phase is 10 nm and the magnetic energy reaches $M_s = 1.6$ T, $M_r = 1.2$ T and $H_{cJ} = 0.3$ T. The magnetic powders containing many soft magnetic phases Fe_3B and α-Fe have good PM properties, which indicates that the isotropic alloys composed of nanoscale soft magnetic phases mixed with PM phases also have strong remanence enhancement effects, i.e., $M_r > 0.5\ M_s$ ($M_r = 0.75\ M_s$), and also have high coercivity. This research result attracted extensive and in-depth research attention, after which researchers in magnetism and magnetic materials conducted extensive and in-depth studies from both theoretical and experimental aspects,[33–45] which led to the development of a new class of PM materials, that is biphasic nanoexchange-coupled PM materials, or "biphasic nanocomposite PMMs".

Nanocomposite PM material is a new type of PM material, which is composed of soft magnetic phase and hard magnetic phase compounded in the nanometer range. To obtain high PM properties, the hard magnetic phase is required to have the highest possible magneto-crystalline anisotropy, while the soft magnetic phase has the highest possible saturation magnetization intensity, and the two phases are compounded in the nanometer range. Under the action of exchange coupling, when there is an external magnetic field, the magnetic moment of the soft magnetic phase has to rotate synchronously with that of the hard magnetic phase, so the magnetization and demagnetization of this magnet have the characteristics of a single ferromagnetic phase. In the remanent state, the magnetic moment of the soft magnetic phase will stay in the average direction of the magnetic moment of the hard magnetic phase, so the isotropic PM has a remanent magnetization enhancement effect. Micromagnetic calculations show that the $(BH)_{max}$ of the $Nd_2Fe_{14}B$/α-Fe nanocomposite magnet can reach 622 kJ/m^3 (83MGOe).

Compared with single-phase nanoPM, the biphasic nanocomposite PM has low rare earth content and thus has the advantages of low raw material cost and good corrosion resistance. Due to the exchange coupling, it has a higher remanent magnetization than single-phase nanoPM, but the coercivity is lower than the latter. Another important characteristic of biphasic nanoexchange-coupled PM is elastic magnetism. It has been shown that the temperature stability of magnetic properties is higher for biphasic nanoexchange-coupled magnetic powders than for non-biphasic nanoexchange-coupled powders.[46]

The preparation method of nanoexchange-coupled magnetic powders is still dominated by the rapid quenching method, such as MQP™-16-7 magnetic powder is a commercial $R_2TM_{14}B/\alpha$-Fe nanoexchange-coupled magnetic powder produced by the fast-quenching method,[47,48] which consists of $R_2TM_{14}B$ and α-Fe phases with a grain size of 20–30 nm. The process points are the same as those of the positively fractionated Nd-Fe-B fast-quenched magnetic powder, and the magnetic properties reached $B_r = 980$ mT, $H_{cJ} = 557$ kA/m and $(BH)_{max} = 127$ kJ/m^3.

Subsequently, researchers have also carried out research work on the preparation of nanoexchange-coupled magnetic powders using mechanical alloying (MA)[48] and HDDR methods.

Regardless of the process method, the alloy composition is the basis, and the composition design is based on the ideal nanocomposite structure that can be obtained under suitable process conditions, i.e., fine grains, not more than 40 nm and uniform distribution. Some principles need to be followed when designing the composition, (i) no or less non-magnetic phase formation, (ii) the added auxiliary elements can effectively inhibit grain growth and (iii) the alloy composition process can be adapted to a wide range, that is, the ideal structure can be obtained under a variety of process conditions.

1.3.4 Sintering Technologies

1.3.4.1 Conventional Preparation Method of Sintered Ferrite PM

The process flow of the conventional solid-phase sintering method for preparing ferrite PM is shown in Figure 1.9. The anisotropic ferrite is pressed in the external magnetic field and sintered at high temperature, of which there are two kinds of pressing and forming methods, wet and dry.

1.3.4.1.1 Wet Forming

Wet forming is a common forming method for manufacturing high-performance anisotropic ferrite PM. Simply put, wet forming is to use water as the lubricant and binder to inject the prepared ferrite PM slurry into the mold cavity by pressure (while applying the orientation magnetic field) and then apply pressure to press the billet with a certain geometry, size, density and mechanical strength.

During the wet-forming process, the easily magnetized axis of the single-domain grains in the die cavity gradually turns to the magnetic field direction

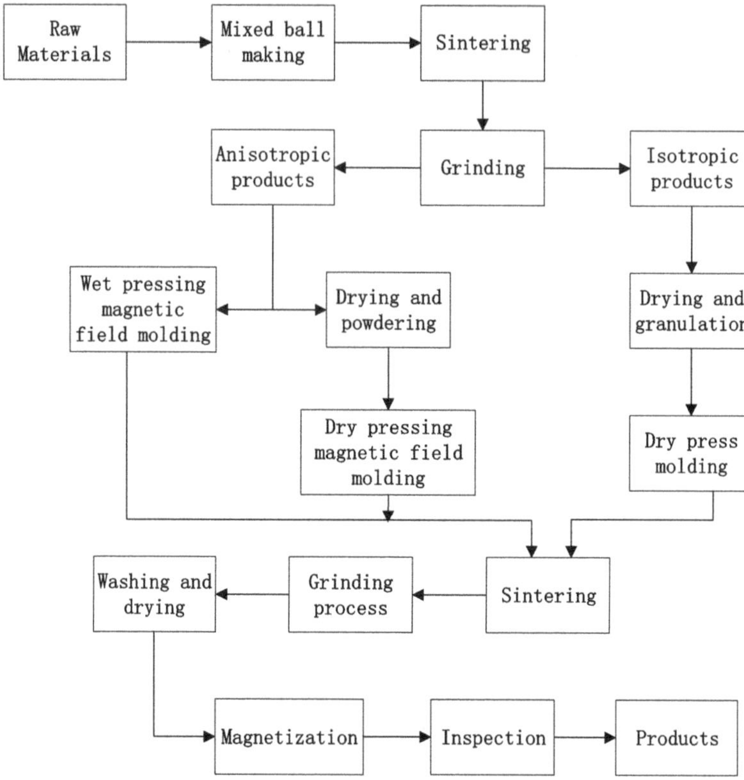

FIGURE 1.9 Process flow diagram of ferrite PM prepared by conventional solid-phase sintering method.[1]

under the good lubrication of water due to the applied magnetic field. Due to the high consistency of grain orientation arrangement, the product has good anisotropy and high magnetic properties. Some factors that affect the forming process are as follows: (i) water content and consistency of the slurry, (ii) injection method and pressure, (iii) forming compaction method and (iv) pressurization pressure and speed.

The density of the raw billet has an important influence on the final properties. Within a certain range, the density d_p of a formed raw billet can be expressed as[49]

$$d_p \infty d_0 P^{t/\mu} \tag{1.7}$$

where d_0 is the loading density, P is the forming pressure, t is the forming time and μ is the internal friction coefficient of the pellet. By analyzing Equation (1.7), we can find some methods to improve the density of raw billets, (i) increase the loading density d_0, (ii) increase the forming pressure P, (iii) control the pressing speed (forming time t) and (iv) reduce the internal friction coefficient μ of the pellet.

The wet production method has better magnetic properties and high and stable product performance due to the high chemical activity of the raw materials and the favorable particle rotation. At the same time, the production process is less dust pollution of the environment but can also make full use of a variety of industrial by-products, to improve quality and reduce costs. However, the productivity and yield of the wet process is low, the wet forming and pressing billets are naturally dried and most of the water will evaporate naturally before sintering in the furnace, so it is vulnerable to environmental factors. Another disadvantage of the wet process is that it requires filter paper and filter cloth in the mold to extract water, which makes the equipment and process complex and makes it difficult to automate mass production.

1.3.4.1.2 Dry Forming

Dry forming is to put the powder without water but mixed with a small amount of binder into the mold, apply sufficient pressure and press the powder into shape under the action of external magnetic field. Due to the high friction between the dry powder, the amount of binder and lubrication effect are limited, and the ferrite particles have difficulty in rotating in the magnetic field, resulting in a lower magnetic property than the wet product. In addition, the dry forming method requires high strength and hardness of the die, and the die life is shorter than that of the wet method. However, dry forming also has more advantages than wet forming, such as

1. The process is simple and the production efficiency is high. The raw material for wet production is a slurry containing about 37% water, so not only do we need to strictly control the water content of the slurry at the powder manufacturing stage but also need to use vacuum pumps to extract the water during the pressing process, which makes the process more complicated. In contrast, the dry method can be used to purchase fine powder directly from the pre-fired material manufacturer for pressing, which is easy to achieve continuous automated production in large quantities.
2. It is suitable for making products with small size, high precision and complex shape. The complex process of wet production causes the processing of the used molds to be difficult, and the products are limited to medium and large sizes, regular ring, square, pie and tile magnets, etc.
3. High density and mechanical strength of pressed billets. Dry forming billets have less moisture and fewer pores, and the density of the pressed billets is relatively high. The high density of the pressed billet and its mechanical strength are correspondingly high, which reduces the breakage rate during the handling of the pressed billet and therefore results in a high yield.
4. Low manufacturing cost. The water content of the pressed billet made by wet forming is 8–10% (mass fraction), so it cannot be directly sintered in the kiln. It needs to be dried naturally for a long time, and only when

the water content of the pressed billet reaches 2–3% by mass can it be loaded into the kiln for sintering. In contrast, dry forming of pressed billets does not need to be dried, so the tendency of cracking of the product is small and the yield is higher. And it saves the water-pumping system, and the mold is easy to manufacture. The formed press billet is smooth and flat, which greatly reduces the amount of grinding process.

1.3.4.1.3 New Sintering Process for Ferrite PM

1. *Low-temperature sintering technology*: The realization of ferrite PM low-temperature sintering is of great significance for energy saving and consumption reduction, the key of which is to refine the powder and add low-temperature sintering additives. The refinement of ferrite powder particles can increase the surface energy of the material, promote the solid-phase chemical reaction and significantly reduce the sintering temperature. The addition of low-temperature sintering additives generally has the effect of mineralization, solubilization and crystal blocking, so as to realize liquid-phase mass transfer sintering and effectively reduce the sintering temperature. Common low-temperature sintering aids include Bi_2O_3, MoO_3, PbO and V_2O_5 and other low-melting oxides or low-melting glass substances. The low-temperature sintering additives have a great impact on the magnetic properties of sintered magnets while lowering the sintering temperature, so the balance between lowering the sintering temperature and ensuring excellent magnetic properties is a real problem to be faced in low-temperature sintering.
2. *Microwave sintering technology*: Conventional sintering is to transfer heat to the specimen by conduction, radiation or convection, and the temperature of the surface is higher while the temperature of the core is lower, forming a temperature gradient from the core to the surface of the specimen. The microwave sintering is through the basic microstructure of the specimen and the special waveband of the microwave coupling effect to generate heat. The specimen undergoes dielectric loss in the electromagnetic field, and the whole specimen is heated to the sintering temperature, thus achieving densification. Microwave sintering has the advantages of fast-heating speed, selective sintering, integral heating, uniform fine crystal structure, energy saving, low sintering temperature, low environmental pollution, etc.
3. *Discharge plasma sintering technology*: Discharge plasma sintering (SPS) is a new technology for material densification that has attracted a large number of researchers because of its low-temperature, rapid and efficient sintering. In the SPS process, a DC pulse current is passed to the electrode, which instantaneously generates discharge plasma, activating the particle surface inside the sintered body and causing each particle to generate heat uniformly by itself, thus achieving sintering by

using the internal heat of the powder itself. In conventional sintering, there is no active force between the powder particles, and the surface of the particles has an inert film, so the sintering time is generally long, while SPS sintering can activate the surface of the powder particles, so it can overcome the disadvantage of long sintering time. In addition, SPS sintering has the advantages of low sintering temperature, the ability to obtain materials with high densities, the ability to sinter gradient materials and complex materials and the ability to obtain uniform and fine tissues.[50]

1.3.4.2 Manufacturing Process of Sintered Nd-Fe-B PM Material

1.3.4.2.1 Manufacturing Processes

Sintered Nd-Fe-B PM materials are produced by powder metallurgy, which includes single-alloy method and double-alloy method. The single-alloy method is to smelt a kind of alloy. The double-alloy method is to smelt two alloys: one is the main alloy, and the other is the auxiliary alloy, or called the grain boundary alloy (phase). The process flow of the single-alloy method is shown in Figure 1.10.

Powder metallurgy method is also called sintering method. Nd-Fe-B PM made by this method is called sintered Nd-Fe-B PM. In some articles or books, the sintered Nd-Fe-B PM is also called Nd-Fe-B PM alloy. There is no clear distinction between sintered Nd-Fe-B PM alloy and sintered Nd-Fe-B PM. Generally speaking, some research units are called alloys because it does not make Nd-Fe-B into magnet components, while companies generally want to make Nd-Fe-B into PM with certain shapes and sizes, and most companies call it sintered Nd-Fe-B PM.

Figure 1.10 shows the manufacturing process flow of S-Nd-Fe-B PM, which has 16 process steps and includes several test analyses in different stages. Strictly speaking, these 16 process steps and testing and analysis are indispensable. The manufacturing process of S-Nd-Fe-B PM is a system engineering, each link has to lay the foundation for the next link, once a process link fails to meet the requirements, the magnets made may not meet the target requirements and become scrap, resulting in a poor yield.

1.3.4.2.2 Method and Process of Making Powder

The purpose of powder making is to crush large alloy ingots into a powder body of a fixed size (3–6 μm). There are traditional powder-making process, improved powder-making process and new powder-making process as shown in Figure 1.11.

The purpose of the 1080°C annealing treatment for ingots is to eliminate α-FeO from the bulk ingot. α-Fe can be avoided only when the Nd content of the bulk ingot is >33% (mass fraction). α-Fe affects the speed of powder making and powder quality. Any alloy ingot with total rare earth metal content <32% should be annealed at 1080°C for 20 hours. For SC flake ingots, alloys with a total rare earth metal content of about 30.0% (mass fraction) or more can suppress the appearance of α-Fe and, therefore, do not need to be annealed for 20 hours at 1080°C. The coarse crushing of ingots to 1–4 mm coarse powder is

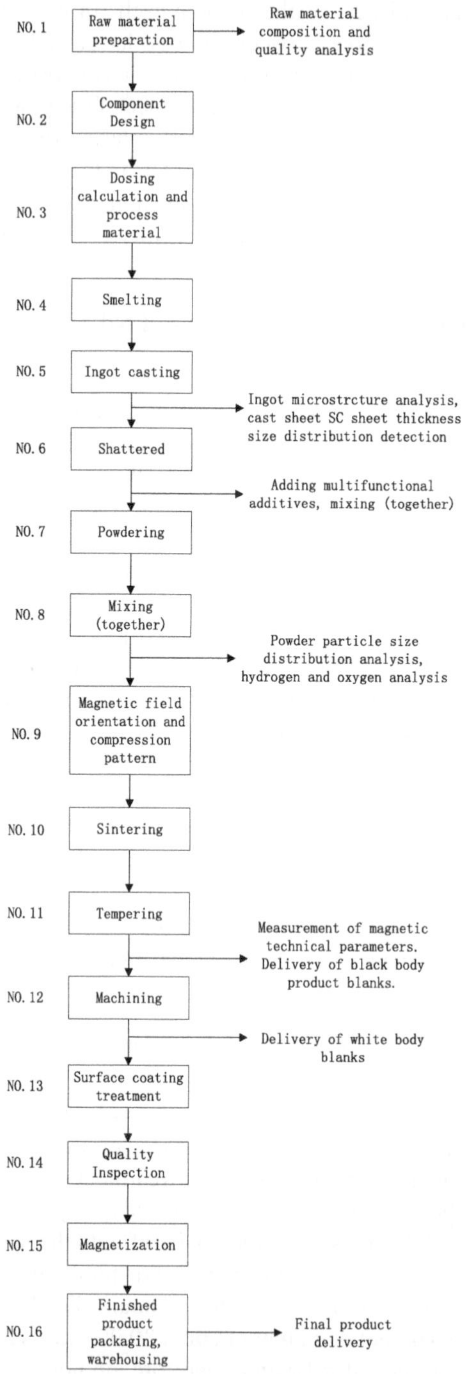

FIGURE 1.10 Manufacturing process of S-Nd-Fe-B PM.[3]

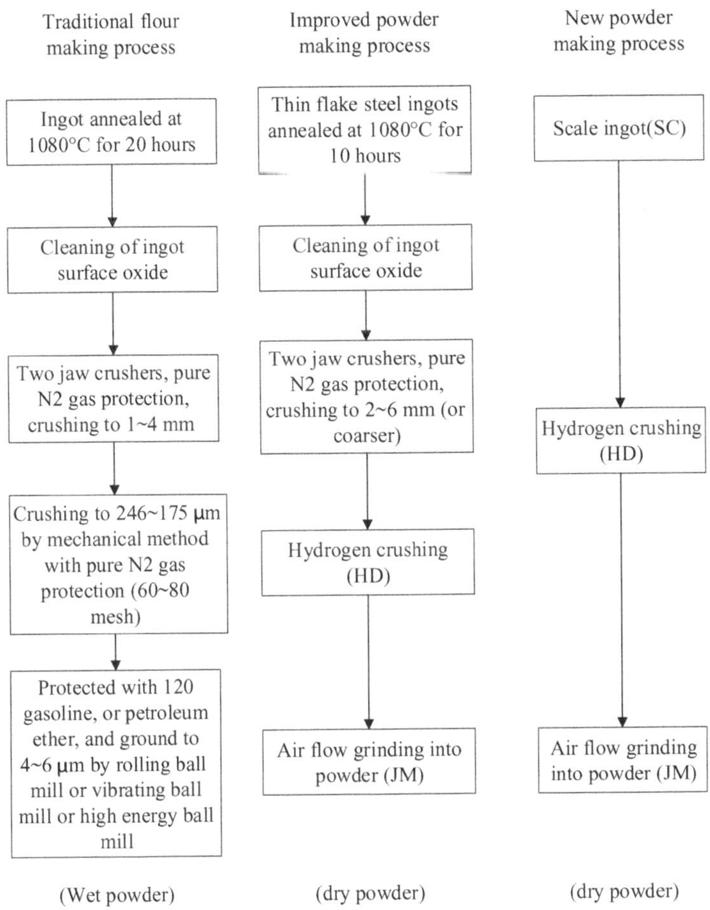

FIGURE 1.11 Method and process of making powder.[3]

generally carried out in a jaw crusher under pure N_2 gas protection with little technical difficulty. In general, the most important process steps for powder making are ball mill powder making (wet), hydrogen crushing and airflow mill powder making steps.

To obtain well-oriented magnets, the magnetic powder has the following requirements: (i) Small size of powder particles (3–4 μm) and narrow size distribution, that is, the requirement of 3–4-μm particles accounted for 95%, do not have <1 and >7 μm particles exist to ensure that all the powder particles are single crystal. (ii) The powder particles are spherical or nearly spherical in shape. (iii) The crystal defects of the powder particles should be as few as possible. (iv) It is better to break along the grain boundary when crushing and try to have Nd-rich phase on the surface of each particle, so as to lay the foundation for the liquid-phase sintering later and prevent the appearance of Class II grain boundary. (v) The impurities and gases adsorbed on the surface of the powder particles should be as little as possible, especially the oxygen content should be less. Therefore, the powder-making process should be carried out under the protection of high purity inert gas (N_2 or Ar) or under vacuum to prevent the magnetic powder from contacting with oxygen or humid air. The above five conditions are necessary for manufacturing high-quality S-Nd-Fe-B PM.

1.3.5 Bonding Technologies

Bonded PM are PM materials made from a homogeneous mixture of magnetic powder and adhesive that is pressed (or injected or calendered) and cured in a mold and are a class of functional composite materials with a binder as the matrix phase.[30] The production of magnets by the bonding process started in the late 1960s and has the following characteristics compared to magnets produced by the initial casting and powder metallurgy (also known as sintering) methods.[51–53]

1. Highly designable, allowing magnetic properties and shapes to be tailored according to the needs of practical applications.
2. High-dimensional accuracy, large freedom of shape and the ability to precisely shape complex magnets in one pass.
3. Uniform and consistent magnet properties.
4. Good mechanical properties, with high toughness, not easy to crack and burst angle, etc.
5. It can realize large-scale automatic production with high economic efficiency.

It is the above advantages that make bonded PM widely used in computer, electronic industry, automobile industry, office automation, household appliances and other fields, becoming the basic materials and devices for high-tech industries. In addition, the density of bonded PM is small, generally equivalent to 50–80% of the density of the dense material of this magnetic powder. At the same time, the bonded Nd-Fe-B has the deficiencies of low magnetic properties and low use temperature due to the influence of Nd-Fe-B magnetic powder itself and polymer binder,[54] but this does not affect it to be a magnet with excellent overall performance. It is currently the fastest growing class of magnetic materials, with a growth rate of more than 20% in recent years.

The factors affecting the performance of bonded Nd-Fe-B should include the following:

1. Types of magnetic powder. The source of displacement of magnetic energy of the magnet is the magnetic powder, so the performance of the magnetic powder directly affects the performance of the magnet.

2. The composition of the magnet. When the binder content is not less than the minimum amount, the smaller the volume fraction of the binder, the higher the magnetic properties of the magnet. The compressive strength of the magnet has an optimal value as the binder content increases.
3. Type of binder. A binder with good mobility and lubrication can effectively eliminate the voids in the magnet and increase the volume fraction of the magnetic powder, thus improving the magnetic properties. The type of binder has a decisive influence on the mechanical properties of the magnet.
4. Forming process parameters. Process measures to reduce cavities in magnets can effectively improve the magnetic properties, corrosion resistance and mechanical properties of magnets.
5. Surface coating. A good surface coating can effectively prevent the oxidation of the magnetic powder in the magnet and maintain the stability of the performance index of the magnet during use.

1.3.5.1 Adhesive for Bonding Nd-Fe-B PM

Since the binder is not magnetic, the magnetic properties of bonded magnets depend mainly on the magnetic properties of the PM powder, but the performance of the binder (including the bond strength, curing temperature, and softening temperature) also has a non-negligible effect on the performance of bonded PM. The basic role of the binder is to increase the fluidity of the magnetic powder particles and the bonding strength between them. There are many types of binders, and the principles of selecting binders are high bonding strength, high bond strength, low water absorption, good dimensional stability and small size shrinkage during curing, making the bonded magnets with high-dimensional accuracy and good thermal stability.

1.3.5.1.1 Thermosetting Binder

Th disadvantge of thermosetting binder is that cannot be used at room temperature, only when heated to a certain temperature or added to the curing agent occurs, after curing cannot be softened again. The outstanding advantage of this binder is the strong bonding force, molding and curing of products with high rigidity, hardness and strength, particularly good dimensional stability, and high heat resistance.

Thermosetting binder is the most commonly used material bonding agent for bonding Nd-Fe-B magnets, which has the advantages of low dosage, high strength, heat resistance, weather resistance, low cost and wide range of process applicability. This type of binder mainly consists of thermosetting epoxy resin, which is mainly used to prepare press molding bonded rare earth PM materials. However, there is still a big difference with thermosetting resins in resin matrix composites. In bonded magnets, the amount of binder should be as little as possible under the premise of ensuring the strength to ensure the best magnetic properties.

1.3.5.1.2 Thermoplastic Binder

At present, most Nd-Fe-B-bonded magnets are bonded magnets with thermosetting resin as the binder, with the expansion of the application of bonded Nd-Fe-B. In some special applications, magnets with special shapes are needed, such as magnets with larger diameters. If thermosetting resin-bonded magnets are still used, a large pressure is required to make the magnets dense and molded, and it is difficult for the current presses to meet such high pressure. At the same time, Nd-Fe-B magnets using thermosetting resin as binder have the problem of environmental pollution when disposed of at the end of life, and there is an urgent need for recyclable bonded magnets. Thermoset resin-bonded magnets have a long production cycle and require curing treatment, and the curing agent is mostly toxic. The bonded magnets with thermoplastic resin as binder can overcome the above-mentioned shortcomings and can be reused, the molding process is simple, the production cycle is short, no curing treatment is required and complex shapes can be manufactured.

Thermoplastic binder also plays an important role in injection-molded bonded magnets. It is mainly used for the flow and bonding between magnetic particles, both as a carrier for the powder flow during injection and as a bridge to connect the powder particles to maintain a specific shape. The dimensional accuracy, shape, processability and mechanical properties of the formed magnet are all importantly related to the binder. The viscosity and mechanical properties of the binder are important factors affecting the performance of the magnet. The viscosity of the thermoplastic binder is too high to lead to poor mobility of the mixture, affecting the orientation of the magnetic powder arrangement and will produce defects, such as under-injection. Viscosity is too low, easy to appear powder and binder of the two-phase separation phenomenon, so that the magnet density distribution is not uniform. Mechanical properties are important factors to ensure that the magnet has a high enough performance at a certain temperature and other conditions.

Injection molding generally uses thermoplastic resin as the binder, and the injection molding process has strict requirements for the binder, which requires not only high strength to maintain the mechanical properties of the magnet but also a suitable viscosity to ensure good fluidity when melting. Melt viscosity is too high, not only poor mobility, affecting the filling of the mold, but also makes the magnetic powder particles in the external field when the orientation of the arrangement is difficult, thus affecting the magnetic properties. Melt viscosity is too low, there is a possibility that the powder and binder two-phase separation, so that injection molding failure or magnet performance is not uniform. At present, the common binders are nylon, polypropylene, polyethylene, polystyrene, polyphenylene sulfide, polytetrafluoroethylene, etc. Table 1.6 lists the characteristics of commonly used thermoplastic resins.

Among these resins, nylon series are the most widely used, especially nylon 12, which has the advantages of high strength, low melt viscosity and good compatibility with magnetic powder. Currently, with the development of the automotive industry, the use of heat-resistant performance of polyphenylene sulfide,

TABLE 1.6
Commonly Used Thermoplastic Binders and Their Advantages and Disadvantages[30]

Resin	Advantages	Disadvantages
Nylon (PA)	Good mechanical properties, good thermal properties, flame retardant, fat resistant and hydrocarbon resistant (gasoline, lubricants, etc.) and good electrical properties	Easy to absorb water and acid to accelerate the rusting of magnetic powder, general weather resistance, not acid resistance and poor repeat processability
Polyester PBT PET	Abrasion resistant, flame retardant, weather resistant, very low moisture absorption, stress cracking resistant, good melt flow and easy to process	Not resistant to acid, alkali and hot water and poor heat resistance and mechanical properties when unmodified
Polyolefin	Easy to process, solvent resistant, low hygroscopicity and low price	Average performance and poor resistance to environmental stress cracking
Polyphenylene sulfide PPS	Creep resistance, wear resistance, flame resistance, heat resistance (use temperature above 180°C), chemical resistance, weather resistance, radiation resistance, good blendability with fillers and other polymers	Low impact strength, high melt viscosity and difficult to mold and process
Polyimide PI	Excellent mechanical properties, creep resistance, heat resistance, solvent resistance and radiation resistance; some varieties can be injected and extruded	Some varieties are not meltable and insoluble, difficult to process and poorly resistant to acids and alkalis and degrade when encountering strong oxidizing agents
Thermotropic Liquid Crystal Polymerization Substances (TLCP)	Excellent physical and mechanical properties, processing fluidity and heat resistance	Expensive
PPO, PEEK, PES, etc.	Good mechanical properties and excellent heat resistance	High melting point, poor flowability, difficult to process

polyimide, liquid crystal resins and other super engineering plastics as a binder for the preparation of high-temperature-bonding Nd-Fe-B is also increasing. In addition, high-temperature-resistant reactive thermoplastic binder is also under development. The characteristic of this kind of binder is to use recyclable low molecular weight and low viscosity oligomer as the initial binder, thereby increasing the loading capacity and reducing the processing temperature and then producing high-temperature-resistant resins like polycarbonate or PBT through catalytic polymerization after injection into the mold.

In the process of injection molding preparation of bonded magnets, the addition of binder is the characteristic and core of its technology, which directly affects the mixing and injection processes, and improper selection of binder may cause various defects. According to the current production process of injection molding bonded magnets, the amount of adhesive for injection molding bonded magnets is usually 25–50% (volume percentage), and the specific choice of which binder to consider the application temperature of the magnet, the process performance of the magnet preparation, the performance of the magnet and price factors.

1.3.5.1.3 Metal and Inorganic Binders

Due to the performance limitations of organic binders, bonded magnets can only be used at lower temperatures. To increase their use temperature, metal and inorganic binders are ideal choices. As early as 1991, Kuhrt et al.[55] studied that Zn is used to bond $m_2Fe_{17}N_x$, but the manufacturability is poor and the production efficiency is low. There is currently no application of work, so there is no successful application yet.

1.3.5.1.4 Other Binders

In addition to the above binders, there are some special binders for improving certain properties of magnets. For example, Ying Liu et al.[10] studied a new bonded PM composite made of dichalcogenide polymer (OPM) compounded with fast-hardened Nd-Fe-B magnetic powder. The magnetic properties of the magnetic polymer-bonded Nd-Fe-B were higher than those of the non-magnetic polymer-bonded Nd-Fe-B with the same volume fraction of magnetic powder, but their temperature stability was also almost the same.

1.3.5.2 Additives for Bonding Nd-Fe-B PM

To improve the processability and stability of the binder and magnetic powder mixture and improve processing efficiency, a small number of additives, such as coupling agents, plasticizers, lubricants and heat stabilizers is usually addded. Coupling agent is mainly used to improve the compatibility between hydrophilic magnetic powder and hydrophobic binder, the main types of silane system and titanate system; plasticizer is mainly used to increase the plasticity of resin and flexibility and improve the orientation of magnetic powder; commonly used plasticizers are phthalates, stearate epoxy compounds, oleic acid lipids, etc.; the role of lubricants is to reduce the mutual friction between magnetic powder and binder, improve the processability of the mixture and make the magnetic powder in the polymer easier to orientation, while reducing the wear and tear of equipment; commonly used lubricants are fatty acid phthalamides, fatty acids and their lipids, metal soaps, hydrocarbons, etc.; and heat stabilizers are mainly used to prevent the resin caused by high-temperature decomposition and deterioration of its performance, the types of metal soaps, organics, etc.

There are many types of additives, and the selection should be based on product requirements, process needs, cost, binder type and other criteria.

1.3.5.2.1 Coupling Agent

The coupling agent is indispensable in the preparation process of magnets. Because the characteristic of magnetic powder is hydrophilic, while that of the general binder is oleophilic. Therefore, the coupling agent is needed to build a "molecular bridge" between the magnetic powder and the binder, connecting the two materials with different properties, thus enhancing the affinity between the magnetic powder and the resin, increasing the amount of resin coverage on the surface of the magnetic powder and improving the magnetic properties.

1.3.5.2.2 Lubricants

According to the mechanism of action of lubricants, they can be divided into two categories: external lubricants and internal lubricants. External lubricants act as an interfacial lubrication. They have little compatibility with the polymer and are easily transferred from the polymer interior to the surface during processing, forming a lubricant molecular layer that reduces the friction between the polymer and the equipment surface. Internal lubricants are somewhat compatible with the polymer, producing a plasticizing effect that weakens the cohesion between polymer molecules and reduces their internal friction. The roles of external and internal lubricants are relative, in fact most lubricants have both lubrication roles.

The molecular structure of general lubricant contains two parts of long-chain non-polar group and polar group, and these two groups show different compatibility in different polymers, thus showing different internal and external lubricating effects. According to the chemical composition, the commonly used lubricants can be divided into the following categories: fatty acids and their esters, fatty acid amides, metal soaps, hydrocarbons, organosilicon compounds and others (including advanced alcohols and compound lubricants). When designing the formulation, the characteristics of the lubricant, the nature of the polymer, the processing process, the processing equipment, the performance of the product and many other factors should be considered. Different applications have different requirements for lubricants, but in general, the ideal lubricant should meet the following requirements: (i) high and long-lasting lubrication efficiency, (ii) moderate size of compatibility with resin, balanced internal and external lubrication, no frosting and not easy to cause scaling, (iii) small surface tension, small viscosity, good diffusion at the interface and easy to form the interface layer, (iv) does not reduce the mechanical strength of the polymer and heat deflection temperature and does not affect the secondary processing performance and printing, bonding and welding of plastic, (v) has good heat resistance and chemical stability, does not decompose and volatilize in high-temperature processing and does not react with resin and other additives and (vi) does not corrode equipment, does not pollute products and has no toxicity. Under normal circumstances, it is difficult for one lubricant to meet all the above requirements, and actually several lubricants are used together; in recent years, compound lubricants have been developing rapidly and mostly from this reason.

1.3.5.2.3 Antioxidants

All polymeric materials, whether synthetic or natural, can actually react with oxygen. Oxidation occurs at every stage of the polymer life cycle, in other words, during the production, storage or addition of the product. The reaction of the binder with oxygen is an autocatalytic process in which the main product at the beginning of the reaction is a peroxide, which decomposes under the right conditions to produce a reactive radical, which in turn can react with a macromolecular hydrocarbon or oxygen to produce a new radical, and so on, so that the oxidation reaction proceeds in a free radical chain process.

Both chain breakage and cross-linking of chains cause deterioration of the binder properties. The typical manifestation of oxidation can be summarized by the typical characteristic of the aging phenomenon. On the one hand, oxidation causes fading (yellowing), loss of transparency, chalking and surface cracking of the polymer, thus affecting the appearance of the product. On the other hand, oxidation causes loss of mechanical properties (impact strength, elongation strength, tensile strength, etc.) of the product. There are several methods to prevent polymer oxidation: (i) polymer structure modification, for example, copolymerization with vinyl-containing antioxidants, (ii) capping of alkyne groups, which is mainly used for polypropylene, (iii) physical stability (tensile) obtained through polymer orientation and (iv) addition of stabilizers, i.e., antioxidants, and addition of antioxidants are the most commonly used stabilization methods. Antioxidant is a chemical additive that can hinder the oxidation and, thus, the aging of the polymer, usually as long as the addition of a tiny concentration of 1–3% antioxidant is very effective. Antioxidants are added as close to the aging cycle as possible, for example, before the polymer dries to a powder. In the processing of bonded magnets, the selection of antioxidants is very important to achieve the desired antioxidant effect. The selected antioxidant should meet the following conditions: (i) do not change the color of the binder, (ii) volatility should be small to ensure that the magnet in the molding does not produce pores due to the volatility of the antioxidant, reducing the mechanical strength and magnetic properties of the magnet, (iii) solubility should be good, the ideal antioxidant in the binder solubility should be high, while in others, its medium solubility should be low, and the solubility of the antioxidant and the binder depends on the antioxidant and binder chemical structure, type, use temperature, etc., (iv) stability should be high, which is to maintain the lasting antioxidant capacity and (v) other additives do not chemically react and (vi) do not affect the processing properties and mechanical properties of bonded magnets.

1.3.5.2.4 Plasticizers

The addition of a certain amount of plasticizer is also very important when bonded magnets are being molded. In the magnetic field injection molding, the heat transfer rate of the melt increases due to the mixing of the binder and the magnetic powder, and when the melt is injected into the mold, the magnetic powder is cooled and cured before it is fully oriented, especially when crystalline polymers, such as nylon resins, are used. To slow down the cooling and curing rate to improve the orientation of the powder and the magnet performance, plasticizers are added. The plasticizer makes the melt viscosity of the polymer decrease, so that the magnetic powder can easily rotate when a magnetic field is applied, and the orientation can be improved as a result. Generally, when using nylon resin, it is more effective to add low molecular weight polyphthalamide or phthalamide compound as plasticizer, which can increase $(BH)_{max}$ by 10–20%.

1.4 APPLICATIONS AND RESEARCH TRENDS

1.4.1 High-Speed PMSM

In AM environment, the drive system units used in the processing equipment are needed to have characteristics of high motion precision and small size. Under the condition of the same power, the high-speed PMSM has compact structure and occupies less equipment space, which is easy to optimize its design and produce high-end manufacturing equipment. The permanent magnets adopted in high-speed PMSM should not only have good magnetic properties but also have high thermal stability to prevent irreversible loss of magnetism caused by rotor overheating.

The overheating of the rotor is caused by the eddy current loss of the rotor, the eddy current loss and hysteresis loss of the stator and the copper loss of the stator coil. The eddy current loss of the rotor can be calculated by the following equation:

$$P_{\text{eddy\_}r} = \frac{L}{T}\int_0^T \iint_S \frac{|J_e|^2}{\sigma} \tag{1.8}$$

where S is the area of the conductive area of the rotor, L is the axial length of the rotor core, and T is the time. Induced eddy current density J_e can be expressed in the following way:

$$J_e = -\left(\frac{\partial A}{\partial t} + \nabla\varphi\right), \tag{1.9}$$

where σ is the conductivity of the material and φ is a scalar magnetic potential.

Stator eddy current loss and hysteresis loss are collectively referred to as core loss; the core loss is given by

$$P_{\text{Fe}} = K_a P_{\text{he}} G_{\text{Fe}}, \tag{1.10}$$

where K_a is the empirical coefficient, G_{Fe} is the core mass and P_{he} is the estimated loss coefficient of the iron core and can be estimated by the following equation:

$$P_{\text{he}} = P_{B/f} B^2 \left(\frac{f}{50}\right)^{1.3}, \tag{1.11}$$

where $P_{B/f}$ is a coefficient related to the alternating frequency f of the magnetic field in the magnetic flux density B in the stator core.

The copper loss in the coil is the resistance loss caused by the excitation current in the stator winding. Due to the relatively high frequency of alternating excitation current required by driving high-speed PMSM, its high-order harmonics are more likely to produce skin effect in the conductor, resulting in the increase of copper loss. To avoid the skin effect, the radius of the designed or selected winding wire should be less than skin depth.

Ignoring the skin effect, the equation to calculate the copper loss of the coil can be simplified as follows:

$$P_{cu} = \sum_{i=1}^{m} I_i^2 R_i, \tag{1.12}$$

where m is the number of motor phases, I_i is the effective value of phase current and R_i is the phase resistance.

Without considering bearing loss and wind resistance loss, the total loss of PMSM converted into thermal energy is estimated by

$$P_{total} = P_{eddy\_r} + P_{Fe} + P_{cu}, \tag{1.13}$$

The above total loss is positively correlated with the temperature rise of the PM, and the relationship is as follows:

$$P_{total} \propto (T_1 - T_0) \tag{1.14}$$

where T_0 is the initial temperature of PM when the PMSM is not running and T_1 is the temperature of the PM when the PMSM runs for a few minutes to reach the internal thermal balance in the PMSM.

It should be emphasized that T_1 shall not exceed the Curie temperature of PM materials. Otherwise, PM will cause irreversible demagnetization due to high temperature, which leads to irreversible functional degradation and damage to the whole driving system.

Since the PM is generally installed on the rotor of the high-speed PMSM, the mechanical strength of the PM must meet certain requirements. Moreover, when the running speed of high-speed PMSM reaches 200,000 rpm or higher, the rotor is usually designed as a sandwich structure with the rotor core on the inner side, the PM in the middle and the protective sleeve on the outer side, as shown in Figure 1.12.

The protective sleeve is usually made by stainless steel or carbon fiber with high tensile strength. The radial and tangential stress of the PM is related to the motor speed, assembly tolerance and the respective thermal expansion characteristics of these three materials; it can be expressed in the following equation:

$$\sigma_{r/\theta}(r) = \mathrm{F}_{(r/\theta)}\Omega^2 + \mathrm{H}_{(r/\theta)}\mathrm{S} + +\Gamma_{(r/\theta)}\nabla\mathrm{T} \tag{1.15}$$

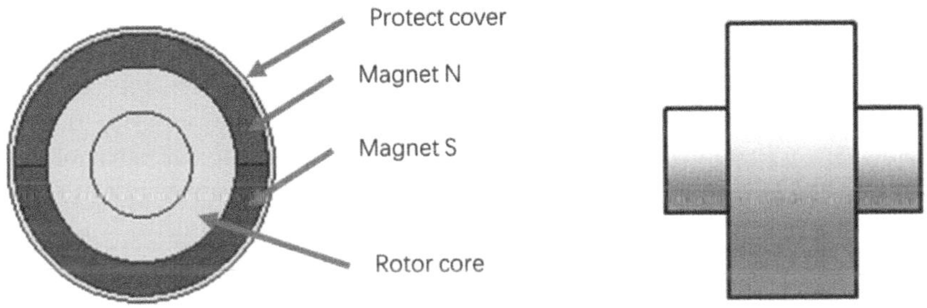

FIGURE 1.12 Rotor structure of high-speed PMSM.

FIGURE 1.13 High-torque PMSM.

where F, F, H and Γ are functions of radius *r*, the external radius and material properties of rotor core, permanent magnet and protection cover.

To summary, the design of a high-speed PMSM and the selection of PM materials should follow the following principles: first, the magnetic flux density in air gap should be strong enough. Second, the selected PM material should ensure the stability of magnetic properties under the specified environmental conditions, working temperature and service conditions. Finally, it should have good machinability and easily be assembled.

1.4.2 High-Torque PMSM

AM includes advanced product manufacturing and advanced production process based on the information and communication technology. Smart manufacturing or Industry 4.0 mainly refers to the latter. In Smart manufacturing environment, industry robot and automated guided vehicle (AGV) are indispensable station operation substitution unit and workshop mobile unit. Different from the above-mentioned high-speed PMSM on the milling machine or grinder used in the AM, the industry robot and AGV can use the low-speed and high-torque direct drive PMSM. To increase the driving torque, the rotor structure usually adopts multi-pole structure, as shown in Figure 1.13.

To get low-speed motion or act with a high torque, as for instance in industrial processes, vehicles or renewable energy generators, commonly a high-speed rotating machine is used together with a mechanical transmission like gearboxes or belts and pulleys. Similarly, a high-speed machine with a ball screw or rack and pinion is used to get linear motion. These transmission elements have the disadvantage of producing losses, being subject to wear and requiring regular maintenance. In addition, they introduce elasticity and non-linear behavior like backlash and stiction, which are all detrimental for motion control.

The alternatives to the mechanical transmission are the direct drives using high-torque machines for rotational motion or linear machines for translational motion. The high-torque, low-speed machines are usually called torque motors, whereas, disregarding how they are called, all of them can operate in motor as well as in generator mode. Several examples of its use as generator can be found in wind, hydraulic or wave power generation.

The electric machines for high torque not only allow for motion control with very high dynamics as required in tooling machines but also result in high reliability and low maintenance as required in wind power generators.

Considering that the size of electric machines is mainly proportional to the torque, and not to the speed or power, the design of such machines is subject to multiple challenges to keep the size as small as possible while maintaining the high efficiency.

Among the usual differences to conventional machines, the torque motor has a higher diameter-to-length ratio, resulting in disk- or ring-shaped machines. In this way, the shear force in the air-gap area acts over a higher radius to produce torque. This is usually combined with an outer rotor construction. As the rotor with permanent magnets is thinner in the radial direction than the stator, placing the rotor outside can increase the radius of the air gap without enlarging the motor.

Also, compared with conventional machines, the torque motors have a higher number of poles and thus reduce the size of the stator and rotor yokes, as well as the windings overhang. This can be achieved with a fractional ratio between pole and slot pitch, which allows non-overlapping concentrated windings. With this kind of windings, the utilization of slot space is the best and the windings overhang the smallest, when compared with the traditional distributed windings used in high-speed motors.

Due to the large radius on the air-gap area, it is not easy to implement small gaps as required for induction or reluctance motors. For this reason, the PMSM is the preferred type for torque motors as it is less sensitive to the air-gap size. This motor type is also very convenient because of its high-torque density when compared with induction or reluctance motors.

To produce high torque, the magnetic flux has to be as high as possible, and this is the most significant limitation. The flux capability can be increased either by extending the cross-section of the magnetic core, which is undesirable as it also raises the motor size, or by increasing the flux density, which is limited by the magnetic material.

Here we have to distinguish between soft magnetic materials required for the stator and rotor core and the hard magnetic materials, i.e., the permanent magnets required for the excitation in the rotor. The soft magnetic material is usually FeSi, where among different variants, the saturation level, i.e., highest flux density, is the parameter to be considered first as mentioned before. The hysteresis and eddy current losses are less important parameters in this case. This is because in torque motors the flux frequency is lower than in conventional motors. Only in special cases, where the weight of the motor is prevalent, as in aeronautics, the CoFe alloy is considered. It saturates at about 30% higher flux density than FeSi but is much more expensive due to the high content of Co.

The preferred choice for the hard magnetic material is Nd-Fe-B in sintered form as it has the highest remanent flux density and provides consequently the highest excitation for the motor. The permanent magnets are usually assembled on the surface of the rotor with glue as in the high-torque low-speed case, the centrifugal forces are not significant. The highest remanent flux density of the permanent magnets is still lower than the saturation level of the used soft magnetic materials. This allows for using flux concentrating magnet configurations like the V-shaped buried magnets. Finally, the eddy current losses in the magnets, which are an important issue in high-speed motors, are not mostly relevant here due to the lower frequency.

Therefore, we can draw the conclusion that the high-torque, low-speed motors are a good alternative to the high-speed counterpart with gearboxes or similar transmission. Nevertheless, all possible measures have to be taken to keep its size (volume and mass) competitive.

1.5 SUMMARY AND PERSPECTIVE

In summary, this chapter presented a comprehensive overview on the various types of hard magnetic materials and various process techniques and highlighted two prominent applications associated with one type of hard magnetic materials and rare earth PM material, in AM environment. Till date, the potentials of rare earth PM material have shown with capabilities to increasing the intrinsic coercivity and maximum magnetic energy product as compared to conventional ferrite materials. These potentials are clearly illustrated for applications such as PMSM, nuclear magnetic resonance equipment, magnetic separation equipment, magnetic machinery and magnetic therapy instruments. With the continual research efforts being put into the developments of new hard magnetic materials and novel process technological advances in topological designs, it should be optimistic that novel hard magnetic materials, such as new type of rare earth PM materials, will be the prospective advanced electrical energy conversion materials of the future.

REFERENCES

1. Zhongwu L. *Basic Principle and Advanced Technology of Permanent Magnet Materials*. Guangdong: South China University of Technology Press, 2017.
2. Fang CM, Kools F, Metselaar R, et al. Magnets and electronic properties of strontium hexaferrite $SrFe_{12}O_{19}$ from first-principles calculations. *Journal of Physics: Condensed Matter*, 2003, 15(36): 6229–6239.

3. Shouzeng, Z, Qingfei, D, Xuexu G. *Sintered Nd-Fe-B Rare Earth Permanent Magnet Materials and Technology*. Beijing: Metallurgical Industry Press, 2011.
4. Knoch KG. Phase relations. In: *Rare-Earth Iron Permanent Magnets* (Ed. Coey JMD). Oxford: Clarendon Press, 1996, pp. 159–177.
5. Boping H. *Rare Earth Permanent Magnet Materials Volume I*. Beijing: Metallurgical Industry Press, 2017.
6. Kim TH, Lee SR, Kim DH, et al. Microstructural evolution of triple junction and grain boundary phases of a Nd-Fe-B sintered magnet by post-sintering annealing. *Journal of Applied Physics*, 2011, 109(7): 07A703.
7. Jin WK, Sun YS, Kim YD. Effect of cyclic sintering process for NdFeB magnet on microstructure and magnetic properties. *Journal of Alloys and Compounds*, 2012, 540(11): 141–144.
8. Sepehri-Amin H, Ohkubo T, Shima T, et al. Grain boundary and interface chemistry of an Nd-Fe-B-based sintered magnet. *Acta Materialia*, 2012, 60(3): 819–830.
9. Liu Q, Xu F, Wang J, et al. An investigation of the microstructure in the grain boundary region of Nd-Fe-B sintered magnet during post-sintering annealing. *Scripta Materialia*, 2013, 68(9): 687–690.
10. Mo W, Zhang L, Liu Q, et al. Dependence of the crystal structure of the Nd-rich phase on oxygen content in an Nd-Fe-B sintered magnet. *Scripta Materialia*, 2008, 59(2): 179–182.
11. Li WF, Ohkubo T, Hono K. Effect of post-sinter annealing on the coercivity and microstructure of Nd-Fe-B permanent magnets. *Acta Materialia*, 2009, 57(5): 1337–1346.
12. Vial F, Joly F, Nevalainen E, et al, Improvement of coercivity of sintered NdFeB permanent magnets by heat treatment. *Journal of Magnetism and Magnetic Materials*, 2002, 245(4): 1329–1334.
13. Park DW, Kim TH, Lee SR, et al. Effect of annealing on microstructural changes of Nd-rich phases and magnetic properties of Nd-Fe-B sintered magnet. *Journal of Applied Physics*, 2010, 107(9): 09A737.
14. Fukagawa T, Hirosawa S, Ohkubo T, et al. The effect of oxygen on the surface coercivity of Nd-coated Nd-Fe-B sintered magnets. *Journal of Applied Physics*, 2009, 105(7): 07A724.
15. Sepehri-Amin H, Ohkubo T, Nagashima S, et al. High-coercivity ultrafine-grained anisotropic Nd-Fe-B magnets processed by hot deformation and the Nd-Cu grain boundary diffusion process. *Acta Materialia*, 2013, 61(17): 6622–6634.
16. Park KT, Hiraga K, Sagawa M. Effect of metal-coating and consecutive heat treatment on coercivity of thin Nd-Fe-B sintered magnets. *The 16th International Workshop on Rare Earth Permanent Magnet and Their Applications*, Sendai, Japan, 2000, pp. 257–264.
17. Hirota K, Nakamura H, Minowa T, et al. Coercivity enhancement by the grain boundary diffusion process to Nd-Fe-B sintered magnets. *IEEE Transactions on Magnetics*, 2006, 42: 2909.
18. Qiong L, Lan Z, Fang X, et al. Dysprosium nitride-modified sintered Nd-Fe-B magnets with increased coercivity and resistivity. *Japanese Journal of Applied Physics*, 2010, 49: 093001.
19. Liu X. Wang X, Liang L, et al. Rapid coercivity increment of Nd-Fe-B sintered magnets by $Dy_{69}Ni_{31}$ grain boundary restructuring. *Journal of Magnetism and Magnetic Materials*, 2014, 370: 76.
20. Suzuki H, Satsu Y, Komuro M, et al. Magnetic properties of a Nd-Fe-B sintered magnet with Dy segregation. *Journal of Applied Physics*, 2009, 105(7): 07A734.

21. Komuro M, Satsu Y, Suzuki H. Increase of coercivity and composition distribution in fluoride-diffused NdFeB sintered magnets treated by fluoride solutions. *IEEE Transactions on Magnetics*, 2010, 46(11): 3831–3833.
22. Sepehri-Amin H, Ohkubo T, Nishiuchi T, et al. Coercivity enhancement of HDDR processed Nd-Fe-B powders by the diffusion of Nd-Cu eutectic alloys. *Scripta Materialia*, 2010, 63(11): 1124–1127.
23. Sepehri-Amin H, Ohkubo T, Nagashima S, et al. High-coercivity ultrafine-grained anisotropic Nd-Fe-B magnets processed by hot deformation and the Nd-Cu grain boundary diffusion process. *Acta Materialia*, 2013, 61(17): 6622–6634.
24. Wan F, Zhang Y, Han T, et al. Coercivity enhancement in Dy-free Nd-Fe-B sintered magnets by using Pr-Cu alloy. *Journal of Applied Physics*, 2014, 115: 203910.
25. Fidler J, Schrefl T. Overview of NdFeB magnets and coercivity. *Journal of Applied Physics*, 1996, 79: 5029.
26. Kim TH, Lee SR, Lee MW, et al. Dependence of magnetic, phase-transformation and microstructural characteristics on the Cu content of NdFeB sintered magnet. *Acta Materialia*, 2014, 66: 12.
27. Chen Z, Yan A, Wang X. Effects of intergranular additions of oxides on the coercivity, thermal stability and microstructure of NdFeB magnets. *Journal of Magnetism and Magnetic Materials*, 1996, 162: 307.
28. Croat JJ, Herbst JF, Lee RW, et al. Pr-Fe and Nd-Fe-based materials: A new class of high- performance permanent magnets(invited). *Journal of Applied Physics*, 1984, 55: 2078–2082.
29. Croat JJ. Current status and future outlook for bonded neodymium permanent magnets (invited). *Journal of Applied Physics*, 1997, 81(8): 4804–4809.
30. Shuxin B. *Manufacturing Principle and Technology of Bonded Nd-Fe-B Permanent Magnet Material.* Beijing: Science Press, 2014.
31. Croat JJ, Herbst JF. Lee RW, et al. High-energy product Nd-Fe-B permanent magnets. *Applied Physics Letters*, 1984, 44(1): 148–150.
32. Coehoorn R, de Mooij DB, Duchateau JPWB. Novel PM materials by rapid quenching. *Journal of Physics D*, 1988, 49(S8): 669–670.
33. Kneller EF, Hawig R. The exchange-spring magnet: A new materials principle permanent magnets. *IEEE Transactions on Magnetics*, 1991, 27(4): 3588–3590.
34. Skomski R, Coey JMD. Giant energy product in nanostructured two phase magnets. *Physical Review B*, 1993, 48: 15812–15816.
35. Manaf A, Buckley RA, Davies HA. New nanocrystalline high-remanence NdFeB by rapid solidification. *Journal of Magnetism and Magnetic Materials*, 1993, 128: 302–306.
36. Kronmüller H, Fischer R, Zerm A. Micromagnetism and microstructure of hard magnetic materials. *Journal of Physics D: Applied Physics*, 1996, 29: 2274–2283.
37. Zuocheng W. Nanocomposite $Pr_2Fe_{14}B/\alpha$-preparation, microstructure and magnetic properties of Fe permanent magnet alloy and magnet. Doctoral Dissertation. Beijing: Beijing University of Science and Technology, 1999.
38. Wang ZC, Zhang MC, Li FB. High coercivity $(NdDy)_2$ $(FeNb)_{14}$ B/α-Fe nanocrystalline Al-loys. *Journal of Applied Physics*, 1997, 81(8): 5097–5099.
39. Sinnema S, Radwauski RT. Magnetic properties of ternary rare-earth compounds of the type $R_2Fe_{14}B$. *Journal of Magnetism and Magnetic Materials*, 1984, 44: 333–341.
40. Inoue A, Takeuchi A. Hard magnetic properties of nanocrystalline Fe-rich Fe-Nd-B alloys prepared by partial crytallisation of amorphous phase. *Materials Transactions JIM*, 1995, 36: 962–971.

41. Manaf A, Buckley RA, Davies HA. Microstructure analysis of nanocrystalline Fe-Nd-B ribbons with enhanced hard magnetic properties. *Journal of Magnetism and Magnetic Materials*, 1993, 128: 307–312.
42. Hadjipanayis GC, Withanawasam L. Nanocomposite $R_2Fe_{14}B/\alpha$-Fe permanent magnets. *IEEE Transactions on Magnetics*, 1995, 33: 3596–3598.
43. Kramer J, Li CP, Dennis KW. Effect of TiC additions to the microstructure and magnetic properties of $Nd_{9.5}Fe_{84.5}B_6$ melt-spun ribbons. *Journal of Applied Physics*, 1998, 83: 6631–6633.
44. Shindo M, Ishizone M, Kato H. Exchange-spring behavior in sputter-deposited α-Fe/NdFeB multilayer magnets. *Journal of Magnetism and Magnetic Materials*, 1996, 161: L1–L5.
45. Shen BG, Yang LY, Zhong JX. Magnetic properties of rapidly quenched NdFeB alloys with lower Nd concentration. *Solid State Communications*, 1990,74(9): 893–897.
46. Folks L, Street R, Woodward RC, et al. Magnetic properties of novel resin-bonded exchange coupled rare-earth magnets, *Journal of Magnetism and Magnetic Materials*, 1995, 147: 360–366.
47. Campbell P, Brown DN, Chen ZM. $R_2Fe_{14}B$-type isotropic powders for bonded magnets. *18th International Workshop on High Performance Magnets and their Application*, Annecy, France, 2004.
48. David NB, Chen ZM, Peter CG, et al. Developments in melt spun poewders for permanent magnets. *Proceedings of 19th International Workshop on Rare Earth Permanent Magnets and Their Applications*, Beijing, China, 2006.
49. Yidong S. *Ferrite Process.* Beijing: Electronic Industry Press, 1984.
50. Ling G, Meiling Z, Meixing Z. Development and application of spark plasma sintering technology. *Powder Metallurgy Technology*, 2002, 3: 128–133.
51. Shouzeng, Z, Qingfei D. *Super Permanent Magnet Rare Earth Iron Series Permanent Magnet Material.* Beijing: Metallurgical Industry Press, 2004.
52. Minbo T. *Magnetic Materials* Beijing: Tsinghua University Press, 2001.
53. Zengfeng L, Ping T, Hanliang Z, et al. Development of bonded rare earth permanent magnets *Rare Metals Letters*, 2005, 24(6): 14–18.
54. Hadjipanyis GC, Tang W, Zhang Y, et al. High temperature 2:17 magnets: Relationship of magnetic properties to microstructure and processing. *IEEE Transactions on Magnetics*, 2000, 36: 3382–3384.
55. Kuhrt C, O'Donnell K, Katter M. Pressure-assisted zinc bonding of microcrystalhe $Sm_2Fe_{17}N_x$ powders. *Applied Physics Letters*, 1992, 60(26): 3316–3318.

2 Advanced Manufacturing of Flexible Piezoelectric Arrays

Han Zhang
Chinese Academy of Sciences

CONTENTS

DOI: 10.1201/9781003182146-2

2.1 INTRODUCTION

From mobile phones and computers to cars, planes and so on, our life and production are inseparable from electronic devices. Traditional electronic devices are based on silicon and other inorganic semiconductor materials, which are hard and brittle (failure strain is about 1%) and not easy to deform. Therefore, traditional electronic devices lack certain flexibility and ductility. Different from traditional electronic devices, flexible electronic devices based on flexible materials or malleable structures have excellent flexibility and ductility. Therefore, flexible electronic devices have become a hot and emerging research direction and show broad application prospects in many fields.

Extensible structures as an important way to achieve bendable rigidity and ductility in flexible electronic devices do not require special flexible materials. Based on traditional inorganic semiconductor materials such as silicon, a large deformation of more than 10% of the entire electronic device can be achieved through clever structural design, and it can be ensured that traditional inorganic semiconductor materials are not destroyed and electrical properties are not affected. Since John Rogers and Yonggang Huang [1] from Northwestern University put forward the concept of malleable structure, the malleable structure has developed rapidly. So far, mechanical structures of extendable structures can be divided into flexion and non-flexion according to deformation modes and corrugated structure, island–bridge structure and paper-cut structure according to shape [2]. Among them, island–bridge structure is the most widely used structure. In addition, to realize the electrical connection of flexible electrons and improve the mechanical properties of the structure, island–bridge structure generally adopts multilayer composite strip structure.

As an important way for flexible electrons to achieve flexibility and extensibility, malleable structures involve complex mechanical deformation. Therefore, to achieve accurate and quantitative deformation of flexible electrons, it is of great significance to the mechanical research of malleable structures.

Piezoelectric materials mainly include piezoelectric single crystal, piezoelectric ceramics, piezoelectric composite materials, piezoelectric semiconductor, piezoelectric polymer, etc. As a kind of high-voltage, dielectric, pyroelectric, ferroelectric piezoelectric material, piezoelectric ceramics have been widely used in the development of ultrasonic transducer, underwater transducer, igniter and other acoustoelectric or electro-acoustic devices. 1–3, 2–2 and 3–3 piezoelectric composites are three kinds of piezoelectric composites which are widely used at present. Type 1–3 piezoelectric composite material is a two-phase piezoelectric composite material formed by the parallel arrangement of one-dimensional connected piezoelectric ceramics in three-dimensional connected polymer. Type 2–2 piezoelectric composite material is a two-phase composite material composed of piezoelectric ceramic phase and polymer spaced to form a lamination, in which the piezoelectric ceramic phase and polymer extend in their respective planes. Type 3–3 piezoelectric composite is a two-phase composite composed of piezoelectric ceramic phase and polymer interwoven in three-dimensional space.

At present, the preparation technologies of piezoelectric composites mainly include cutting and filling method [3,4], demolding method [5], injection molding method [6], laser cutting method [7], and flow stretch-stacking method [8].

Around the advanced manufacturing of flexible piezoelectric arrays, this chapter will be developed around the following three aspects to the work.

Section 2.1 mainly introduces the research of ultrasonic phased array transducers now.

Section 2.2 introduces flexible piezoelectric arrays, mainly analyzing the types of flexible piezoelectric arrays and the characteristics of each type of piezoelectric array.

Section 2.3 introduces the design and manufacturing process of flexible piezoelectric arrays, mainly from the design of mechanical properties, electrical properties, the design and manufacturing of scalable coils and the design of electrodes, i.e., the process.

2.2 FLEXIBLE PIEZOELECTRIC ARRAYS

2.2.1 Types of Flexible Piezoelectric Arrays

Integration of mechanical flexibility into energy-harvesting materials presents an emerging development trajectory toward a wide range of applications [9], such as sensory skins for robotics, artificial muscles, self-powered sensors and wearable devices [10–12]. Piezoelectric materials are particularly attractive because they offer an efficient route to scavenge energies from the living environment, such as human activities, to power mobile consumer electronics and nanosystems [13–15]. In addition to piezoelectricity, mechanical flexibility and cost-effective scalability are key matrices for these energy-harvesting materials [16–18]. However, mechanical flexibility and piezoelectricity appear to be two conflicting properties, as improving one often compromises the other [10]. Ferroelectric polymers, such as poly(vinylidene fluoride) and its copolymers, are mechanically flexible and easily processable [19–23], but their piezoelectric coefficients are over one order of magnitude less than their ceramic counterparts [24]. Piezoceramics and single crystals with high piezoelectric coefficients, on the other hand, are very fragile, prone to fracture upon mechanical deformation and difficult to be processed into mechanically compliant structures [25]. To improve mechanical flexibility, high-density nanowire arrays [26–29] and thin films [30–35] of piezoelectric ceramics and piezoelectric semiconductor oxides [36–38] have been extensively explored. However, these one- and two-dimensional (1-D and 2-D) nanostructures are cost-ineffective for scalable production. Their low dimensions also limit the loading directions with considerable piezoelectric effects.

An intuitive yet logical approach toward flexible piezoelectric materials is the development of ceramic-polymer composites with piezoelectric ceramic nanostructures dispersed in a polymer matrix [39–45]. Such composites are expected to combine excellent piezoelectricity of electroceramics with mechanical flexibility of polymers. Various competitive attributes and parameters, including the volume fraction and the morphology of the piezoelectric ceramics, dictate the combined mechano-piezoelectric responses of the composites. For example, increasing the piezoelectric ceramic fraction improves the piezoelectric effect [46], but necessarily compromises mechanical flexibility of the resulting composites. The currently available piezoelectric-polymer nanocomposites are commonly based on the low-dimensional (0-D nanoparticles and 1D nanowires) ceramic fillers [12,46,47]. However, the large stiffness disparity of the material constituents and the spatial discontinuity of the ceramic phase in these composites render fairly poor load transfer from the surrounding polymer matrix to the active ceramic fillers, thus significantly limiting their piezoelectricity [46,48–52].

Flexible piezoelectric materials are the key to wearable electronic devices and medical sensing devices. Current fine ceramic-polymer hybrid materials embedded with low-dimensional ceramic fillers have low piezoelectric coefficients due to very poor load-transfer efficiency, which is proportional to the stiffness ratio of the polymer substrate and the ceramic filler, despite their mechanical flexibility. Therefore, the following three flexible piezoelectric types are presented in this section: flexible film piezoelectric element materials, island–bridge structures for flexible electronics and scalable piezoelectric arrays.

2.2.2 Piezoelectric Component Materials of Soft Films

The most important materials used in flexible film piezoelectric components are poly(vinylidene fluoride) and its copolymers. Poly(vinylidene fluoride), abbreviated as PVDF, is a semi-crystalline polymer compound. PVDF is usually made by polymerizing vinylidene fluoride with the structural formula $-(CF_2CH_2)n-$. PVDF is an excellent polymer piezoelectric PVDF piezoelectric film that was introduced in Japan in the 1970s and has been widely used in various fields of national economy, and it will have a very broad market prospect in the future [53]. Therefore, it is of great importance to conduct research on high-performance composite piezoelectric films. In this section, PVDF and its copolymers are introduced, and the existing ways to prepare PVDF piezoelectric films are carefully summarized [54].

In the polymerization process of PVDF, when the monomer molecules expand into molecular chains, the carbon atoms form the main carbon chain, while the hydrogen and fluorine atoms can rotate around the chemical bonds to form different spatial configurations. In PVDF molecules, there are two common conformations: adjacent substitutes approximately 180° to each other, called trans conformation (TTTT), and adjacent substitutes approximately 60° to each other, called torsional conformation (TGTG) [55].

There are at least five crystal structures in PVDF, and the common ones are α, β, δ, γ and ε crystalline phases. As shown in Figure 2.1, α crystal phase is a monoclinic crystal system with a molecular chain configuration of TGTG′; β crystal phase belongs to the orthorhombic crystal system with a molecular chain configuration of TTT; and γ crystal phase is a monoclinic crystal system with a molecular chain structure similar to that of α crystal phase [56].

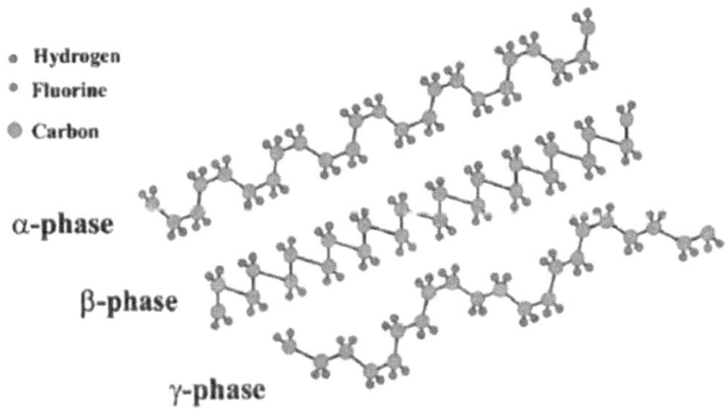

FIGURE 2.1 Molecular chain structure of α, β and γ crystalline phases [57].

Among all crystalline phases, α crystalline phase is a nonpolar molecule and does not possess piezoelectricity. β crystalline phase is a polar molecule with better piezoelectricity and is the focus of research and preparation of high-performance piezoelectric materials [58].

PVDF has a variety of copolymers. Among them, polyvinylidene fluoride-hexafluoropropylene and polyvinylidene fluoride-trifluoroethylene are two copolymers that are used more in piezoelectric materials.

1. PVDF-HFP can be obtained from vinylidene fluoride (VDF) and hexafluoropropylene (HFP) by emulsion polymerization. The reaction equation is as follows [53].

   Compared with PVDF, PVDF-HFP has lower crystallinity, but PVDF-HFP has higher F content and it has better hydrophobicity than PVDF. PVDF-HFP also has better flexibility and solidification properties than PVDF, which makes it easy to prepare and use. In terms of piezoelectric properties, PVDF-HFP can reach a piezoelectric coefficient d_{31} of 30 pC/N when the mixing ratio of HFP is 5%, which is higher than that of PVDF [59].
2. PVDF-TrFE. PVDF-TrFE is produced by the polymerization of VDF and trifluoroethylene (TrFE) with the reaction equation [53].

PVDF-TrFE is an excellent performing piezoelectric, ferroelectric polymer that exhibits piezoelectric ferroelectric effects without stretching. The β phase can be obtained when the PVDF content in PVDF-TrFE is between 50% and 80% due to

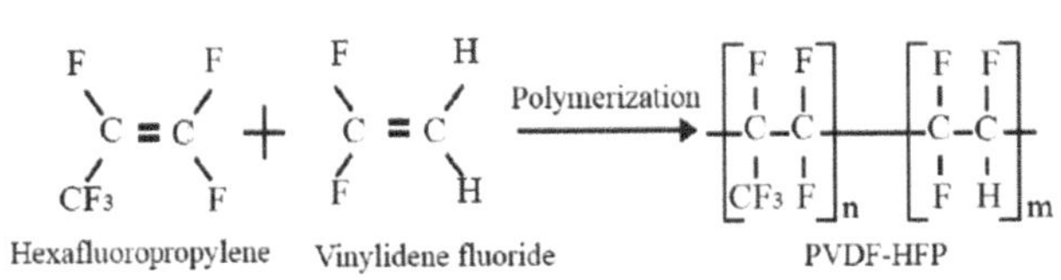

FIGURE 2.2 Polymerization equation of PVDF-HFP.

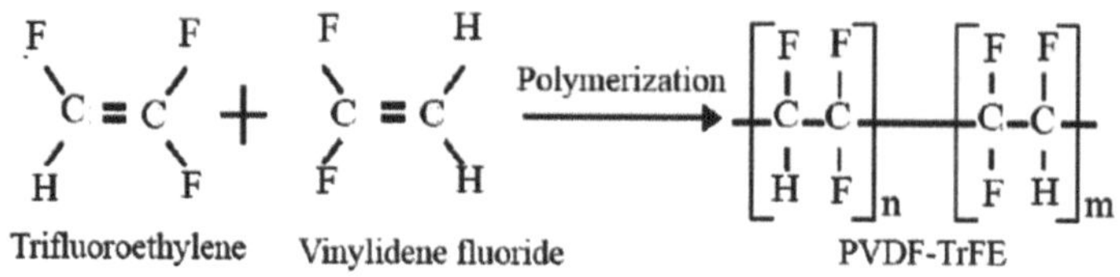

FIGURE 2.3 Polymerization reaction equation of PVDF-TrFE.

the TrFE spatial site resistance in the main chain that prevents the formation of the parafluoroethylene chain segment and promotes the formation of the β phase in an all-trans conformation [59]. Compared to PVDF, PVDF-TrFE contains only one crystalline structure of β phase at well below Curie temperature and the well-defined Curie temperature gives it better piezoelectric ferroelectric properties with higher electromechanical coupling coefficient [60]. The crystallinity of PVDF-TrFE is usually distributed between 30% and 80%, and its crystallinity is mainly affected by the TrFE content with a maximum crystallinity of up to 90% [61] (Figure 2.3).

In PVDF materials usually prepared by solution cooling or solution casting method, the crystalline phase composition is mainly α crystalline phase, because the molecular chain of α crystalline phase is nonpolar and does not have piezoelectricity, so it is necessary to use certain processing means to transform the α crystalline phase to β crystalline phase with excellent piezoelectricity.

The methods commonly used to increase the β crystalline phase content are uniaxial stretching, high-pressure crystallization, heat treatment and addition of nanoparticles. Figure 2.4 shows the transition conditions between various crystalline phases of PVDF [62].

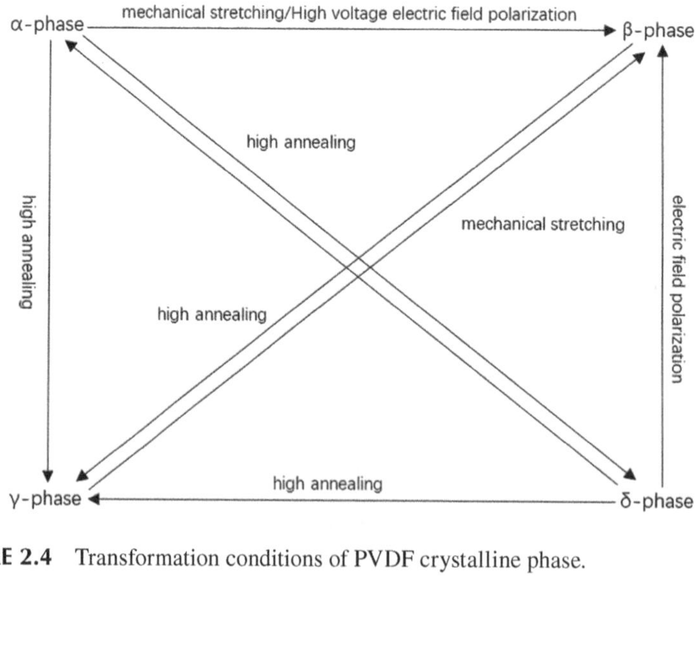

FIGURE 2.4 Transformation conditions of PVDF crystalline phase.

Uniaxial stretching is the most commonly used method to transform the α crystalline phase to the β crystalline phase. During axial stretching, the stress acting on the PVDF film causes the molecular chain of the substrate to rotate in the direction perpendicular to the axial force, transforming the TGTG′ conformation into the TTTT all-trans conformation. It has been shown that the parameters of axial stretching, such as elongation of stretching, stretching speed and stretching temperature, affect the transformation rate of the β crystalline phase [63].

The effect of stretching conditions on the morphology and crystalline phase structure of PVDF was investigated by Sencadas et al. It was shown that during the stretching process, the molecular chains of the polymers rotate inside the crystals, resulting in an all-trans structure. The highest β phase content can be obtained when the stretching temperature is 80°C and the elongation is 5. Also, the higher the stretching rate, the higher the orientation of the polymer. Increasing the stretching temperature also increases the orientation of the molecular chains but at the same time decreases the efficiency of the β crystalline phase transition. However, the transition from α to β crystalline phases is accompanied by a decrease in crystallinity, which decreases by 10% at the highest β crystalline phase content [64].

Salimi and Yousefi found that the highest β crystalline phase content was obtained at a stretching temperature of 90°C and a stretching elongation of 4.5–5.0. The experimental results showed that the effect of elongation on the β crystalline phase content was higher than that of the stretching temperature. It was concluded that the main reason for the increase in β crystalline phase content could be the orientation of molecular chains in the crystalline region and the movement of molecular chains in the amorphous region that occurred during the stretching process [65].

Gomes et al. investigated the effect of elongation from 1 to 5 and stretching temperature from 80°C to 140°C on the ferroelectric and piezoelectric properties of PVDF films. It was found that elongation and stretching temperature have a very important effect on the transition from α to β crystalline phases [66].

It was also found that increasing the temperature during stretching leads to a decrease in the polarization of the β crystalline phase, even though the temperature at this point is lower than the melting point and Curie temperature of PVDF. The main reason for this is that the grains of the β crystalline phase obtained by stretching are very small, along with the presence of structural defects [67].

High-pressure crystallization is also an effective method to obtain the high content of β crystalline phase. It has been shown that β crystalline phase can be obtained from PVDF in the molten state by crystallization at pressure higher than 4000 kg/cm [68]. Scheinbeim et al. investigated the effect of high-pressure quenching on the β crystalline phase content of PVDF. The experimental results showed that the content of β crystalline phase increased from 0% to 85% when the quenching pressure was increased from 200 to 700 MPa. When the initial quenching pressure was 100 MPa and the quenching pressure at the end was 800 MPa, the β crystalline phase content in the specimens could reach 100% [69].

Hattori et al. investigated the effect of high pressure on the piezoelectric and ferroelectric properties of PVDF and showed that piezoelectric films containing β crystalline phase could be obtained by high-temperature and high-pressure crystallization. The experimental results showed that the optimal crystallization conditions were an initial melting pressure of 100–200 MPa, followed by an increase in crystallization pressure to 500–600 MPa and a temperature maintained at 24°C–300°C. The study also showed that although high-pressure crystallization could improve the piezoelectric properties of PVDF films, the electromechanical coupling coefficient of the films remained low, which was mainly caused by a decrease in the thermal stability of the films [70].

Satish et al. investigated the effect of hot pressing on the properties of PZT-PVDF composite films. It was shown that the polarization process of the composite films treated with high pressure was more effective and thus had a higher piezoelectric coefficient. It was also shown that the dielectric coefficient of the composite films decreased with increasing pressure, mainly because the increase in pressure increased the effect of polarization and brought the number of dipole moments close to the saturation value. On the other hand, the dielectric constant of the composite films increases with the increase in temperature during hot pressing, mainly because the increase in temperature enhances the mobility of the polymer molecular chains, thus improving the effect of polarization orientation [71].

Adding nanoparticles to PVDF to prepare composite films can likewise increase the β crystalline phase content in PVDF films, thus obtaining composite films with excellent properties. The main principles of adding nanoparticles to enhance the piezoelectric properties of PVDF films are as follows: (i) using the piezoelectric effect of the particles themselves, such as adding $BaTiO_3$ particles with piezoelectric effect; (ii) using the nucleation effect of the added particles to reduce the size of the grains in the initial crystallization stage; (iii) in the axial stretching stage, the stress concentration between the interface of the particles and the matrix, promoting the stretching of molecular chains and thus increasing the β crystalline phase; and (iv) formation of microcapacitance in PVDF film by adding particles to improve the orientation effect of dipole moment during film polarization. The commonly used nanoparticles are $BaTiO_3$, graphene, carbon nanotube (CNT), carbon black (CB), etc.

Siddiqui et al. added $BaTiO_3$ particles to PVDF-TrFE matrix using spin-coating method. The experimental results showed that the output voltage and current of the piezoelectric films increased with the increase of $BaTiO_3$ particle content, and the output voltage of the piezoelectric films reached 9.8 V, the output current was 0.69 μA and the output power density reached 13.5 μW/cm under cyclic bending conditions when the $BaTiO_3$ particle content reached 40 wt.%. It was concluded that the $BaTiO_3$/PVDF-TrFE piezoelectric film performance enhancement is mainly attributed to the high crystallinity of PVDF-TrFE, while the nature of $BaTiO_3$ itself enhances the piezoelectric properties of the film [72].

Kim et al. investigated the effect of adding multiwalled carbon nanotubes (MWCNT) on the crystal structure of PVDF films. It was found that for the stretch-treated films, the addition of MWCN could increase the content of β crystalline phase, mainly because MWCNT increased the crystallization rate of the films, which was beneficial to the formation of β crystalline phase. When the MWCNT content was lower than 0.2 wt.%, polarization of the stretched films also increased the β crystalline phase content, up to nearly 100% of the β crystalline phase. However, when the MWCNT content is further increased, it leads to a decrease in the polarization effect and hinders the formation of the β crystalline phase [58].

Wu et al. prepared PVDF-HFP/CB composite films by adding CB to PVDF-HFP using the solution casting method. At the CB content of 0.5 wt.%, the open-loop voltage and AC recovery power were simultaneously maximized, 104% and 364% higher than those of the pure film, respectively. It was found that in the initial crystallization stage, the CB particles mainly played the role of nucleation, and the matrix was pure in a large number of arbitrarily distributed spherical and sheet crystals. In contrast, during the stretching stage, the molecular chains of the films with the addition of CB particles were more fully stretched and the degree of orientation of the dipole moment was better during polarization [73].

Wu et al. investigated the effect of CNT on the properties of PVDF/CNT films using electrostatic weaving, focusing on the changes in their crystal structure and piezoelectric properties. It was found that when the content of CNT was 1.0 wt.%, the content of β crystalline phase increased from 79% to 89% and the piezoelectric coefficient d_{33} increased from 16.8 to 31.3 pC/N [74].

2.2.3 Island–Bridge Structures of Flexible Electronics

Island–bridge structures [75] are a typical structure for flexible electronic devices, which can integrate both functional components and achieve stretchable deformation capability. As the most widely used structure, island–bridge structures can be further divided into straight interconnected island–bridge structures, curved interconnected island–bridge structures and fractal interconnected island–bridge structures, and the following section will focus on the principles, characteristics and applications of these three island–bridge structures.

2.2.3.1 Straight Bridge Connected Island–Bridge Structure

Ko et al. [76] transferred a pre-made planar island–bridge structure (functional electronic components, such as semiconductor p–n junctions on top of a rigid "island" and "island" to "island" circuit connections) to a hemispherical PDMS substrate stretched radially into a plane and then released the pre-strain on the PDMS substrate. The straight "bridge" between the "island" and "island" is connected to the hemispherical PDMS substrate stretched in the radial direction, and then the pre-strain of the PDMS substrate is released and the straight bridge is flexed under the compression of the PDMS (Figure 2.5a). On the one hand, the functional electronics on top of the "island" are protected by the "island", and on the other hand, the overall structure can be stretched in any direction (the flexed "bridge" is gradually flattened during the stretching process).

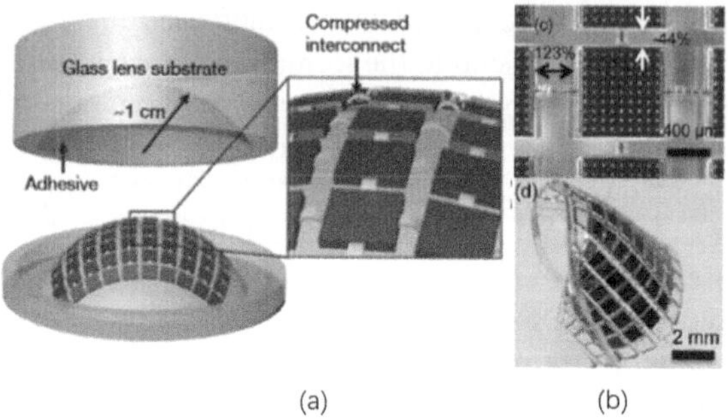

FIGURE 2.5 Straight bridge interconnect island–bridge structure [76,77]. (a) Flexible curved electronics; (b) Stretchable solar cells.

Lee et al. [77] prepared a stretchable GaAs solar cell using an island–bridge structure (Figure 2.5b), and in addition, this work treated the multilayer strip structure as a multilayer composite beam and obtained the effective bending stiffness (per unit width) of the multilayer composite strip structure as:

$$\overline{EI} = \sum_{i=1}^{3} \overline{E}_i h_i \left[\left(\sum_{j=1}^{i} h_i \right)^2 - \left(\sum_{j=1}^{i} h_i \right) h_i + \frac{h_i^2}{3} \right] - \frac{\left[\sum_{i=1}^{3} \overline{E}_i h_i \left(\sum_{j=1}^{i} h_j - \frac{h_i}{2} \right) \right]}{\sum_{i=1}^{3} \overline{E}_i h_i} \tag{2.1}$$

where h_i is the thickness of the ith layer (from the top to the bottom) and $\overline{E_i} = E_i / \left(1 - v_i^2\right)$ is plane strain modulus of elasticity (E_i and v_i are the modulus of elasticity and Poisson's ratio of the ith layer, respectively).

2.2.3.2 Curved Bridge Interconnected Island–Bridge Structure

To further improve the tensile strain of flexible electronic devices and to eliminate pre-stretching as an operation, researchers have improved the straight bridge for island-to-island interconnection to a curved bridge. Kim et al. [78] used a curved bridge to achieve a mechanical and electrical connection between islands and prepared a high-performance CMOS circuit that could achieve a tensile strain of 120% and a small strain in the curved bridge material (lower than the material breaking strain) of a high-performance CMOS circuit (Figure 2.6a). This high tensile rate CMOS circuit can be laminated to any bendable/stretchable/twistable object surface, such as fabric, leather and paper. On the other hand, this curved bridge interconnect island–bridge structure is inherently stretchable and compressible, so the substrate does not require pre-stretching and releasing pre-strain to achieve the stretchability of the bridge, which makes the processing easier. Hu et al. [80] proposed a process to connect commercial electronics with stretchable serpentine wires using a flip-chip bonding process to fabricate stretchable electronics. A stretchable light-emitting diode (LED) array containing 50 LEDs was

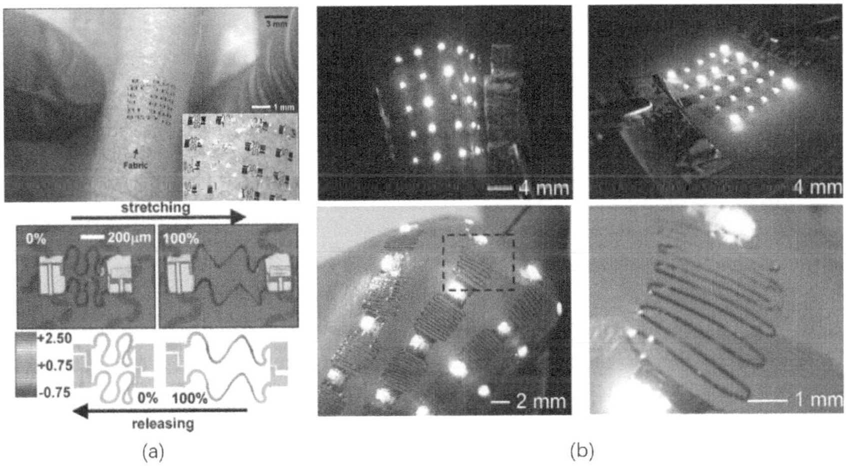

FIGURE 2.6 Curved bridge interconnect island–bridge structure [78,79]. (a) High stretch CMOS circuits; (b) Stretchable LED arrays.

fabricated (Figure 2.6c). After 90,000 cycles of reciprocal stretching of the array (100% stretching strain), all LEDs still functioned properly, demonstrating the reliability of the curved bridge interconnect structure.

2.2.3.3 Fractal Interconnected Island–Bridge Structure

To further improve the ductility of the island–bridge structure, Xu et al. [81] prepared a lithium battery with 300% stretchability using silicone elastomer as the base material and a self-similar interconnection structure (Figure 2.7a). Based on the interconnected island–bridge structure of the curved bridge, the interconnected curved bridge is designed as a fractal structure, the graded structure of the first level is still a curved bridge and the second level and above is obtained according to the self-similarity theory (Figure 2.7b). According to the finite element simulations and experiments, the self-similar structure will unfold in an orderly manner from the largest to the smallest level when stretching the graded structure. Furthermore, Zhang et al. [80] proposed a hierarchical computational model (HCM) for post-buckling analysis of graded structures based on the mechanism of graded structure unfolding by bases. Compared with the traditional finite element calculation method, this computational model can greatly reduce the computational volume, and at the same time, it can ensure sufficient computational accuracy. According to this computational model (HCM), the ductility of the graded conductor increases rapidly with the number of stages, starting from stage 2, and the ductility of the graded conductor increases by a factor of 3 for each increase of 1 in the number of stages (Figure 2.7c). The system was cyclically stretched at 30% tensile strain and 0.6 Hz for 6000 times without structural and functional damage, demonstrating the reliability of the fractal interconnected island–bridge structure.

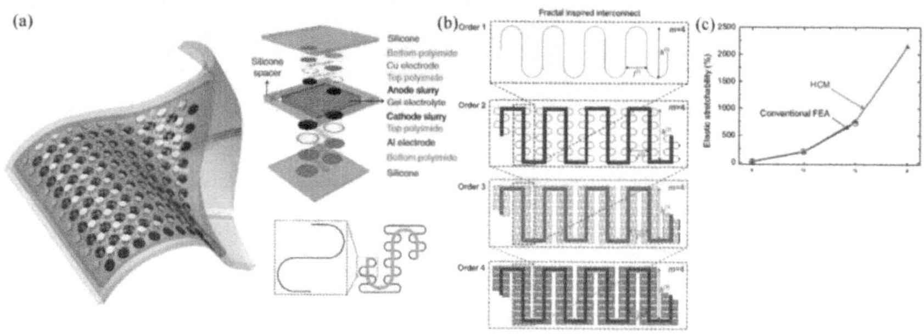

FIGURE 2.7 Fractal interconnect island–bridge structure [80–82]. (a) Stretchable lithium battery; (b) Fractal wire; (c) Relationship between fractal grade and ductility of conductors.

2.2.4 Stretchable Piezoelectric Arrays

Detection methods for object interface and internal defects in the existing technology mainly include ultrasonic detection method, eddy current detection method, radiographic detection method, magnetic particle detection method and so on. Among them, ultrasonic inspection method is the most widely used at home and abroad, the highest frequency and the rapid development of a non-destructive testing technology.

Currently used ultrasonic transducers for the detection of complex surfaces require the use of specially tailored transducers and wedge coupling and are mechanically scanned to achieve large areas of detection. Such detection devices and labor and time costs are expensive and extremely inefficient. Although there are detection devices that use the bottom surface of the wedge to be designed as a concave surface according to the diameter of the pipe, thus realizing the coupling between the wedge and the pipe, the design can only be limited to a specific diameter of the pipe for large spherical structures, such as natural gas storage tanks, and some other irregular structures are not applicable at all.

To solve the problem that existing rigid transducers cannot detect complex surfaces, this section provides a stretchable ultrasonic transducer array that can be dynamically affixed to an arbitrarily shaped object under test during service. In particular, this section provides a stretchable ultrasound transducer array that can fit any curved surface. It includes the following parts.

A detection array provided with a plurality of transducers arranged in a predetermined array shape, said transducers being connected to each other by upper and lower stretchable wires to form a scanning circuit; a signal interface for connecting said detection array and providing a signal output interface; a signal interface for connecting said detection array and providing a signal output interface; and cladding layer, formed by a resilient material covering said detection array.

The transducer in the stretchable ultrasonic transducer array is one of the lead zirconate titanate piezoelectric ceramic (PZT), sodium potassium niobate-based ceramic, BNT-based piezoelectric ceramic, $BaTiO_3$-based piezoelectric ceramic and bismuth laminate-structured piezoelectric ceramic. The transducers are filled with polymers that prevent signal crosstalk between each other. The lower surface of the detection array is provided with a backing layer composed of epoxy resin, and the upper surface is provided with a matching layer to realize the impedance matching of acoustic characteristics between the transducer and the sound transmission medium. The said upper and lower layers of conductors between the transducers, respectively, as positive and negative layers, are unidirectional island–bridge structures, with the bridge being serpentine. The upper and lower layers between the transducers are laser cut from flexible copper laminates.

The stretchable ultrasonic transducer array is an essential improvement over existing conventional rigid probes in the way it couples to the object under test, which greatly improves detection effectiveness and efficiency. At the same time, the stretchable feature enables it to be dynamically coupled to any shape of the object under test (including both expandable and non-expandable surfaces), widening the range of objects to be inspected, and thus can be applied to debonding the inspection of complex laminar structures and stress detection of complex structures. In addition, the array structure can be used to achieve high-efficiency real-time inspection over a large area, completely eliminating the dependence of ultrasonic inspection on mechanical walking/scanning and its automation devices.

As shown in Figures 2.8–2.12, the stretchable ultrasonic transducer array for 100 generally includes a detection array for 20, a signal interface and a cladding layer 10. Layer 11 in the figure indicates a matching layer, 12 indicates a backing layer, 13 indicates a transducer, 14 indicates a polymer and 15 indicates a wire.

The detection array includes a plurality of transducers arranged in a predetermined array shape, and the transducers are connected to each other by wires. The transducer is an energy conversion device, using piezoelectric crystals made of ultrasonic detection equipment, the shape of the transducer can be thin plate shape, circular sheet shape, circular ring shape, round tube shape, round rod shape, thin shell spherical, piezoelectric film and so on. The shape of the array can be set according to the actual test equipment, where the settings include the test shape of general-purpose equipment, that is, the array arrangement that can be applied to most of the equipment testing, and also include the shape of the test for specialized equipment, that is, the display shape of a specific arrangement set for the shape of a specific device to obtain more accurate measurement values. The signal interface is used to connect the detection array to provide an excitation signal or to collect the

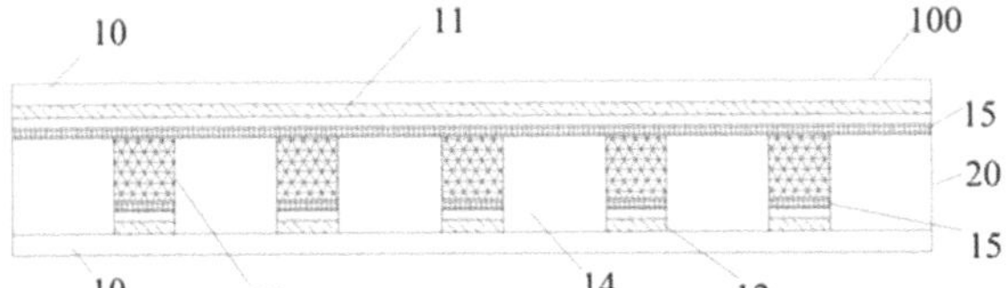

FIGURE 2.8 Schematic diagram of the cross-sectional structure of the stretchable ultrasonic transducer array [83].

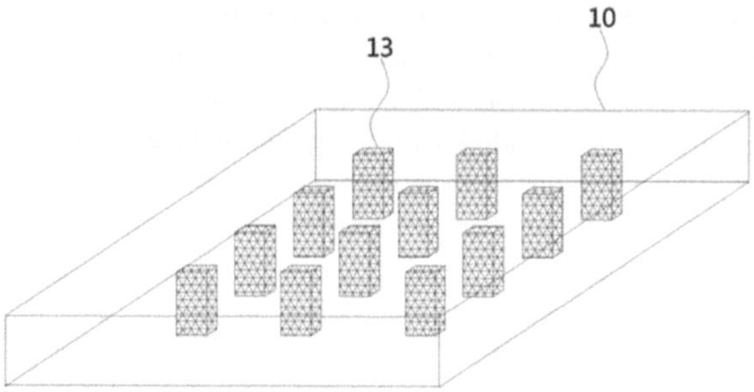

FIGURE 2.9 Schematic diagram of transducer arrangement structure [83].

FIGURE 2.10 Schematic diagram of electric wire structure [83].

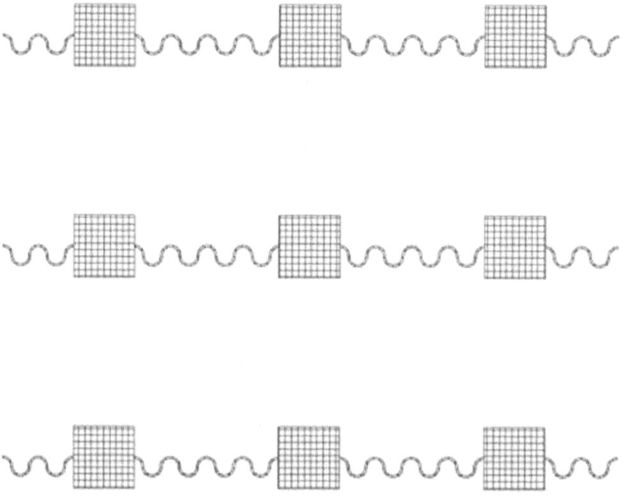

FIGURE 2.11 Schematic diagram of negative layer wire arrangement [83].

detection signal from each transducer and to provide a signal output interface for outputting the detection signal to an external device. The cladding layer is used to cover the detection array and can be made of a resilient material, e.g., this stretchable ultrasonic transducer array uses a resilient silicone film as the cladding layer.

The stretchable ultrasonic transducer array adopts an integrated structure, the shape of which can be a plate structure made according to the test requirements and the overall shape can be round or rectangular. The stretchable ultrasonic transducer array through the arbitrary deformation of the cladding layer so that

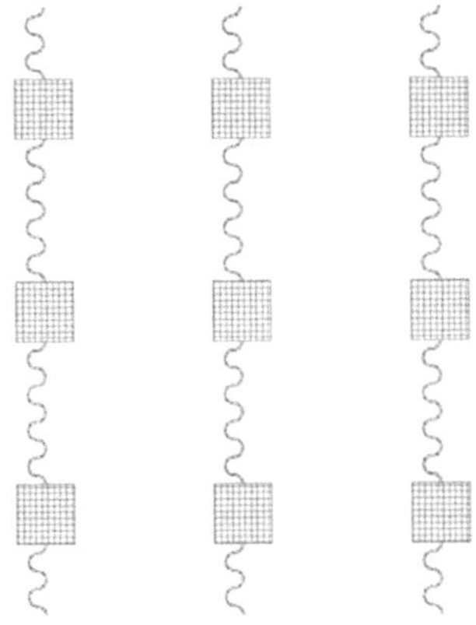

FIGURE 2.12 Schematic diagram of positive layer wire arrangement [83].

the detection array can be completely fit with the surface shape of the device under test, the cladding layer in the detection array of the transducer that can be the closest way to detect the parts under test. Then, the detection signal is transmitted to the connected processing equipment through the signal interface.

The stretchable ultrasonic transducer array can improve the contact effect with the device under test by arbitrary deformation during the testing process. When deforming, the whole cladding layer can be deformed according to the shape of the surface of the object under test, the position of the internal transducer changes and the wire connecting the transducer adopts a stretchable arrangement, which makes the distance between transducers longer, while it can be stretched simultaneously to ensure the signal normal transmission. The structure of the specific wire can be a unidirectional island–bridge structure with a serpentine bridge (Figure 2.10), so that the length of the wire between adjacent transducers under normal conditions is greater than the distance between adjacent transducers. When the deformation of the transducer array within a certain range, the bending part of the serpentine wire is unfolded, mainly through the rigid displacement of the wire to coordinate the deformation of the transducer array. However, the straight wire is mainly through the stretching or compression of the wire to coordinate the deformation. Due to the local larger deformation will lead to local deformation of the wire is also larger, the straight wire stretching or compression degree is greater, so that the resistance changes more or even the wire is pulled off, and then affect the voltage signal. Therefore, this implementation uses a serpentine wire compared to the straight wire that has good stretchability, for the transducer array dynamically to fit the complex surface to provide the conditions.

The stretchable ultrasonic transducer array substrate replaces the traditional rigid substrate, can be bent in any direction and is pioneered for real-time and accurate detection of interface bonding conditions of complex surfaces and stress detection of complex structures, which can be repeatedly bent and deformed during the measurement process, greatly improving the measurement results. In addition, it can be rolled and folded when not in use, making it easy to carry.

Compared with the existing conventional rigid probe, this implementation has an essential improvement in the coupling method with the object under test, which greatly improves the detection effect and efficiency. At the same time, the stretchable feature enables it to be dynamically coupled with any shape of the object under test (including both expandable and non-expandable surfaces), widening the range of detection objects, and thus can be applied to the debonding detection of complex laminar structures and stress detection of complex structures. In addition, the array structure can be used to achieve high-efficiency real-time inspection over a large area, completely eliminating the dependence of ultrasonic inspection on mechanical walking/scanning and its automation devices.

In the stretchable ultrasound transducer array, the specific transducer is a PZT layer, and the PZT blocks can be filled with polymer between each PZT block to achieve isolation between adjacent transducers and reduce mutual crosstalk between adjacent PZT blocks. The specific polymer can be silicone with ductile properties. The lower surface of the detection array is provided with a backing layer composed of epoxy resin, and the upper surface is provided with a matching layer to realize the impedance matching of acoustic characteristics between the transducer and the sound transmission medium. The matching layer is a layer of acoustic material laid on the radiating surface of the probe wafer to realize the matching of acoustic characteristic impedance between the transducer wafer and the sound transmission medium so that the acoustic energy can be well transmitted. The cladding layer contains all of the above components when covered.

In the stretchable ultrasonic transducer array, a method of making a stretchable ultrasonic transducer array is given in the following steps: (i) plating a layer of copper foil on the PI film; (ii) soldering a PZT block that has been polarized and plated with silver on top of the copper foil; (iii) spin coating a layer of epoxy resin damping backing layer on the other side of the PI film; (iv) soldering a flexible copper-clad plate on the other side of the PZT array; (v) epoxy resin matching layer; (vi) cutting the two copper foil-PI-epoxy resin composite layers on the top and bottom, respectively, by laser to form a unidirectional island–bridge structure, that is, to obtain the transducer array; and (vii) encapsulating the whole array with elastic silicone film.

2.3 DESIGN AND MANUFACTURING PROCESS

2.3.1 High-Performance Design of Mechanical Properties

Stretchable electronics [84–86] is a field of technology that has recently attracted much research interest, partly due to its wide-ranging prospects for applications, from sensory skins for robotics [87] to photovoltaics [88,89], wearable communication devices [90], transient energy harvester [91] and other systems that require lightweight, rugged construction in thin, conformal formats [92] to bio-integrated devices for surgical and diagnostic implements that naturally integrate with the human body to provide advanced therapeutic capabilities [93], and further to cameras that use biologically inspired designs to achieve superior performance [94].

Success of stretchable electronics depends on the availability of electronic materials and structures (such as the shape of serpentine interconnects and transition area of island interconnect [95]) that can be highly bent, stretched, compressed and twisted [96,97], in onetime stretching and cyclic stretching conditions [98].

One challenge in the development of stretchable electronics is to achieve large stretchability [99,100] without significantly sacrificing the aerial coverage of active devices. Fractal-inspired geometric designs for electrical interconnects between islands of active devices provide an effective approach, with demonstrated utility in stretchable lithium-ion batteries [101], epidermal electronics and radio frequency (RF) antennas [102]. Figure 2.13 illustrates zigzag, sinusoidal and serpentine shapes, produced using a fractal design approach.

This chapter determines analytically the elasticity of fractal-inspired interconnects. For an order-n fractal interconnect of arbitrary shape, the tensile stiffness is obtained analytically. The key parameters governing the tensile stiffness are clearly identified. Experiments are carried out to verify the elastic stiffness for the order-2 fractal interconnects.

Figure 2.14a shows an order-n fractal interconnect. At each fractal order, the interconnect has the same, centrosymmetric shape, but the size increases by a factor η as the fractal order increases by 1. Figure 2.14b shows a representative

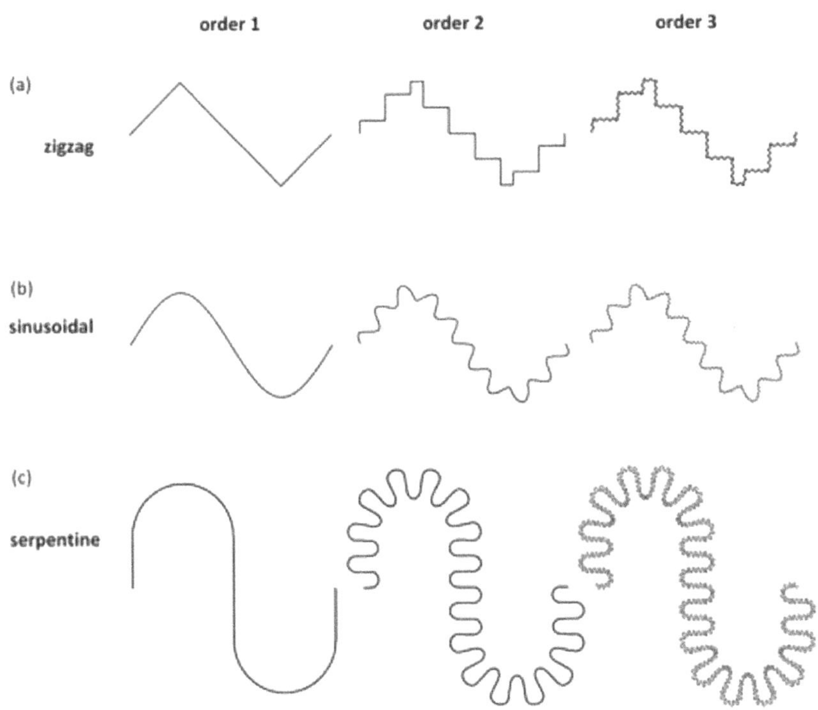

FIGURE 2.13 Schematic illustrations of the order-1, -2 and -3 fractal interconnects with the (a) zigzag, (b) sinusoidal and (c) serpentine shapes [103].

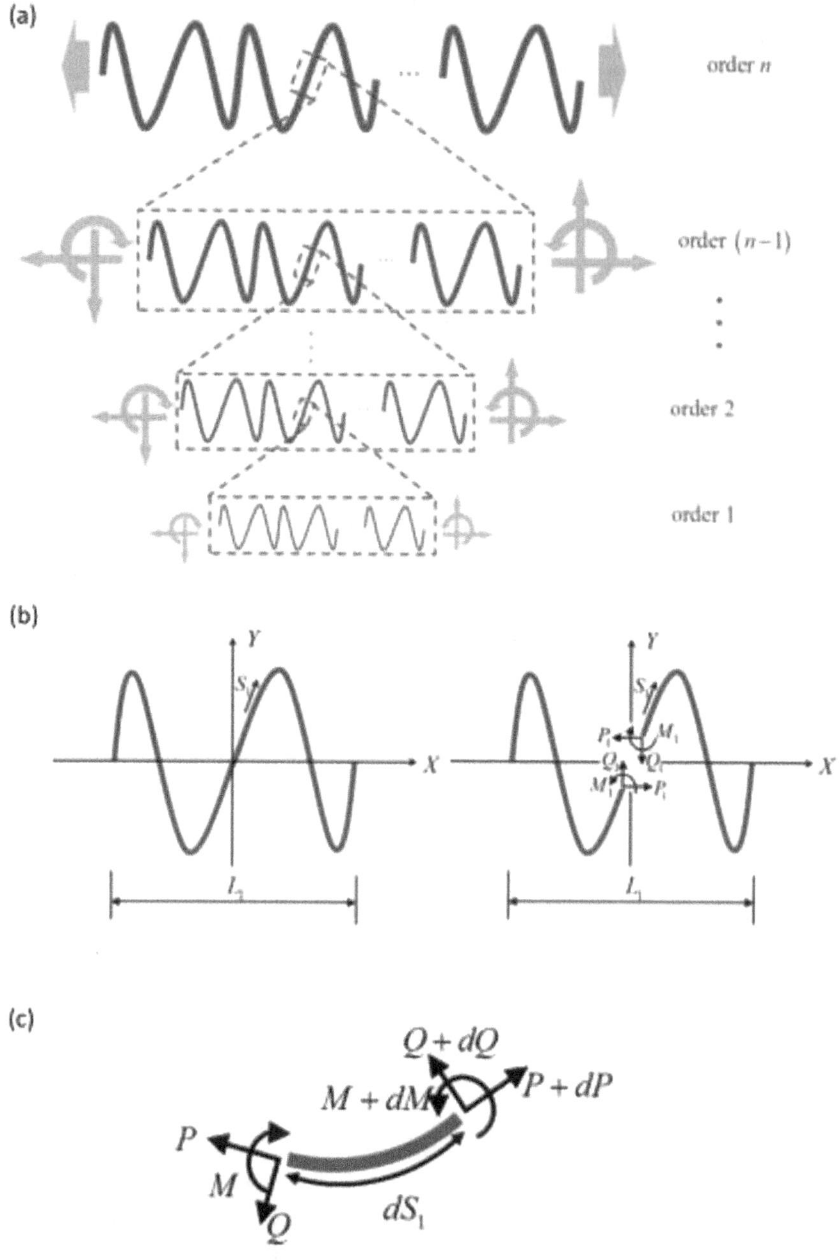

FIGURE 2.14 (a) Schematic illustrations of the fractal interconnects with any centro-symmetric shape in a representative element; (b) the curvilinear coordinate S_1 along the arc length and the internal forces) P_1 and Q_1 and bending moment M_1 at the center $S_1=0$ for the order-1 structure; and (c) the sign convention for the internal axial force P, shear force Q and bending moment M [103].

order-1 element, and its shape is described by the local Cartesian coordinates (X, Y) (Figure 2.14b).

$$X = X(S_1), \quad Y = Y(S_1) \tag{2.2}$$

where the origin of (X, Y) is at the center of this centrosymmetric structure and the curvilinear coordinate S_1 is along the arc length of the order-1 structure, with $S_1 = 0$ at the center of antisymmetry (Figure 2.14b) such that $X(-S_1) = -X(S_1)$ and $Y(-S_1) = -Y(S_1)$. The apparent length of the representative element (Figure 2.14) is $L_1 = X(S_{1t}/2) - X(-S_{1t}/2) = 2X(S_{1t}/2)$, where S_{1t} is the total length of the representative element. At the two ends of the representative element $Y(-S_{1t}/2) = -Y(S_{1t}/2) = 0$.

The order-1 fractal interconnect is modeled as a beam. Figure 2.14c shows the sign convention for the positive internal axial force *P*, shear force *Q* and bending moment *M*. Let P_1 and Q_1 denote the internal forces along the local coordinates *X* and *Y*, respectively, at the center $S_1 = 0$ and M_1 bending moment for the order-1 structure (Figure 2.14b). The axial and shear forces and bending moment (at any S_1) are then obtained from force equilibrium as

$$\begin{vmatrix} M(S_1) = M_1 + P_1 Y(S_1) - Q_1 X(S_1) \\ P(S_1) = P_1 X'(S_1) + Q_1 Y'(S_1) \\ Q(S_1) = -P_1 Y'(S_1) + Q_1 X'(S_1) \end{vmatrix} \tag{2.3}$$

where $X'(S_1) = dX(S_1)/dS_1$ and $Y'(S_1) = dY(S_1)/dS_1$. The membrane energy is negligible as compared to the bending energy such that the energy density is dominated by the bending energy per apparent length and is given by

$$U_1(M_1, P_1, Q_1) = \frac{1}{L_1}\frac{1}{2EI} \int_{-S_1,/2}^{S_1/2} [M(S_1)]^2 \, dS_1 = \frac{1}{EIL_1} \int_{0}^{S_{1/2}} [M(S_1)]^2 \, dS_1 \tag{2.4}$$

where *EI* is the bending stiffness of the order-1 fractal interconnect. Substitution of Equation (2.3) into Equation (2.4) gives the energy density

$$U_1(M_1, P_1, Q_1) = \frac{1}{2EI}\left[\beta_1 M_1^2 + \beta_2 (P_1 L_1)^2 + \beta_3 (Q_1 L_1)^2 + \beta_4 (P_1 L_1)(Q_1 L_1)\right] \tag{2.5}$$

where

$$\begin{cases} \beta_1 = \dfrac{S_{1t}}{L_1}, \quad \beta_2 = \dfrac{2}{L_1^3} \displaystyle\int_0^{S_1/2} [Y(S_1)]^2 \, dS_1 \\ \beta_3 = \dfrac{2}{L_1^3} \displaystyle\int_0^{S_1/2} [X(S_1)]^2 \, dS_1 \\ \beta_4 = \dfrac{-4}{L_1^3} \displaystyle\int_0^{S_1/2} X(S_1)Y(S_1)dS_1 \end{cases} \tag{2.6}$$

are dimensionless parameters that depend only on the shape of the order-1 structure, not on its size, and β_1 represents the ratio of total length to the apparent length.

The shape of the order n is the same as that of the order 1 and therefore is represented by $X = X(S_n)$ and $Y = Y(S_n)$ in its local coordinate, where S_n is the curvilinear coordinate along the arc length of the order-n structure, and the functions X and Y are the same as those in Equation (2.2). Let P_n and Q_n denote the internal forces along the local coordinates and M_n bending moment at the center $S_n = 0$. Similar to Equation (2.5), the energy density for the order n can be generally written as

$$U_n(M_n, P_n, Q_n) = \frac{1}{2EI}\left[\alpha_1^{(n)} M_n^2 + \alpha_2^{(n)}(P_n L_n)^2 + \alpha_3^{(n)}(Q_n L_n)^2 + \alpha_4^{(n)}(P_n L_n)(Q_n L_n)\right] \tag{2.7}$$

where the apparent length L_n for order n is related to L_1 for order 1 by $L_n = \eta^{n-1} L_1$, the dimensionless coefficients $\alpha_i^{(n)}$ $(i = 1,2,3,4)$ are to be determined, and $\alpha_1^{(n)} = \beta_i$. The axial force, shear force and bending moment in the order-n structure are actually the internal forces P_{n-1} and Q_{n-1} and bending moment M_{n-1} at the center $(S_{n-1} = 0)$ for order $n - 1$. Similar to Equation (2.3), force equilibrium then gives

$$\begin{cases} M_{n-1} = M_n + P_n Y(S_n) - Q_n X(S_n) \\ P_{n-1} = P_n X'(S_n) + Q_n Y'(S_n) \\ Q_{n-1} = -P_n Y'(S_n) + Q_n X'(S_n) \end{cases} \tag{2.8}$$

The energy density for order n is the integration of that of the order $n - 1$,

$$U_n(M_n, P_n, Q_n) = \frac{1}{L_n} \int_{-S_{W}/2}^{S_{\alpha i}/2} U_{n-1}(M_{n-1}, P_{n-1}, Q_{n-1}) dS_n \tag{2.9}$$

where $S_{nt} = \eta^{n-1} S_{1t}$. Substitution of Equations (2.7) and (2.8) into Equation (2.9) gives the recursive relation

$$\begin{Bmatrix} \alpha_1^{(n)} \\ \alpha_2^{(n)} \\ \alpha_3^{(n)} \\ \alpha_4^{(n)} \end{Bmatrix} = \begin{bmatrix} \beta_1 & 0 & 0 & 0 \\ \beta_2 & (\beta_1 - \beta_5)/\eta^2 & \beta_5/\eta^2 & -\beta_6/\eta^2 \\ \beta_3 & \beta_5/\eta^2 & (\beta_1 - \beta_5)/\eta^2 & \beta_6/\eta^2 \\ \beta_4 & 2\beta_6/\eta^2 & -2\beta_6/\eta^2 & (\beta_1 - 2\beta_5)/\eta^2 \end{bmatrix} \begin{Bmatrix} \alpha_1^{(n-1)} \\ \alpha_2^{(n-1)} \\ \alpha_3^{(n-1)} \\ \alpha_4^{(n-1)} \end{Bmatrix} u \tag{2.10}$$

where the additional dimensionless shape parameters are

$$\begin{aligned} \beta_5 &= \frac{2}{L_1} \int_0^{S_1/2} [Y'(S_1)]^2 \, dS_1 \\ \beta_6 &= \frac{2}{L_1} \int_0^{s_1/2} Y'(S_1) X'(S_1) dS_1 \end{aligned} \tag{2.11}$$

Equation (2.10) can be rearranged to an explicit (non-recursive) relation

$$\alpha_1^{(n)} = (\beta_1)^n \cdot \begin{Bmatrix} \alpha_2^{(n)} \\ \alpha_3^{(n)} \\ \alpha_4^{(n)} \end{Bmatrix} = \left[\sum_{k=1} (\beta_1)^{n-k} \eta^{2k-2} \begin{Bmatrix} \beta_1 - \beta_5 & \beta_5 & -\beta_6 \\ \beta_5 & \beta_1 - \beta_5 & \beta_6 \\ 2\beta_6 & -2\beta_6 & \beta_1 - 2\beta_5 \end{Bmatrix}^{k-1} \right] \begin{Bmatrix} \beta_2 \\ \beta_3 \\ \beta_4 \end{Bmatrix} \tag{2.12}$$

For relatively large size increase $\eta \gg 1$ at each fractal order, the above equation can be significantly simplified to

$$\left\{ \alpha_1^{(n)}, \alpha_2^{(n)}, \alpha_3^{(n)}, \alpha_4^{(n)} \right\} \simeq (\beta_1)^{n-1} \{\beta_1, \beta_2, \beta_3, \beta_4\} \tag{2.13}$$

A fractal shape at each order can be given in the Cartesian coordinates $Y = Y(X)$, such as $Y = -(H_1/2)\sin(2\pi X/L_1)(-L_1/2 \le X \le L_1/2)$ for the sinusoidal shape in Figure 2.13b, where H_1 is the apparent height of the order-1 representative element (Figure 2.14b). The dimensionless shape parameters are expressed as

$$\begin{aligned} \beta_1 &= \frac{2}{L_1} \int_0^{L_1/2} \sqrt{1 + (dY/dX)^2} \, dX \\ \beta_2 &= \frac{2}{L_1} \int_0^{L/2} [Y(X)]^2 \sqrt{1 + (dY/dX)^2} \, dX \end{aligned} \tag{2.14}$$

$$\beta_3 = \frac{2}{L_1^3}\int_0^{L_1/2} X^2\sqrt{1+(dY/dX)^2}\,dX$$

$$\beta_4 = \frac{-4}{L_1^3}\int_0^{L/2} XY(X)\sqrt{1+(dY/dX)^2}\,dX$$

$$\beta_5 = \frac{2}{L_1}\int_0^{L_0/2} \frac{(dY/dX)^2}{\sqrt{1+(dY/dX)^2}}\,dX$$

$$\beta_6 = \frac{2}{L_1}\int_0^{L_1/2} \frac{dY/dX}{\sqrt{1+(dY/dX)^2}}\,dX$$

where β_1 is the ratio of total length to apparent length, and all other shape parameters depend only on β_1 for each given fractal shape, for example, for the zigzag shape in Figure 2.13a.

For an order-n fractal interconnect between two active device islands with the spacing L, the tensile stiffness for a representative element of apparent length L is given by

$$K^{(n)} = \frac{\overline{EI}}{L^3}\frac{4\alpha_3^{(n)}}{4\alpha_2^{(n)}\alpha_3^{(n)} - \left[\alpha_4^{(n)}\right]^2} \tag{2.15}$$

For relatively large size increase $\eta \gg 1$ at each fractal order, substitution of Equation (2.13) gives

$$K^{(n)} \approx \frac{\overline{EI}}{L^3}\frac{1}{\beta_1^{n-1}\left(\beta_2 - \dfrac{\beta_4^2}{4\beta_3}\right)} \tag{2.16}$$

to be independent of n. It decreases as the fractal order n increase (because $\beta_1 > 1$) and depends on the fractal shape through $\beta_1^{n-1}\left[\beta_2 - \beta_4^2/(4\beta_3)\right]$. Here, β_2, β_3 and β_4 all depend only on the β_1 for each fractal shape. For example, $\beta_1^{n-1}\left[\beta_2 - \beta_4^2/(4\beta_3)\right]$ becomes $7\beta_1^n\left[\beta_1^2 - 1\right]/768$ for the zigzag shape.

Experiments were performed to examine the deformation in a representative fractal interconnect structure. Figure 2.15 shows micrographs collected at the original and stretched states of a Cu trace cut by laser milling; the cross-sectional dimensions of the trace are 60 μm × 60 μm and the apparent length is 4 mm. The interconnect is not adhered to a substrate except its ends, sometimes referred to as "free-standing". Stretching of each fractal shape was repeated three to five times and all the data points are plotted together to compare with theoretical predictions. Figure 2.15 also shows the measured force versus nominal strain for zigzag, sinusoidal and serpentine

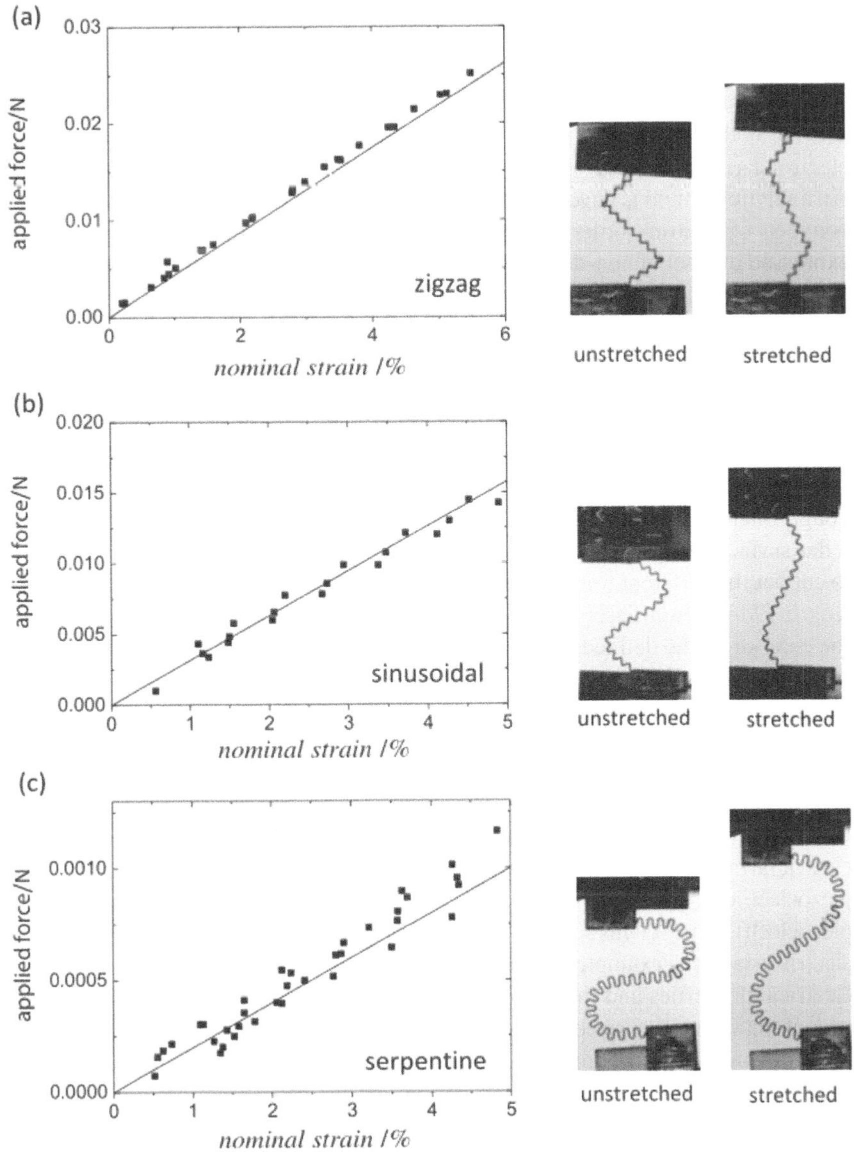

FIGURE 2.15 Comparison between the experimental (repeated three to five times for each fractal shape) and analytic force-nominal strain curves and the optical microscope images for undeformed and deformed interconnects for the (a) zigzag, (b) sinusoidal and (c) serpentine fractal shapes of fractal order $n = 2$ [103].

fractal shapes, which agrees well with the analytic model without any parameter fitting. Here, the nominal strain is defined as the ratio of the stretched displacement to the apparent length L of the interconnect. The small deviations between experiments and analytical predictions may result from the equipment's limited capability to accurately dictate very small forces (<10 mN). The repeating measurement of stretching verified the repeatability and robustness of the measurements.

In summary, the fractal-inspired geometric designs presented in this study provide both key features that are contradicting in most of the other stretchable electronics designs: large stretchability and high aerial coverage of active devices. An analytic method is developed to determine the tensile stiffness of fractal interconnects of arbitrary order n and in arbitrary shape; the tensile stiffness is clearly expressed by a set of non-dimensional parameters; experimental measurements in tensile stiffness agree very well with the analytic predictions.

The fractal interconnects, in general, are much longer than the straight one, which leads to undesired increase in the electric resistance. This can be overcome by using a larger cross-section of the interconnects, for which the elastic stiffness is particularly important because their deformation modes tend to be in-plane bending instead of buckling. It is helpful to avoid interfacial peeling under stretching as well [7,104].

Fabrication of the Fractal Structures: copper foils (Annealed Alloy 110, Electro Tough Pitch, smooth finish on both sides, All Foils, Inc.) were laminated against a flat surface using water-soluble tape (3 M). A laser milling machine was used to cut out the different traces precisely defined by a CAD file. An infrared (wavelength 790–820 nm) laser beam with a power of 0.10 W runs three passes along the exterior of the defined edges at a speed of 2.0 mm/s to cut vertically through the copper foil. Excessive material was removed after the laser milling and the desired traces were left on the flat water-soluble tape for ease of handling.

2.3.2 High-Performance Design of Electrical Properties

Not only the mechanical structure affects the performance of flexible piezoelectric arrays, but also the rational design of the electrical properties of flexible piezoelectric arrays is one of the methods to improve the performance of flexible piezoelectric arrays. This section introduces four methods to improve flexible piezoelectric arrays and examples of applications: (i) seeking materials with high-voltage electrical properties and high flexibility; (ii) designing the microform/microstructure of the material surface; (iii) adding dopants; and (iv) using piezoelectric techniques to prepare strain/pressure-driven gate-controlled sensing transistors.

2.3.2.1 Seeking Materials with High-Voltage Electrical Properties and High Flexibility

Seeking for materials with high piezoelectricity and flexibility to improve the performance of piezoelectric sensors. Piezoelectric materials play a decisive role in determining the performance of piezoelectric sensors. The sensitivity and sensing ranges of piezoelectric sensors fabricated with different piezoelectric materials vary greatly. To date, ZnO, PZT, $BaTiO_3$ and PVDF are the major functional piezoelectric materials that have been fully researched and widely applied (Figure 2.16) [105–109].

ZnO is the earliest studied piezoelectric semiconductive material, due to its relatively simple chemical composition and crystal structure and easy control

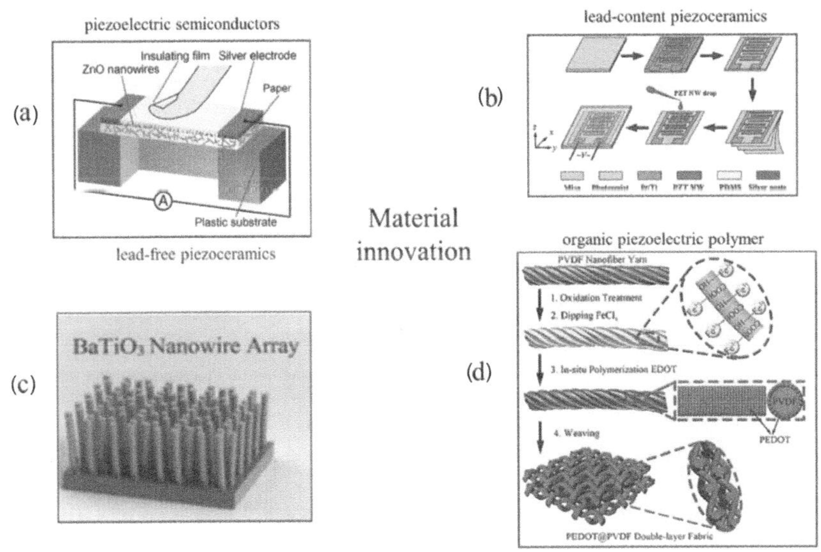

FIGURE 2.16 Material innovation: piezoelectric semiconductors, inorganic piezoceramics and organic piezoelectric polymer [103]. (a) Structure of a sensor based on ZnO. (b) Structure of a sensor based on $Pb_xZr_{1-x}TiO_3$ (PZT). (c) Structure of a sensor based on $BaTiO_3$. (d) Structure of a sensor based on PVDF.

of purity, size and surface morphology [111]. Meanwhile, the semiconducting, desirable piezoelectric property, biosafety and easy fabrication make ZnO a great candidate for piezoelectric sensors. ZnO nanowires can generate electrical signals in response to small mechanical force for the sake of its large aspect ratio. Therefore, the electrical response of a single ZnO nanowire under small forces was reported in 2006 [112]. As shown in Figure 2.17a, Zhao et al. fabricated a strain sensor based on a single ZnO nanowire. Compared with a commercial strain sensor, its gauge factor is 200 times higher (above 400). However, the strain sensor only withstand ultra-small mechanical strain on the order of 10^{-6} [113].

Considering that the single nanowire will break under large force and generates a relatively poor electrical response, in practical applications, it is essential to improve sensitivity and broaden the sensing range of piezoelectric sensors by integrating large numbers of ZnO nanowires on a flexible platform. For example, the field-limited ZnO nanorods were grown on a flexible Kapton substrate by using a low-temperature hydrothermal method, which was then packaged with PDMS to fabricate a one-piece flexible tactile sensor with good linearity and sensitivity of 403 mVm/Pa at around 0.7 mPa [114]. Meanwhile, the diameter, height and density of the micropillars/microrods can influence the performance of piezoelectric sensors, including flexibility, sensitivity and sensing range [115], which deserves to be further researched by analyzing the coupling effects of the mechanical and piezoelectric properties under different diameters, heights and density. As shown in Figure 2.16a, a study-based ZnO nanowires touch pad was

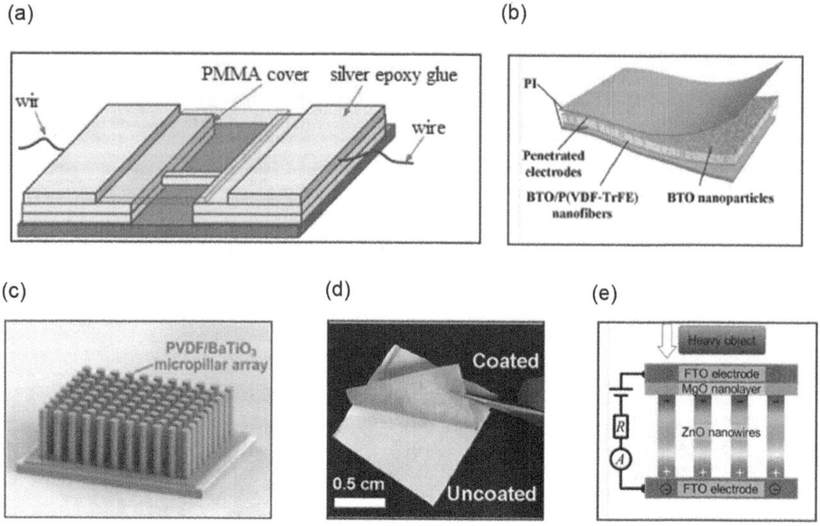

FIGURE 2.17 Different strategies facilitating high-performance flexible piezoelectric sensors [110]: (a) piezoelectric semiconductor, (b) nanocomposite materials, (c) microstructure, (d) addition of dopants and (e) piezotronic effect.

designed [116], which exhibited a sensitivity of 0.57 nA/N and a linear relationship between the applied force and the response current. In addition, semiconductive piezoelectric materials combined with piezotronic, ZnO for instance, can be used to fabricated amperometric piezoelectric sensors.

However, the relatively poor piezoelectric characteristics of ZnO lead to the low sensitivity of ZnO-based piezoelectric sensors. Therefore, alternative materials with higher piezoelectric coefficient need to be found to fabricate high-performance piezoelectric sensors. Lead zirconate titanate (PZT) is a solid solution of two materials including a ferroelectric lead titanate ($PbTiO_3$) and an anti-ferroelectric lead zirconate ($PbZrO_3$) [117]. It owns high piezoelectric coefficient, outstanding electromechanical coupling effect, high dielectric constants and good crystallinity [118–120]. PZT has been vastly researched and used for piezoelectric sensors. For instance, as shown in Figure 2.16b, a flexible semi-transparent sensor based on laterally aligned PZT nanowires and interdigital Pt/Ti electrodes was assembled using a simple spin-coating method, which exhibits high pressure sensitivity (0.14 Vk/Pa) [121]. A PZT diaphragm strain sensor with a diameter of 1.35 cm and a thickness of 300 μm was fabricated, which generated a voltage magnitude of 7.3 mV at the maximum displacement of 250 μm and can sense load forces between 89 and 539 kPa [122]. Owing to the of PZT-based ceramic materials, these piezoelectric sensors exhibit higher performance, which makes them more competitive in microelectronics. However, PZT contains more than 60% lead (Pb) by weight, which has an adverse effect on human beings and ecosystems. Therefore, regulators around the world have begun severely restricting the use of lead [82,120]. Compared with PZT, lead-free piezoelectric materials not only have no side effect on environment but also exhibit comparable or even superior piezoelectric properties [109,111]. Therefore, they have attracted increasing attention in recent years.

Lead-free piezoceramics, such as barium titanate, bismuth potassium titanate (BKT), potassium sodium niobate (KNN) and bismuth sodium titanate (BNT), without PbO in its composition, are expected to replace lead-based piezoceramics in the future due to their environmental friendliness and biocompatibility [120,123]. Although BNT–BKT lead-free piezoceramics have a high Curie temperature, they have many disadvantages that cannot be ignored, such as highly corrosive due to the presence of alkalis, low piezoelectric properties and volatile [120]. For KNN-based lead-free ceramics with a high piezoelectric coefficient, their drawbacks, such as volatility of alkali components at high temperature and instability of KNN phase at high temperature, make it difficult to fabricate [117]. Therefore, BKT, BNT and KNN materials have not been widely investigated and developed.

The piezoelectric coefficient (d_{33}) of pure $BaTiO_3$ through solid state processing is 190 pC/N [124], which is higher than that of pure ZnO. Owing to the excellent ferroelectric properties, piezoelectric properties and biocompatibility, the performance of $BaTiO_3$-based sensors has been further improved. For example, as shown in Figure 2.16c, Zhou et al. prepared vertically aligned $BaTiO_3$ nanowires on a conductive substrate based on a two-step hydrothermal reaction, which exhibited a high piezoelectric response (d_{33}=43±2 pm/V). Besides, ultra-long vertically aligned $BaTiO_3$ nanowire arrays were fabricated by a two-step hydrothermal reaction, of which the diameter and length, respectively, were about 600–630 nm and 40 μm [125]. When applied a sinusoidal base acceleration of 0.25 g rms at lower resonant frequency (170 Hz), the device outputted a high peak to peak voltage (345 mV). In 2017, a self-powered wearable sensor based on $BaTiO_3$/alginate spherical composite beads was fabricated through the ionotropic gelation method [126]. In addition, Wu et al. fabricated a piezoelectric strain sensor based on field effect transistors with a high sensitivity of 1.608 × 107/N [127].

Although $BaTiO_3$ has many outstanding advantages such as outstanding ferroelectricity, piezoelectricity, nontoxicity and easy synthesis, the high polarization temperature and inherently poor flexibility make it difficult to satisfy the increasing need of flexible piezoelectric sensors. PVDF and its derived copolymers have fabulous properties, such as high deformability, low permittivity, high thermal stability and outstanding chemical durability [128], which make them the most promising materials for piezoelectric sensors. Besides, their low acoustic impedance close to human tissues and water improves the competitiveness in the field of medical diagnosis, health monitoring and wearable devices [129]. As shown in Figure 2.16d, a sensor with sensitivity (18.376 k/Pa at ~100 Pa), fast response (15 ms), wide pressure range (0.002–10 kPa) and high durability (7500 cycles) was designed and fabricated by weaving PVDF electrospun yarns coated with poly(3,4-ethylene dioxythiophene) [130]. In 2016, a highly flexible self-powered sensor was fabricated through printed circuit board and near-field electrospinning technology, which provided a possible industrial preparation process for low-cost production [131]. Lu et al. adopted spin coating and mask etching to fabricate a flexible 4 × 4 PVDF strain sensor array with high sensitivity (12 mVk/Pa) and fast recover time (<25 μs) [132]. However, compared with traditional piezoceramics like PZT and $BaTiO_3$, the d_{33} of PVDF is relatively lower, which limits the performance of PVDF-based piezoelectric sensors. However, the β phase content, the strongest polar moment contributing to piezoelectricity of PVDF among all the crystalline phases (α,β,γ,δ and ε phases) [133], could be increased by doping [134,135] or improving processing technology [128,136] to enhance the piezoelectricity of PVDF. Electrospinning, as the most straightforward and versatile technique to prepare PVDF nanofibers with large length to diameter ratio, has been widely studied and optimized [130,137–139]. For example, Zheng et al. increased the β phase content by adjusting the technical parameters during electrospinning [140]. The technical parameters could be potentially modified by adding a low boiling point solvent like acetone, decreasing the flow rate of the tip, shortening the distance between the tip and the collector and decreasing the environment temperature [141,142]. However, the β phase semi-crystalline PVDF with high piezoelectric coefficients cannot be produced in large scale due to the difficult synthesis [143], which still needs a great effort to find more facile, efficient and stable methods to improve the β phase content.

In summary, flexibility and piezoelectricity of piezoelectric materials are the key factors that determine whether they could be widely used to fabricate piezoelectric sensors in the future. Nevertheless, for the most well-researched piezoelectric materials, such as ZnO, PZT, $BaTiO_3$ and PVDF, none of them have both high flexibility and good piezoelectricity. Although ZnO has semiconductor properties and piezoelectricity, its flexibility and piezoelectric properties are relatively poor. Similarly, inorganic ceramic materials with high ferroelectric and piezoelectric properties, such as PZT and $BaTiO_3$, have inherently poor flexibility. Moreover, PZT-based piezoelectric materials will be gradually abandoned due to its toxicity to environment and living things. Despite high flexibility, PVDF and its derived copolymers have poor piezoelectricity. With the increasing need for high sensing performance and flexibility, piezoelectric sensors need to be fabricated with highly piezoelectric and flexible materials. Therefore, new piezoelectric materials with high piezoelectricity and flexibility need to be explored and designed.

Composed of organic polymers of high flexibility and inorganic particles of high piezoelectricity, flexible nanocomposite materials provide a new solution to fabricate piezoelectric sensors with high sensing performance and flexibility. Hence, designing nanocomposite materials of high piezoelectricity and flexibility is another research hotspot in piezoelectric sensors' field. The concept of piezoelectric ceramic/polymer composites was first proposed in 1978 [144]. Composite materials, the mixture of two or more materials, are formed by dispersing inorganic nanoparticles into an elastomeric matrix without a chemical reaction occurring between them [145]. Piezoelectric sensors based on composite materials are sensitive to the applied force due to high piezoelectricity of inorganic materials in composite materials. In addition, they show high conformity and flexibility owing to high elasticity of the elastomeric matrix in composite materials.

Besides high flexibility and piezoelectricity, composite materials have many other advantages, such as eco-friendliness, cost-effectiveness and ease of large-scale fabrication. To date, the most commonly used inorganic materials are ZnO and $BaTiO_3$. The most used elastomeric matrixes are non-piezoelectric organic materials [i.e., polydimethylsiloxane (PDMS), polyvinyl alcohol, polystyrene (PS)] and piezoelectric organic materials, such as PVDF and its derived polymers [115,146–155]. As shown in Figure 2.17b, Hu et al. improved sensing performance of the P(VDF-TrFE) nanofibers-based sensor by utilizing $BaTiO_3$ NPs. The $BaTiO_3$/P(VDF-TrFE) composite material exhibits the maximum β phase crystallinity and piezoelectricity when adding 5 wt.% $BaTiO_3$ NPs. The sensor could detect the creep force of an ant (~1 mg). Figure 2.18 shows the micro-morphology, device design and sensing performance of piezoelectric sensors based on composite thin-film materials.

As shown in Figure 2.18a, a high-strain sensor based on ZnO NWs and PS nanofiber was fabricated by electrospinning [149], which can measure and withstand strains up to ~50%. However, due to the poor piezoelectricity of ZnO nanowire, its gauge factor is only ~116. To further improve the sensing performance of piezoelectric sensors, inorganic materials with higher piezoelectricity need to be used. Therefore, Bian et al. adopted dip-coating method to fabricate a biomimetic force sensor by combining $BaTiO_3$ nanoparticles with PDMS based on 3D polypyrrole network (Figure 2.18b), which showed an excellent stretchability of up

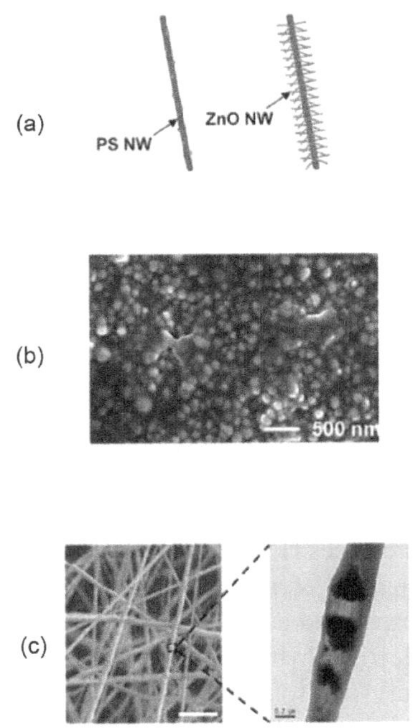

FIGURE 2.18 Material innovation: nanocomposite materials [110]: (a) piezoelectric semiconductors/non-piezoelectric organic materials, (b) piezoceramics/non-piezoelectric organic materials and (c) piezoelectric semiconductors/piezoelectric organic materials,

to 310%. Due to high piezoelectricity of $BaTiO_3$ and three-dimensional porous network, the sensor could detect acceleration with sensitivity of 1.03 μAs^2/m in a range from 1.7 to 16.2 m/s^2, static pressure with sensitivity of 12.6 pFk/Pa in a range from 0.6 to 5 kPa and dynamic pressure with sensitivity of 57.58 μAM/Pa in a range from 31.25 to 187.5 kPa [153]. In these composite thin-film materials, the non-piezoelectric polymer materials, such as PDMS and PS, only work as supporting cross-linker materials to hold inorganic nanoparticles to form a non-directional polymer structure, which cannot improve piezoelectric properties or the electrical energy response of composite materials. In contrast, piezoelectric organic polymers, such as PVDF and P(VDF-TrFE), act as not only a support layer but also a piezoelectric active layer to enhance piezoelectric properties of composite materials. Moreover, the addition of inorganic materials can improve the β phase contents in piezoelectric organic polymers [147]. The synergistic coupling effect between inorganic materials and piezoelectric organic polymer materials makes piezoelectric polymers promising materials for composite materials. For example, a flexible self-powered sensor (Figure 2.18c) on the base of the cowpea-structured PVDF/ZnO nanofibers with a fantastic bending sensitivity (4.4 mV/deg), a wide range (from 44° to 122°), and ultrahigh pressing sensitivity (0.33 Vk/Pa) was fabricated, which was used for remote control of gestures in human–machine interactive system [108]. One of the major problems in the preparation of composite materials is to make inorganic nanoparticles homogeneously dispersed in organic polymers. However, traditional methods including hot embossing, spin coating and solvent casting lead to heterogeneous dispersion of inorganic nanoparticles in polymer materials owing to the higher density of inorganic nanoparticles compared with polymer materials and agglomeration during fabrication process, which weakens mechanical and functional properties of composite materials. Therefore, Kim et al. adopted filament extrusion and 3D printing to fabricate homogeneous composite materials composed of $BaTiO_3$ nanoparticles and PVDF [156]. The result showed that the piezoelectric response of nanocomposites fabricated with 3D printing exhibited three times higher than that of nanocomposites fabricated with solvent casting.

In addition to investigating and exploring nanocomposite materials, researchers have been seeking flexible materials with both high piezoelectricity and flexibility. Due to their light weight, flexibility, low acoustical impedance, ease of low-temperature processing, structural tenability and biocompatibility, molecular ferroelectrics are potential promising materials for the exploration of next generation piezoelectric devices [157]. To date, various molecular ferroelectrics have been designed through different screen strategies. As shown in Figure 2.19a, Liao et al. found and researched a molecular perovskite $(TMFM)_x(TMCM)_{1-x}$-$CdCl_3(0 \leq x \leq 1)$ (TMFM, trimethylfluoromethyl ammonium; TMCM, trimethylchloromethyl ammonium) solid solution [158]. Different piezoelectric materials can be synthesized by changing the value of x. When x equals 0.26, the synthesized piezoelectric material has a d_{33} of ~1540 pC/N, which is two times than that of PZT. In addition, Ye et al. designed the family of metal-free ABX_3-type (where A is a divalent organic cation and X is Cl, Br or I) 3D perovskites

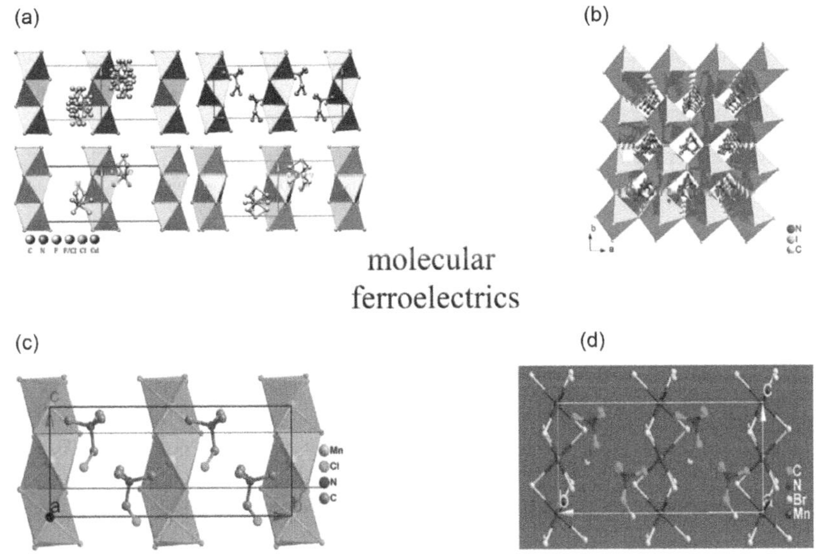

FIGURE 2.19 Material innovation: molecular ferroelectrics [110]: (a) $(TMFM)_x(TMCM)_{1-x}CdCl_3(0 \leq x \leq 1)$, (b) MDABCO–$NH_4H_3$, (c) TMCM–$MnCl_3$ and (d) TMBM–$MnBr_3$.

(Figure 2.19b), among which MDABCO (*N*-methyl-*N*′-diazabicyclo [2.2.2] octonium) – NH_4I_3 has a large spontaneous polarization ($Ps = 22$ μC/cm^2) and a high phase transition temperature ($T_0 = 448$ K) [159].

Unlike inorganic ferroelectrics whose polarization can be switched among multiple directions easily in each grain, the random grains of uniaxial molecular ferroelectrics are only polarized between two opposite polarization directions under the electric field, which leads to a weak macroscopic polarization and low piezoelectricity [160]. Therefore, multiaxial molecular ferroelectrics are proposed and researched. As shown in Figure 2.21c, You et al. discovered an organic–inorganic perovskite ferroelectric material of TMCM–$MnCl_3$with a d_{33}value of 190 pC/N and a high transition temperature of 406 K. At the same year, a molecular ferroelectric thin film [161] based on TMBM–$MnBr_3$($Me_3NCH_2BrMnBr_3$) was fabricated through dip coating, which has a large piezoelectric coefficient ($d_{33} = 112$ pC/N) (the crystal structure model of TMBM–$MnBr_3$ is shown in Figure 2.19d).

Despite the high piezoelectricity and flexibility, strict design and screening strategies and manufacturing conditions limit the development of molecular ferroelectrics [160]. Therefore, there are still many research and engineering work to be conducted. On the other hand, although nanocomposite materials have been researched widely, many challenges need to be tackled, such as the refined control of surface roughness, developing composite materials with higher piezoelectricity and the best optimization of weight ratio of inorganic materials and organic polymer materials [162].

2.3.2.2 Designing the Microform/Microstructure of the Material Surface

The performance of piezoelectric sensors can be improved by designing the micromorphology/microstructure of the material surface. Apart from high-performance piezoelectric materials, precisely controlling the micro-morphology on the surface of materials can enhance the performance of sensors as well, which usually adopts different preparation technologies under rational reaction temperature. Micro-morphologies/microstructures on the surface of materials could increase contact areas and redistribute the applied force uniformly. For example, a self-powered piezoelectric sensor (Figure 2.17c) based on vertically aligned P(VDF-TrFE)/$BaTiO_3$ micropillars was fabricated through nanoimprinting process, which could be used for detecting air pressure working in a noncontact mode [115]. The performance of piezoelectric sensors with different micro-morphologies is summarized in Table 2.1.

To detect the larger strain, a laterally aligned, patterned and highly stretchable polyaniline/poly (vinylidene fluoride) (PANI/PVDF) strain sensor was developed by electrospinning and in situ polymerization [139]. Compared with common nonwoven PANI/PVDF mat, it could detect a strain up to 110% and exhibited an excellent linear response to a wide strain range (0%–85%). Besides, various sensors based on laterally aligned nanofibers were fabricated via electrospinning [163,164], which exhibit excellent stretchability and sensing performance. However, the response to normal force of these sensors is nonideal due to its planarization structure. Therefore, vertically aligned structures were designed to detect normal force efficiently. In vertically aligned structures, traditional one-dimensional nanostructures include nanowires, nanorods, nanopillars, nanotubes

TABLE 2.1
Comparison of Piezoelectric Sensors with Different Microstructures [110]

Microstructures	Sensitivity	Detect Limit	Sensing Range	Response Time	Stability (Cycles)
Laterally aligned P(VDF-TrFE) nanofibers	1.1/kPa	0.1 Pa	0.4–2 kPa	–	–
PANI/PVDF nanofiber webs	1.84/kPa	–	0%–110% (strain)	–	10,000
P(VDF-TrFE)/graphene nanofiber webs	15.6/kPa	1.2 Pa	–	5 ms	100,000
Vertically aligned P(VDF-TrFE) nanowires	0.046/kPa	–	–	–	36,000
Vertically aligned PVDF/$BaTiO_3$ nanopillars	0.0264/kPa	–	50–600 kPa	–	12,000
Vertically aligned P(VDF-TrFE) nanopyramids	0.005/N (sensor size not specified)	–	–	–	5000
Single-layer interlocked rGO/PVDF nanohemispheres	35/kPa below 2.45 kPa	0.6 Pa	0.6 Pa to 49.5 kPa	–	5000

and nanoribbons. Besides, Chen et al. fabricated a vertically well-aligned (PVDF-TrFE) nanowires-based piezoelectric sensor with a high sensitivity (458.2 mV/N) [165], which were applied as wearable pressure sensors to detect some weak human activities, such as breath, low magnitude sound wave and heartbeat pulse. To survive, living things in nature have evolved their micro- and nanostructures to efficiently response to environmental changes. Therefore, researches sought inspiration from those unique structures of biologic systems to further improve the performance of piezoelectric sensors. Inspired by the structure of human fingertips, Park et al. fabricated an interlocked and hierarchical microstructured ferroelectric sensor based on reduced graphene (rGO) and PVDF composite films [166], which could detect various stimuli ranging from vibration, static and dynamic pressure to temperature. By adopting this interlocked structure, small spots between interlocked microdomes could concentrate and amplify stress, which then greatly deforms the interlocked microdomes and increases contact areas. Consequently, this kind of sensors are more sensitive to small pressure.

Designing different micro-morphologies on material surface could increase contact area and distribute the applied force more uniformly to improve the performance of piezoelectric sensors. However, due to a nominal enhancement of piezoelectric properties from micro-morphologies, the performance of sensors based on such materials is restricted by their inherently piezoelectricity.

2.3.2.3 Adding Dopants

Adding Dopants to Improve the Performance of Piezoelectric Sensors. To date, piezoelectric sensors based on ZnO, PZT, $BaTiO_3$ and PVDF exhibit good performance, but they are restricted by their piezoelectric properties. Meanwhile, designing different micro-morphologies on material surface is not essential to improve the piezoelectric properties. In addition, although a series of achievements were reported in molecular ferroelectrics, there are many theoretical and engineering problems need to be researched and optimized before its large-scale applications. On the other hand, composite materials are produced by dispersing inorganic nanoparticles into an elastomeric matrix without a chemical reaction occurring between them, of which piezoelectric properties are mainly decided by inorganic piezoelectric materials.

To further improve the performance of piezoelectric sensors based on ZnO, PZT, $BaTiO_3$ and PVDF, various dopants (i.e., metallic elements, graphene) were researched and added into piezoelectric materials to increase the piezoelectric coefficient (d_{33}) [167–171]. For example, as shown in Figure 2.17d, graphene was used to induce the structural units of $-CH_2-$ and $-CF_2-$ of PVDF nanofiber chains to arrange directionally during phase separation to enrich the electroactive phase. The flexible fabrics-based piezoelectric sensor coated with PVDF/graphene exhibited high sensitivity of 34 V/N and low detecting limit of 0.6 mN [172]. Piezoelectric sensors based on dopant-modified ZnO, PZT and PVDF are shown in Figure 2.20.

Adding different chemical dopants into piezoelectric materials could modify their piezoelectric coefficients and dielectric constants, therefore improving their piezoelectric properties. Due to the semiconductor properties, the doping

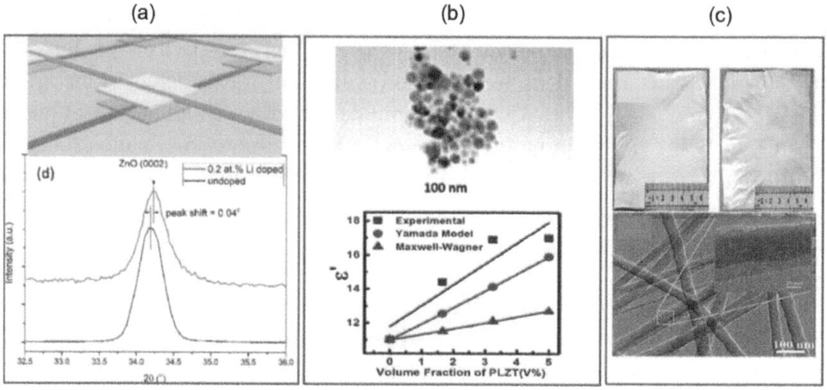

FIGURE 2.20 Addition of dopants [110]: (a) adding dopants in piezoelectric semiconductors, (b) adding dopants in piezoceramics and (c) adding dopants in organic piezoelectric polymer.

methods for ZnO-based piezoelectric materials are classified into n-doping and p-doping. For n-doping, dopants-induced strain of the lattice along the polar *c*-axis can improve the piezoelectric coefficient of ZnO crystal. Nevertheless, the addition of dopants with large ionic radius or high doping concentration would lead to many lattice defects, which increases resistance to the flow of electrons and decreases the performance of ZnO-based piezoelectric sensors. For p-doping, the dopant-generated shielding effect can be reduced in ZnO nanowires [143]. TO date, dopants that have been researched to improve the piezoelectric properties of ZnO-based piezoelectric materials include halogen ions (F^{1-}, Cl^{1-}, Br^{1-}, I^{1-}) [173,174] and metallic elements (Y^{3+}, Fe^{3+}, Cr^{3+}, V^{5+}, La^{3+}, Mn^{2+}, Mg^{2+}, Gd^{2+}, Eu^{3+}, Ce^{3+}, Cu^{3+}, Fe, Ag, Li, V, Ga) [167,175–188]. For halogen doping, the radius of F^{1-} is slightly smaller than that of O^{2-}, which induces a compressive strain within the crystal structure and, then, decreases the piezoelectric performance [175]. In contrast, the radius of Cl^{1-}, Br^{1-} and I^{1-} is larger than that of O^{2-}, which expands the strain along the *c*-axis of the lattice and then increases the piezoelectric performance. However, the radius of I^{1-} is much larger than that of O^{2-}, which leads to the flaw and distortion of the ZnO lattice and enhancement of the screening effect generated by free electrons and then leads to a decrease in piezoelectric performance. Generally, owing to polarization rotation generated under external electric fields, the ZnO nanowires doped by metal ions exhibit higher piezoelectric coefficients compared with pure ZnO nanowires. From the microscopic origin, the Zn^{2+} ion on the ZnO crystal site would be substituted by the doped metal ion (Mn^{+}), leading to the form of an M–O bond. It is noted that the radius of metallic ion influences the piezoelectric properties. When the size of the doped metal ion is small, substitution of Mn^{+} ion for Zn^{2+} results in easier rotation of non-collinear M–O bonds along the *c*-axis under the applied field, which produces large piezoelectric displacement and corresponding piezoelectric response enhancement. When the size of the doped ion is large, it is difficult to rotate the non-collinear M–O bond, leading to a decrease in the piezoelectric response of the ZnO. That is, doping ZnO with a small ion produces enhanced piezoelectric response, whereas doping ZnO with a larger ion results in decreased piezoelectric response. However, when the size of the doped ion is large, the orientation of the ZnO nanowires along the *c*-axis will be enhanced, and the lattice constant along the *c*-axis in the ZnO crystals will decrease. Therefore, the synergistic effect between enhanced orientation and decreased lattice constant will enhance the piezoelectric response of ZnO nanowires [143]. For example, as shown in Figure 2.20a, an Li-doped ZnO in-plane strain mapping sensor was fabricated by RF sputtering. The X-ray diffraction patterns of the un-doped and Li-doped ZnO thin films demonstrated that the un-doped film has a (0002) peak corresponding to ZnO single crystal at 34.20°, which reveals that the *c*-axis of ZnO thin films is perpendicular to the substrate. Meanwhile, the corresponding peak of Li-doped ZnO film is at 34.24°. When the Li substitutes the Zn site, the lattice constant of ZnO will reduce a little bit, which induces this tiny peak shift. Furthermore, Song et al. conducted the test of carrier type and conductivity by the hall effect, which demonstrated that doped ZnO film is still an n-type semiconductor. The in-plane strain mapping sensor achieved a high gauge factor (~199) through adapting nanowire arrays and doping Li element [171].

Overall, chemical doping could be regarded as practical routine for enhancing the performance of piezoelectric sensors on conditions of reasonably controlling the ionic size of dopants and doping concentration.

The piezoelectric properties of inorganic ceramic PZT- and $BaTiO_3$-based piezoelectric materials could also be improved by doping chemical elements. For example, due to the coexistence of polymorphic phase boundaries in Zr/$BaTiO_3$ hybrid piezoelectric thin films [189], the piezoelectric coefficients of $BaTiO_3$ thin films doped with Zr were enhanced greatly. The Ti^{4+} was substituted by Zr^{4+} and $BaTi_{(1-x)}Zr_xO_3$ was formed, which led to the increase of the interplanar spacing of crystals and the shift of the charge center inside the crystal. These two factors synergistically increased the piezoelectric coefficients (285–290 pC/cm^2) of $BaTi_{(1-x)}Zr_xO_3$ ($0.02 \leq_x \leq 0.08$) hybrid piezoelectric materials [190]. In 2017, Zhuang et al. fabricated a flexible tactile sensor based on Ce-doped $BaTiO_3$ nanofibers and analyzed the doping mechanism and working process [191]. In addition, a nanocomposite film force sensor based on La-modified PZT/paint was developed with brushing technique [192], which exhibited high performance with the output voltage of 51.7 mV and the power of 0.38 μW when working in the –31– piezoelectric mode (Figure 2.20b).

Not only the inorganic piezoelectric materials but also organic piezoelectric materials like PVDF and its derived copolymers could be doped with chemical elements, such as Ag, Ce^{3+} and Eu^{3+}, to improve the piezoelectric properties [168,169,193]. For example, the addition of Eu^{3+} could prevent external polarization conditions from damaging the β phase of (PVDF-HFP) and induce self-polarization of its electroactive β phase, which maintains the β phase content and leads to higher piezoelectric coefficients compared to the pure P(VDF-HFP) [168]. As shown in Figure 2.18c, the Ce^{3+}-doped PVDF/graphene nanofibers for nanopressure sensor and acoustic nanogenerator were developed via in situ poled fabrication strategy [169], which exhibited a well-behaved, linear response to the applied pressure and could detect a low pressure of 2 Pa. In addition, for PVDF organic piezoelectric materials, rGO or graphene are sometimes used to enhance piezoelectricity by transforming the PVDF phase from the nonpolar α phase to the polar β phase [194,195]. However, too high doping concentration of rGO would result in the decrease in piezoelectric performance. This behavior can be attributed to the enhanced screening of piezoelectric polarization with the increase of film conductivity due to higher rGO doping [166,196].

Although the performance of piezoelectric sensors could be improved by elaborately choosing the radius of dopants and dispersing dopants uniformly in matrix, all the reports only make qualitative analysis from semiconductor physics (lattice), mechanism (stress redistribution) and electricity (dielectric constant). Therefore, the detailed principle behind doping should be further researched and analyzed quantitatively to establish effective guidelines.

2.3.2.4 Using Piezoelectric Techniques to Prepare Strain/Pressure-Driven Gate-Controlled Sensing Transistors

Utilizing piezotronics to fabricate strain-/pressure-driven gated-controlled sensing transistors to improves the performance of piezoelectric sensors. The coupling effect of piezoelectricity and semiconductor properties of piezoelectric semiconductor materials brings about some impressive characteristics and provides theoretical basis for piezotronics [197–199]. In 2006 and 2007, Wang proposed the concept of piezotronics, which aimed to study the coupling mechanism between piezoelectricity and conductivity of piezoelectric semiconductor materials [200]. Since then, this field has attracted increasing research interests and many piezoelectric sensors based on piezoelectric transistors have been reported.

When an external force is applied to the piezoelectric semiconductor materials, piezoelectric potential is created due to the strain-/pressure-induced polarization charges. If the piezoelectric material is connected to a metal [metal–semiconductor (M–S) contact formed] or other semiconductor (p–n junction formed), polarization charges at interface modulate the local interfacial band structure and then influence the concentration/distribution of free carriers and interfacial electronic charged states in a M–S contact or p–n junction. Consequently, the current in piezoelectric semiconductor devices is changed [201,202].

In recent years, many piezoelectric sensors based on piezotronic transistors by replacing the external gating voltage with the piezoelectric polarization have been reported, in which the current of transistors is modulated by the applied strain on piezoelectric semiconductor materials. As shown in Figure 2.21a, a strain sensor based on individual ZnO piezoelectric fine wires was fabricated with a high gauge factor of 1250 at a strain of 0.99%, which is much higher than that of sensors discussed above [203]. Although many techniques have been reported to synthesize one-dimensional semiconductor nanostructures [204,205], they are either cumbersome or incompatible with state-of-the-art microfabrication technologies. Moreover, the single nanowire cannot withstand large force, which hinders their applications.

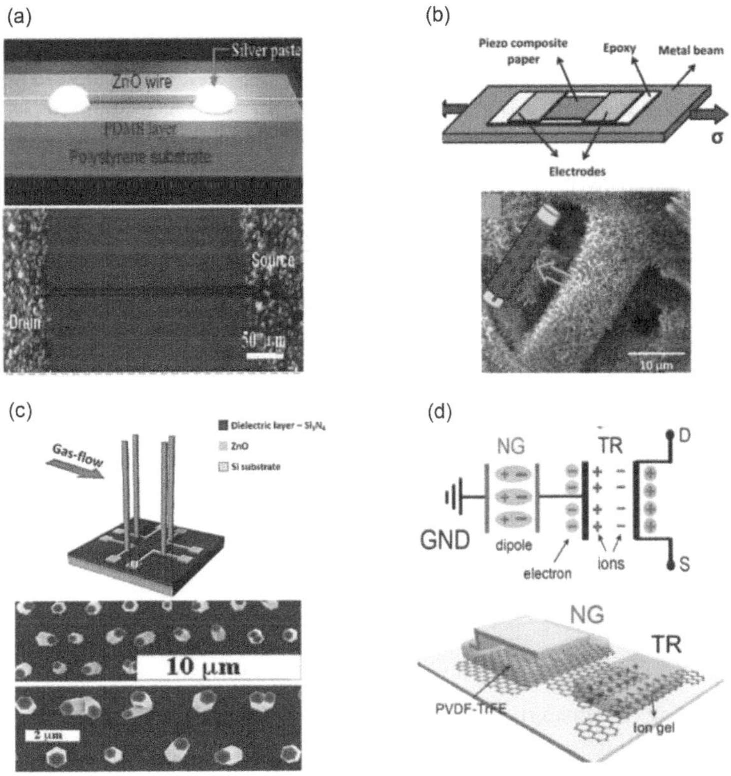

FIGURE 2.21 Sensors based on piezoelectric effect [110]. (a) Strain sensor based on individual ZnO piezoelectric fine-wires: structural schematic diagram (upper part) and SEM image (lower part). (b) Strain sensor based on ZnO-paper: structural schematic diagram (upper part) and SEM image (lower part). (c) Sensor based on individually contacted vertical piezoelectric nanowires: structural schematic diagram (upper part) and SEM image (lower part). (d) Transistor with (PVDF-TrFE) as active materials and graphene as gate electrodes: circuit diagram (upper part) and structural schematic diagram (lower part).

Therefore, piezotronic thin films were researched and fabricated by the state-of-the-art microfabrication technologies and overcame the limitations existing in 1D nanostructures. For example, a strain sensor based on In-doped ZnO nanobelts was developed by the chemical vapor deposition technique, which exhibited high gauge factor of 4036 (compressive strain) and 135 (tensile strain) [206]. As shown in Figure 2.21b, a strain sensor based on ZnO–paper nanocomposite materials was fabricated to monitor the real-time current response under both static and dynamic mechanical loading. Besides, a micro-strain sensor based on ZnO microwire network structure was developed under low temperature, of which the highest gauge factor was ~900 [207]. However, *c*-axes of crystal in thin films cannot be aligned well, which reduces the macroscopic piezoelectricity of piezotronic thin film and decreases the performance of piezoelectric sensors. Meanwhile, the response of the piezotronic thin films to normal force is nonideal due to its plane structure. Therefore, vertically well-aligned ZnO nanowires were fabricated to detect small normal pressure. For example, a device based on individually contacted vertical piezoelectric nanowires was fabricated by ultraviolet and e-beam lithography techniques [208], which exhibited a high linearity with flow rate (Figure 2.21c). In addition, Liao et al. reported a pressure sensor based on vertically well-aligned ZnO nanorods [209] (Figure 2.17e). Due to the electron-tunneling modulation effect of MgO nanolayer, the sensor had a "on/off" ratio up to 10^5 and exhibited a high sensitivity (7.1×10^4g/f) and a fast response time (128 ms). As shown in Figure 2.21d, a transistor with PVDF-TrFE as active materials and graphene as gate electrodes was fabricated to form an active-matrix strain sensor, which exhibited a high gauge factor (389) and a low detectable strain (0.008%) [210].

Different from Ohmic-contacted sensors, of which the sensitivity is decided by bulk channel modulation under strain or pressure, the sensitivity of piezoelectric sensors based on Schottky contact is mainly decided by tuning local SBH at interface. Therefore, how to choose piezoelectric semiconductor and metal whose SBH at interface is highly sensitive to strain is still challenge in the future. In addition, various piezoelectric semiconductor materials with more excellent piezoelectricity are waiting to be investigated and developed.

2.3.3 Stretchable Coil Design and Fabrication

The receiving or transmitting coils consisted of two overlapping stretchable windings [211]. The winding was designed to be a square with the side length of 73 mm (d_{out}) and six turns in total. The winding consisted of treble strand serpentine copper traces with the width of 0.3 mm (w) for one of the treble strand copper traces, the 0.1-mm thickness (t), the space between the two serpentine copper traces of 0.2 mm (s), the radian of serpentine copper traces of 150° (θ), the length of tangential path in serpentine copper traces of 1 mm (l), the arc radius of one winding of 1.25 mm (r) and the distance between two windings of 0.1 mm (d). Finally, there was a square with the side length of 23 mm (d_{in}) in the center of coil. The coil was prepared by laser cutting the copper foil according to the dimension parameter. The device was fabricated starting from the substrate that was coating a layer Ecoflex of 0.1-mm thickness on a clean glass panels (Figure 2.22I and II) and then putting the prepared coil of copper on it (Figure 2.22III). A new layer Ecoflex was coated on the first coil to insulate the second coil (Figure 2.22IV and V). The entire device was peeled off the glass after coating the Ecoflex at the top as encapsulation layer (Figure 2.22VI–VIII).

2.3.4 Electrode Design and Process

Current instruments for detecting ECG signals or EMG signals require electrodes to be attached to the surface of human skin, and the material commonly used for electrodes is currently conductive rubber. In addition, metal sheets and conductive cloth can be used as electrodes in principle. Among them, the metal sheet is not

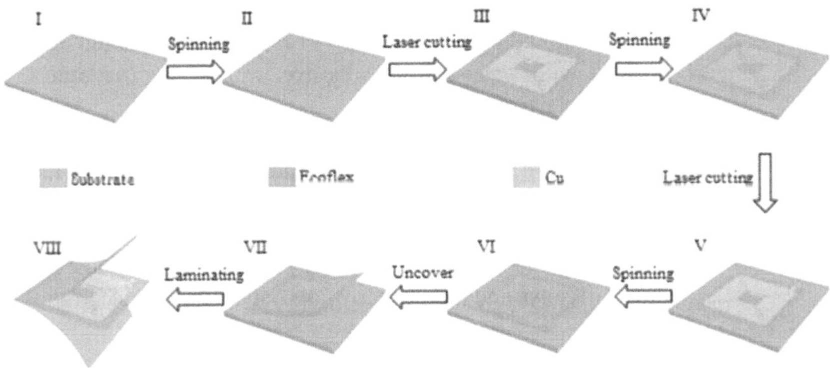

FIGURE 2.22 Fabrication of the inductance coil, involving (I–III) putting the wingdings on an Ecoflex encapsulation layer, (IV, V) putting another same wingdings on a new Ecoflex layer to insulate the two layers of windings, (VI, VII) encapsulating the upper wingdings with Ecoflex and peeling the complete coil off and (VIII) exploded-view schematic illustration of the reductive coil [211].

stretchable, and the fit with the skin is very poor, and some metals are easy to rust, so it is not suitable for contacting the skin directly as electrode; the conductive cloth has good fit with the skin, and the stretchability is related to the weaving method, so it can have good stretchability, but the metal particles in the outer layer of the fiber are easy to wash off after many times, so it is not particularly suitable as electrode; the conductive rubber sheet has certain stretchability, but it generally does not meet the requirements. And the conductive rubber sheet will significantly increase its resistance after stretching, and even lose its conductivity, so the existing conductive rubber sheet directly used as an electrode cannot play its stretching properties.

A method for preparing an extensible flexible electrode comprises the following steps:

a. Hollowing out of the conductive rubber sheet using high-precision mechanical cutting or laser cutting to give it an island–bridge structure.
b. Bonding one side of the skeletonized conductive rubber sheet to the metal sheet with leads using conductive adhesive, bonding only a small area of the edge of the conductive rubber sheet to maintain stretch.
c. Filling the gaps of the hollow structure and encapsulation with elastic sealant with good flowability, high elongation at break, low modulus of elasticity and low hardness.

Furthermore, said step (c) is specified as:

1. will be hollowed out and has been bonded to the metal conductive rubber sheet into the rectangular container without a lid, the conductive rubber sheet is not bonded to the surface of the metal

sheet to be water-soluble adhesive on the inner bottom of the container and the other inner walls of the container should also be coated with water-soluble adhesive, waiting for the water-soluble adhesive curing;

2. then inject the elastic sealant with good fluidity, high elongation at break, low modulus of elasticity and low hardness into the container; and
3. leave the container for a period of time, wait for the elastic sealant to solidify, then soak the container into the water to make the hydrosol dissolve and then directly take out the flexible electrode.

Embodiments of the present invention are described in detail in this section in conjunction with the accompanying drawings. It is to be noted that the embodiments and the features in the embodiments in this section can be combined with each other in any combination, provided there is no conflict.

2.3.4.1 Example of Implementation I

This section provides an extendable flexible electrode as shown in Figure 2.23, where the electrode is a skeletonized island–bridge structure, where island 1 and bridge 2 are connected, with an island on the right side having an enlarged area for affixing a metal sheet, and the skeletonized island–bridge structure has a similar bridge width and thickness dimensions to ensure that the structure does not flex out of plane. Bridge 2 is curved in shape. The island–bridge structure is periodically arranged. The material of electrodes is conductive rubber. The hollowing is processed by high-precision mechanical cutting or laser cutting. The diameter of the island in this embodiment is 2 mm (except for the enlarged one) and the width of the bridge is 0.5 mm.

As shown in Figure 2.24, this chapter also provides a method for preparing a flexible electrode, comprising the following steps:

a. Hollowing out of the conductive rubber sheet using high-precision mechanical cutting or laser cutting to give it an island–bridge structure;

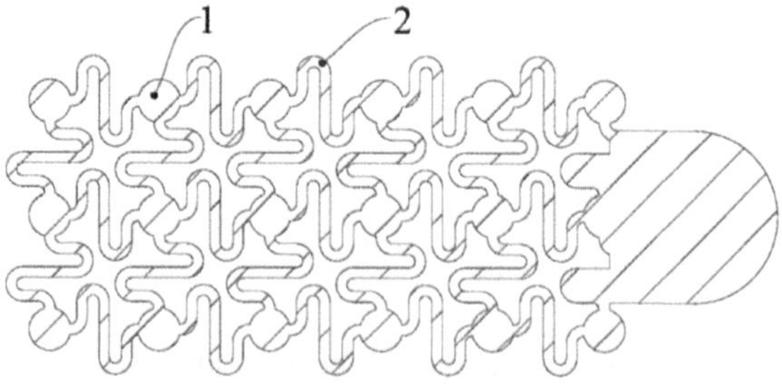

FIGURE 2.23 An extendable flexible electrode [212].

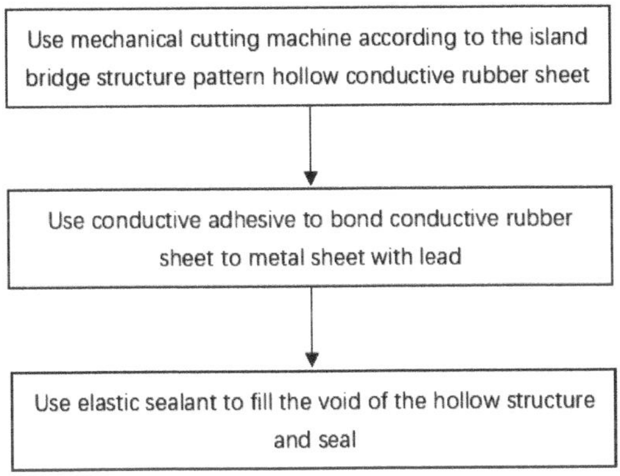

FIGURE 2.24 A flexible electrode preparation method [212].

b. Bonding one side of the skeletonized conductive rubber sheet to the metal sheet with leads using conductive adhesive, bonding only a small area of the edge of the conductive rubber sheet to maintain stretch; and
c. Filling the gaps in the hollow structure as well as the encapsulation with WACKER E41 elastomeric sealant:
    1. Will be hollowed out and has been bonded to the metal conductive rubber sheet into the rectangular container without a lid, the conductive rubber sheet is not bonded to the surface of the metal sheet to be water-soluble adhesive on the inner bottom of the container and the other inner walls of the container should also be coated with water-soluble adhesive, waiting for the water-soluble adhesive curing;
    2. Then inject WACKER E41 elastomeric sealant into the container; and
    3. Leave the container for a period of time, wait for the elastic sealant to solidify, then soak the container into the water to make the hydrosol dissolve and then directly take out the flexible electrode.

The method in this chapter significantly increases the stretchability compared to the direct use of conductive rubber sheets as electrodes. Compared to using metal sheets as electrodes, the method in this chapter has excellent stretchability and adhesion and does not rust. The method in this chapter has a higher service life than using conductive cloth as the electrode.

2.4 SUMMARY AND PERSPECTIVE

The future development direction of information technology is to achieve comprehensive and efficient utilization of physical world, information data and human social resources through the interaction and fusion of human and external information. To achieve this goal, the biggest bottleneck lies in the traditional rigid inorganic integrated devices cannot be highly closely fitted and integrated with the flexible tissue of human body. It is an effective way to break through this bottleneck by designing the micro–nano structure of inorganic components at the physical bottom and integrating it with the flexible stretchable substrate to achieve the acquisition, processing and wireless transmission of human body information. This chapter reviews the advanced manufacturing technologies of flexible piezoelectric arrays, which can break through technical barriers when traditional manufacturing methods are faced with complex surfaces and multifunctional integrations, especially for large-scale efficient or high-precision micro-operations. This chapter introduces the research background, design emphasis and fabrication process of flexible piezoelectric arrays. The latest research trends, achievements, challenges and future prospects in these fields are analyzed and discussed in detail.

REFERENCES

1. Khang, D.-Y. A stretchable form of single-crystal silicon for high-performance electronics on rubber substrates. *Science*, 2006, 311(5758), 208–212.
2. Zhang, Y.; Yan, Z.; Nan, K., et al. A mechanically driven form of Kirigami as a route to 3D mesostructures in micro/nanomembranes. *Proceedings of the National Academy of Sciences of the United States of America*, 2015, 112(38), 11757.
3. Howarth, T.R.; Ting, R.Y. Electroacoustic evaluations of 1-3 piezocomposite SonoPanel™ materials. *IEEE Transactions on Ultrasonics Ferroelectrics & Frequency Control*, 2000, 47(4), 886–894.
4. Zhengkang, L. Preparation and structure and properties of 3-3 type piezoelectric composites. *Xi'an University of Technology*, 2010.
5. Li, L. 1-3 Series of Piezoelectric Composites and Hydroacoustic Transducers. *Beijing University of Posts and Telecommunications*, 2008.
6. Lous, G.M.; Cornejo, I.A.; Mcnulty, T.F., et al. Fabrication of piezoelectric ceramic/polymer composite transducers using fused deposition of ceramics. *Journal of the American Ceramic Society*, 2008, 83(1), 124–128.
7. Ohara, Y.; Miyayama, M.; Koumoto, K., et al. PZT-polymer composites fabricated with YAG laser cutter. *Sensors & Actuators a Physical*, 1994, 40(3), 187–190.
8. Janas, V.F.; Mcnulty, T.F.; Walker, F.R., et al. Processing of 1-3 piezoelectric ceramic/polymer composites. *Journal of the American Ceramic Society*, 1995, 78(9), 2425–2430.
9. Zhang, G. et al. Flexible three-dimensional interconnected piezoelectric ceramic foam based composites for highly effcient concurrent mechanical and thermal energy harvesting. *Energy & Environmental Science*, 2018, 11(8), 2046–2056.
10. Fan, F.R.; Tang, W.; Wang, Z.L. *Advanced Materials*, 2016, 28, 4283.
11. Park, S.; Kim, H.; Vosgueritchian, M.; Cheon, S.; Kim, H.; Koo, J.H.; Kim, T.R.; Lee, S.; Schwartz, G.; Chang, H.; Bao, Z. *Advanced Materials*, 2014, 26, 7324.
12. Wu, H.; Huang, Y.A.; Xu, F.; Duan, Y.Q.; Yin, Z.P. *Advanced Materials*, 2016, 28, 9881.
13. Erturk, A.; Inman, D.J. *Piezoelectric Energy Harvesting*. John Wiley & Sons, Ltd., Chichester, 2011.
14. Bowen, C.R.; Kim, H.A.; Weaver, P.M.; Dunn, S. Piezoelectric and ferroelectric materials and structures for energy harvesting applications. *Energy & Environmental Science*, 2014, 7, 25.
15. Briscoe, J.; Dunn, S. Piezoelectric nanogenerators – a review of nanostructured piezoelectric energy harvesters. *Nano Energy*, 2015, 14, 15.

16. Dagdeviren, C.; Li, Z.; Wang, Z.L. Energy harvesting from the animal/human body for self-powered electronics. *Annual Review of Biomedical Engineering*, 2017, 19, 85.
17. Qi, Y.; Mcalpine, M.C. Nanotechnology-enabled flexible and biocompatible energy harvesting. *Energy & Environmental Science*, 2010, 3, 1275.
18. Trung, T.Q.; Lee, N.E. Flexible and stretchable physical sensor integrated platforms for wearable human-activity monitoringand personal healthcare. *Advanced Materials*, 2016, 28, 1338.
19. Persano, L.; Dagdeviren, C.; Su, Y.; Zhang, Y.; Girardo, S.; Pisignano, D.; Huang, Y.; Rogers, J.A. High performance piezoelectric devices based on aligned arrays of nanofibers of poly (vinylidenefluoride-co-trifluoroethylene). *Nature Communications*, 2013, 4, 1633.
20. Lee, J.; Lee, K.Y.; Gupta, M.K.; Kim, T.Y.; Lee, D.-Y.; Oh, J.; Ryu, C.; Yoo, W.J.; Kang, C.-Y.; Yoon, S.-J.; Yoo, J.-B.; Kim, S.-W. Highly stretchable piezoelectric-pyroelectric hybrid nanogenerator. *Advanced Materials*, 2014, 26, 765.
21. Sun, C.L.; Shi, J.; Bayerl, D.J.; and Wang, X.D. PVDF microbelts for harvesting energy from respiration. *Energy & Environmental Science*, 2011, 4, 4508.
22. Zhang, Z.; Yao, C.; Yu, Y.; Hong, Z.; Zhi, M.; Wang, X. Mesoporous piezoelectric polymer composite films with tunable mechanical modulus for harvesting energy from liquid pressure fluctuation. *Advanced Functional Materials*, 2016, 26, 6760.
23. Zabek, D.; Taylor, J.; Boulbar, E.L.; Bowen, C.R. Micropatterning of flexible and free standing polyvinylidene difluoride (PVDF) films for enhanced pyroelectric energy transformation. *Advanced Energy Materials*, 2015, 5, 1401891.
24. Horiuchi, S.; Tokura, Y. Organic ferroelectrics. *Nature Materials*, 2008, 7, 357.
25. Damjanovic, D. Contributions to the piezoelectric effect in ferroelectric single crystals and ceramics. *Journal of the American Ceramic Society*, 2005, 88, 2663.
26. Koka, A.; Zhou, Z.; Sodano, H.A. Vertically aligned $BaTiO_3$ nanowire arrays for energy harvesting. *Energy & Environmental Science*, 2014, 7, 288.
27. Xu, S.; Hansen, B.J.; Wang, Z.L. Piezoelectric-nanowire-enabled power source for driving wireless microelectronics. *Nature Communications*, 2010, 1, 93.
28. Koka, A.; Sodano, H.A. High-sensitivity accelerometer composed of ultra-long vertically aligned barium titanate nanowire arrays. *Nature Communications*, 2013, 4, 2682.
29. Zhang, G.; Zhang, X.; Huang, H.; Wang, J.; Li, Q.; Chen, L.-Q.; Wang, Q. Toward wearable cooling devices: highly flexible electrocaloric $Ba_{0.67}Sr_{0.33}TiO_3$ nanowire arrays. *Advanced Materials*, 2016, 28, 4811.
30. Dagdeviren, C.; Yang, B.D.; Su, Y.; Tran, P.L.; Joe, P.; Anderson, E.; Xia, J.; Doraiswamy, V.; Dehdashti, B.; Feng, X.; Lu, B.; Poston, R.; Khalpey, Z.; Ghaffari, R.; Huang, Y.; Slepian, M.J.; Rogers, J.A. Conformal piezoelectric energy harvesting and storage from motions of the heart, lung, and diaphragm. *Proceedings of the National Academy of Sciences of the United States of America*, 2014, 111, 1927.
31. Hwang, G.-T.; Park, H.; Lee, J.-H.; Oh, S.; Park, K.-I.; Byun, M.; Park, H.; Ahn, G.; Jeong, C.K.; No, K.; Kwon, H.; Lee, S.-G.; Joung, B.; Jae, K. Self-powered cardiac pacemaker enabled by flexible single crystalline PMN-PT piezoelectric energy harvester. *Advanced Materials*, 2014, 26, 4880.
32. Qi, Y.; Kim, J.; Nguyen, T.D.; Lisko, B.; Purohit, P.K.; McAlpine, M.C. Enhanced piezoelectricity and stretchability in energy harvesting devices fabricated from buckled PZT ribbons. *Nano Letters*, 2011, 11, 1331.
33. Seo, M.-H.; Yoo, J.-Y.; Choi, S.-Y.; Lee, J.-S.; Choi, K.-W.; Jeong, C.K.; Lee, K.J.; Yoon, J.-B. Versatile transfer of an ultralong and seamless nanowire array crystallized at high temperature for use in high-performance flexible devices. *ACS Nano*, 2017, 11, 1520.

34. Hwang, G.-T.; Annapureddy, V.; Han, J.H.; Joe, D.J.; Baek, C.; Park, D.Y.; Kim, D.H.; Park, J.H.; Jeong, C.K.; Park, K.-I.; Choi, J.-J.; Kim, D.K.; Ryu, J.; Lee, K.J. Self-powered wireless sensor node enabled by an aerosol-deposited PZT flexible energy harvester. *Advanced Energy Materials*, 2016, 6, 1600237.
35. Jeong, C.K.; Cho, S.B.; Han, J.H.; Park, D.Y.; Yang, S.; Park, K.-I.; Ryu, J.; Sohn, H.; Chung, Y.-C.; Lee, K.J. Flexible highly-effective energy harvester via crystallographic and computational control of nanointerfacial morphotropic piezoelectric thin film. *Nano Reserarch*, 2017, 10, 437.
36. Yang, Y.Y.; Zhang, H.; Zhu, G.; Lee, S.; Lin, Z.H.; Wang, Z.L. Flexible hybrid energy cell for simultaneously harvesting thermal, mechanical, and solar energies. *ACS Nano*, 2013, 7, 785.
37. Bhang, S.H.; Jang, W.S.; Han, J.; Yoon, J.K.; La, W.-G.; Lee, E.; Kim, Y.S.; Shin, J.-Y.; Lee, T.-J.; Baik, H.K.; Kim, B.-S. Zinc oxide nanorod-based piezoelectric dermal patch for wound healing. *Advanced Functional Materials*, 2017, 27, 1603497.
38. Ha, M.; Lim, S.; Park, J.; Um, D.S.; Lee, Y.; Ko, H. Bioinspired interlocked and hierarchical design of ZnO nanowire arrays for static and dynamic pressure-sensitive electronic skins. *Advanced Functional Materials*, 2015, 25, 2841.
39. Furukawa, T.; Ishida, K.; Fukada, E. Erratum: Piezoelectric properties in the composite systems of polymers and PZT ceramics. *Journal of Applied Physics*, 1980, 51, 4561.
40. Furukawa, T.; Fujino, K.; Fukada, E. Electromechanical properties in the composites of epoxy resin and PZT ceramics. *Japanese Journal of Applied Physics*, 1976, 15, 2119.
41. Jeong, C.K.; Lee, J.; Han, S.; Ryu, J.; Hwang, G.T.; Park, D.Y.; Park, J.H.; Lee, S.S.; Byun, M.; Ko, S.H.; Le, K.J. A hyper-stretchable elastic-composite energy harvester. *Advanced Materials*, 2015, 27, 2866.
42. Yang, Y.; Jung, J.H.; Yun, B.K.; Zhang, F.; Pradel, K.C.; Guo, W.; Wang, Z.L. Flexible pyroelectric nanogenerators using a composite structure of lead-free $KNbO_3$ nanowires. *Advanced Materials*, 24, 5357.
43. Chen, X.; Xu, S.Y.; Yao, N.; Shi, Y. 1.6 V nanogenerator for mechanical energy harvesting using PZT nanofibers. *Nano Letter*, 2010, 10, 2133.
44. Jung, J.J.H.; Lee, M.; Hong, J.I.; Ding, Y.; Chen, C.Y.; Chou, L.J.; Wang, Z.L. Lead-free $NaNbO_3$ nanowires for a high output piezoelectric nanogenerator. *ACS Nano*, 2011, 5, 10041.
45. Ploss, B.; Ploss, B.; Shin, F.G.; Chan, H.L.W.; Choy, C.L. Pyroelectric or piezoelectric compensated ferroelectric composites. *Applied Physics Letters*, 2000, 76, 2776.
46. Park, K.; Lee, M.; Liu, Y.; Moon, S.; Hwang, G.T.; Zhu, G.; Kim, J.E.; Kim, S.O.; Kim, D.K.; Wang, Z.L.; Lee, K.J. Flexible nanocomposite generator made of $BaTiO_3$ nanoparticles and graphitic carbons. *Advanced Materials*, 2012, 24, 2999.
47. Zhou, Z.; Tang, H.; Sodano, H.A. Scalable synthesis of morphotropic phase boundary lead zirconium titanate nanowires for energy harvesting. *Advanced Materials*, 2014, 26, 7547.
48. Naskar, A.K.; Keum, J.K.; Boeman, R.G. Polymer matrix nanocomposites for automotive structural components. *Nature Nanotechnology*, 2016, 11, 1026.
49. Kim, K.; Zhu, W.; Qu, X.; Aaronson, C.; McCall, W.R.; Chen, S.; Sirbuly, D.J. 3D optical printing of piezoelectric nanoparticle–polymer composite materials. *ACS Nano*, 2014, 8, 9799.
50. Meza, L.R.; Das, S.; Greer, J.R. Strong, lightweight, and recoverable three-dimensional ceramic nanolattices. *Science*, 2014, 345, 1322.

51. Shin, S.H.; Kim, Y.H.; Lee, M.H.; Jung, J.Y.; Nah, J. Hemispherically aggregated BaTiO3 nanoparticle composite thin film for high-performance flexible piezoelectric nanogenerator. *ACS Nano*, 2014, 8, 2766.
52. Xu, S.; Yeh, Y.; Poirier, G.; McAlpine, M.C.; Register, R.A.; Yao, N. Flexible piezoelectric PMN–PT nanowire-based nanocomposite and device. *Nano Letter*, 2013, 13, 2393.
53. Liangke, W. Piezoelectric properties of PVDF-HFP composites. *Zhejiang University*, 2016.
54. Jing, C. Research on high performance piezoelectric composites. *Chongqing University*, 2019.
55. Bowen, C.; Wan, C. Multiscale-structuring of polyvinylidene fluoride for energy harvesting: the impact of molecular-, micro- and macro-structure. *Journal of Materials Chemistry A*, 2017, 5(7), 3091–3128.
56. Chunlei, W.; Jichao, L.; Minglei, Z. Piezoelectric Ferroelectric Physics. *Science Publishers,* 2009.
57. Martins, P.; Lopes, A.C.; Lanceros-Mendez, S. Electroactive phases of poly (vinylidene fluoride): Determination, processing and applications. *Progress in Polymer Science*, 2014, 39(4), 683–706.
58. Kim, G.H.; Hong, S.M.; Seo, Y. Piezoelectric properties of poly (vinylidene fluoride) and carbon nanotube blends: β-phase development. *Physical Chemistry Chemical Physics*, 2009, 11(44), 10506–10512.
59. Lingling, Z.; Shishang, G. Progress in the study of ferroelectric polymer PVDF and its copolymers. *Advances in Physics,* 2016, 36(2), 35–45.
60. Jian, J. Polymer physics. *Northwestern Polytechnic University Press*, 2015.
61. Zhang, Q.M.; Xu, H.; Fei, F., et al. Critical thickness of crystallization and discontinuous change in ferroelectric behavior with thickness in ferroelectric polymer thin films. *Journal of Applied Physics*, 2001, 89(5), 2613–2616.
62. Tashiro, K.; Tadokoro, H.; Kobayashi, M. Structure and piezoelectricity of poly(vinylidene fluoride). *Ferroelectrics*, 1981, 32(1), 167–175.
63. Wu, L.; Huang, G.; Hu, N. et al. Improvement of the piezoelectric properties of PVDF-HFP using AgNWs. *RSC Advances*, 2014, 4(68), 35896–35903.
64. Sencadas, V.; Gregoriojr, R.; Lanceros-Mendez, S. α to β phase transformation and microestructural changes of PVDF films induced by uniaxial stretch. *Journal of Macromolecular Science Part B*, 2009, 48(3), 514–525.
65. Salimi, A.; Yousefi, A.A. Analysis method: FTIR studies of β-phase crystal formation in stretched PVDF films. *Polymer Testing*, 2003, 22(6), 699–704.
66. Gomes, J.; Nunes, J.S.; Sencadas, V., et al. Influence of the β-phase content and degree of crystallinity on the piezo- and ferroelectric properties of poly(vinylidene fluoride). *Smart Materials & Structures*, 2010, 19(6), 065010.
67. Hattori, T.; Hikosaka, M.; Ohigashi, H. The crystallization behaviour and phase diagram of extended-chain crystals of poly(vinylidene fluoride) under high pressure. *Polymer*, 1996, 37(1), 85–91.
68. Matsushige, K.; Nagata, K.; Takemura, T. Direct observation of crystal transformation process of poly (vinylidene fluoride) under high pressure by PSPC X-ray system. *Japanese Journal of Applied Physics*, 1978, 17(3), 467–472.
69. Scheinbeim, J.; Nakafuku, C.; Newman, B.A., et al. High-pressure crystallization of poly(vinylidene fluoride). *Journal of Applied Physics*, 1979, 50(6), 4399–4405.
70. Hattori, T.; Kanaoka, M.; Ohigashi, H. Improved piezoelectricity in thick lamellar β-form crystals of poly(vinylidene fluoride) crystallized under high pressure. *Journal of Applied Physics*, 1996, 79(4), 2016–2022.

71. Satish, B.; Sridevi, K.; Vijaya, M.S. Study of piezoelectric and dielectric properties of ferroelectric PZT-polymer composites prepared by hot-press technique. *Journal of Physics D Applied Physics*, 2002, 35(35), 2048.
72. Siddiqui, S.; Kim, D.I.; Le, T.D., et al. High-performance flexible lead-free nanocomposite piezoelectric nanogenerator for biomechanical energy harvesting and storage. *Nano Energy*, 2015, 15, 177–185.
73. Wu, L.; Yuan, W.; Hu, N., et al. Improved piezoelectricity of PVDF-HFP/carbon black composite films. *Journal of Physics D Applied Physics*, 2014, 47(13), 394–401.
74. Wu, C.M.; Chou, M.H.; Zeng, W.Y. Piezoelectric response of aligned electrospun polyvinylidene fluoride/carbon nanotube nanofibrous membranes. *Nanomaterials*, 2018, 8(6), 420.
75. Shizhen, Y. A study on the theory and application of flexible electronic multilayer island bridge structure in monitoring and inspection of high speed railway structures. *University of Chinese Academy of Sciences*, 2019.
76. Ko, H.C.; Stoykovich, M.P.; Song, J., et al. A hemispherical electronic eye camera based on compressible silicon optoelectronics. *Nature*, 2008, 454(7205), 748–753.
77. Lee, J.; Wu, J.; Shi, M., et al. Stretchable GaAs photovoltaics with designs that enable high areal coverage. *Advanced Materials*, 2011, 23(8), 986–991.
78. Kim, D.-H.; Kim, Y.S.; Wu, J., et al. Ultrathin silicon circuits with strain-isolation layers and mesh layouts for high-performance electronics on fabric, vinyl, leather, and paper. *Advanced Materials*, 2009, 21(36), 3703–3707.
79. Hu, X.; Krull, P.; De Graff, B., et al. Stretchable inorganic-semiconductor electronic systems. *Advanced Materials*, 2011, 23(26), 2933–2936.
80. Zhang, Y.; Fu, H.; Xu, S., et al. A hierarchical computational model for stretchable interconnects with fractal-inspired designs. *Journal of the Mechanics and Physics of Solids*, 2014, 72, 115–130.
81. Kim, D.H.; Kim, H.S.; Kim, R.H., et al. Stretchable batteries with self-similar serpentine interconnects and integrated wireless recharging systems. *Nature Communications*, 2013, 4, 1543.
82. Aksel, E.; Jones, J.L. Advances in lead-free piezoelectric materials for sensors and actuators. *Sensors*, 2010, 10, 1935–1954.
83. Yewang, S.; Han, Z.; Shizhen, L., et al. A stretchable ultrasound transducer array. *Institute of Mechanics, Chinese Academy of Sciences*, 2017.
84. Someya, T. *Stretchable Electronics*, Wiley-VCH, Weinheim Germany, 2013.
85. Rogers, J.A.; Someya, T.; Huang, Y. Materials and mechanics for stretchable electronics. *Science*, 2010, 327(5973), 1603–1607.
86. Khang, D.Y.; Jiang, H.Q.; Huang, Y.; Rogers, J.A. A stretchable form of single-crystal silicon for high-performance electronics on rubber substrates. *Science*, 2006, 311(5758), 208–212.
87. Someya, T., Sekitani, T., Iba, S., Kato, Y., Kawaguchi, H., Sakurai, T. A large-area, flexible pressure sensor matrix with organic field-effect transistors for artificial skin applications. *Proceedings of the National Academy of Sciences of the United States of America*, 2004, 101(27), 9966–9970.
88. Lee, J.; Wu, J.; Shi, M.; Yoon, J.; Park, S.I.; Li, M.; Liu, Z.; Huang, Y.; Rogers, J.A. Stretchable GaAs photovoltaics with designs that enable high areal coverage. *Advanced Materials*, 2011, 23(8), 986–991.
89. Tang, R.; Huang, H.; Tu, H.; Liang, H.; Liang, M.; Song, Z.; Xu, Y.; Jiang, H.; Yu, H. Origami-enabled deformable silicon solar cells. *Applied Physics Letters*, 2014, 104(8), 083501.

90. Axisa, F.; Schmitt, P.M.; Gehin, C.; Delhomme, G.; McAdams, E.; Dittmar, A. Flexible technologies and smart clothing for citizen medicine, home healthcare, and disease prevention. *IEEE Transactions on Information Technology B,* 2005, 9(3), 325–336.
91. Dagdeviren, C.; Hwang, S.-W.; Su, Y.; Kim, S.; Cheng, H.; Gur, O.; Haney, R.; Omenetto, F.G.; Huang, Y.; Rogers, J.A. Transient, biocompatible electronics and energy harvesters based on ZnO. *Small,* 2013, 9(20), 3398–3404
92. Kim, D.H.; Lu, N.S.; Ma, R.; Kim, Y.S.; Kim, R.H.; Wang, S.D.; Wu, J.; Won, S.M.; Tao, H.; Islam, A.; Yu, K.J.; Kim, T.I.; Chowdhury, R.; Ying, M.; Xu, L.Z.; Li, M.; Chung, H.J.; Keum, H.; McCormick, M.; Liu, P.; Zhang, Y.W.; Omenetto, F.G.; Huang, Y.G.; Coleman, T.; Rogers, J.A. Epidermal electronics. *Science,* 2011, 333(6044), 838–843.
93. Viventi, J.; Kim, D.H.; Moss, J.D.; Kim, Y.S.; Blanco, J.A.; Annetta, N.; Hicks, A.; Xiao, J.L.; Huang, Y.G.; Callans, D.J.; Rogers, J.A.; Litt, B. A Conformal, Bio-Interfaced Class of Silicon Electronics for Mapping Cardiac Electrophysiology. *Science Translational Medicine,* 2010, 2(24), 24ra22.
94. Ko, H.C.; Stoykovich, M.P.; Song, J.Z.; Malyarchuk, V.; Choi, W.M.; Yu, C.J.; Geddes, J.B.; Xiao, J.L.; Wang, S.D.; Huang, Y.G.; Rogers, J.A. A hemispherical electronic eye camera based on compressible silicon optoelectronics. *Nature,* 2008, 454(7205), 748–753.
95. Vanfl eteren, J.; Gonzalez, M.; Bossuyt, F.; Hsu, Y.Y.; Vervust, T.; De Wolf, I.; Jablonski, M. Printed circuit board technology inspired stretchable circuits. *MRS Bulletin,* 2012, 37(3), 254–260.
96. Rogers, J.A.; Someya, T.; Huang, Y.G. Materials and mechanics for stretchable electronics. *Science,* 2010, 327(5973), 1603–1607.
97. Lazarus, A.; Miller, J.T.; Reis, P.M. Continuation of equilibria and stability of slender elastic rods using an asymptotic numerical method. *Journal of the Mechanics and Physics of Solids,* 2013, 61(8), 1712–1736.
98. Hsu, Y.-Y.; Gonzalez, M.; Bossuyt, F.; Vanfl eteren, J.; De Wolf, I. Polyimide-enhanced stretchable interconnects: design, fabrication, and characterization. *IEEE Transactions on Electron Devices,* 2011, 58(8), 2680–2688.
99. Lacour, S.P.; Jones, J.; Wagner, S.; Li, T.; Suo, Z.G. Stretchable interconnects for elastic electronic surfaces. *Proceedings of IEEE,* 2005, 93(8), 1459–1467.
100. Li, T.; Huang, Z.Y.; Suo, Z.; Lacour, S.P.; Wagner, S. Stretchability of thin metal films on elastomer substrates. *Applied Physics Letters*, 2004, 85(16), 3435–3437.
101. Xu, S.; Zhang, Y.; Cho, J.; Lee, J.; Huang, X.; Jia, L.; Fan, J.A.; Su, Y.; Su, J.; Zhang, H.; Cheng, H.; Lu, B.; Yu, C.; Chuang, C.; Kim, T.-I.; Song, T.; Shigeta, K.; Kang, S.; Dagdeviren, C.; Petrov, I.; Braun, P.V.; Huang, Y.; Paik, U.; Rogers, J.A. Stretchable batteries with self-similar serpentine interconnects and integrated wireless recharging systems. *Nature Communications,* 2013, 4, 1543.
102. Fan, J.A.; Yeo, W.-H.; Su, Y.; Hattori, Y.; Lee, W.; Jung, S.-Y.; Zhang, Y.; Liu, Z.; Cheng, H.; Falgout, L.; Bajema, M.; Coleman, T.; Gregoire, D.; Larsen, R.J.; Huang, Y.; Rogers, J.A. Fractal design concepts for stretchable electronics. *Nature Communications,* 2014, 5, 3266.
103. Su, Y.; Wang, S.; Huang, Y.; Luan, H.; Dong, W.; Fan, J.A.; Yang, Q.; Rogers, J.A.; Huang, Y. Elasticity of fractal inspired interconnects. *Small*, 2015, 11, 367–373.
104. Hsu, Y.-Y.; Gonzalez, M.; Bossuyt, F.; Axisa, F.; Vanfl eteren, J.; De Wolf, I. In situ observations on deformation behavior and stretching-induced failure of fine pitch stretchable interconnect. *Journal of Materials Research*, 2009, 24(12), 3573–3582.

105. Shin, S.; Kim, Y.H.; Lee, M.H.; Jung, J.-Y.; Nah, J.-H. Hemispherically aggregated $BaTiO_3$ nanoparticle composite thin film for high-performance flexible piezoelectric nanogenerator. *ACS Nano*, 2014, 8, 2766–2773.
106. Park, K.-I.; Lee, M.; Liu, Y.; Moon, S.; Hwang, G.-T.; Zhu, G.; Kim, J.E.; Kim, S.O.; Kim, D.K.; Wang, Z.L.; et al. Flexible nanocomposite generator made of $BaTiO_3$ nanoparticles and graphitic carbons. *Advanced Materials*, 2012, 24, 2999–3004.
107. Yu, R.; Pan, C.; Wang, Z.L. High performance of ZnO nanowire protein sensors enhanced by the piezotronics effect. *Energy & Environmental Science*, 2013, 6, 494–499.
108. Deng, W.; Yang, T.; Jin, L.; Yan, C.; Huang, H.; Chu, X.; Wang, Z.; Xiong, D.; Tian, G.; Gao, Y.; et al. Cowpea-structured PVDF/ZnO nanofibers based flexible self-powered piezoelectric bending motion sensor towards remote control of gestures. *Nano Energy*, 2019, 55, 516–525.
109. Choi, Y.-Y.; Yun, T.G.; Qaiser, N.; Paik, H.; Roh, H.S.; Hong, J.; Hong, S.; Han, S.M.; No, K. Vertically aligned P(VDF-TrFE) core-shell structures on flexible pillar arrays. *Scientific Reports*, 2015, 5, 10728.
110. Duan, S., et al. Innovation strategy selection facilitates high-performance flexible piezoelectric sensors. Sensors, 2020, 20(10), 2820.
111. Wang, Z.L. Piezopotential gated nanowire devices: Piezotronics and piezo-phototronics. *Nano Today,* 2010, 5, 540–552.
112. Wang, X.; Zhou, J.; Song, J.; Liu, J.; Xu, N.; Wang, Z.L. Piezoelectric field effect transistor and nanoforce sensor based on a single ZnO nanowire. *Nano Letter*, 2006, 6, 2768–2772.
113. Zhao, H.; Liu, X.; Chen, L.; Chang, S.; Chang, M. Fabrication of ultra-high-sensitivity flexible strain sensor based on single ZnO nanowire. *Microsystem Technologies*, 2016, 23, 1703–1707.
114. Deng, W.; Jin, L.; Zhang, B.; Chen, Y.; Mao, L.; Zhang, H.; Yang, W. A flexible field-limited ordered ZnO nanorod-based self-powered tactile sensor array for electronic skin. *Nanoscale*, 2016, 8, 16302–16306.
115. Chen, X.; Li, X.; Shao, J.; An, N.; Tian, H.; Wang, C.; Han, T.; Wang, L.; Lu, B. High-performance piezoelectric nanogenerators with imprinted P(VDF-TrFE)/$BaTiO_3$ nanocomposite micropillars for self-powered flexible sensors. *Small*, 2017, 13, 1604245.
116. Li, X.; Wang, Y.-H.; Zhao, C.; Liu, X. Paper-based piezoelectric touch pads with hydrothermally grown zinc oxide nanowires. *ACS Applied Materials & Interfaces*, 2014, 6, 22004–22012.
117. Panda, P.K.; Sahoo, B. PZT to lead free piezo ceramics: A review. *Ferroelectrics*, 2015, 474, 128–143.
118. Zhou, Q.; Lam, K.H.; Zheng, H.; Qiu, W.; Shung, K.K. Piezoelectric single crystal ultrasonic transducers for biomedical applications. *Progress in Materials Science*, 2014, 66, 87–111.
119. Kang, M.-G.; Jung, W.-S.; Kang, C.-Y.; Yoon, S.-J. Recent progress on PZT based piezoelectric energy harvesting technologies. *Actuators*, 2016, 5, 5.
120. Wei, H.; Wang, H.; Xia, Y.; Cui, D.P.; Shi, Y.; Dong, M.; Liu, C.; Ding, T.; Zhang, J.-X.; Ma, Y.; et al. An overview of lead-free piezoelectric materials and devices. *Journal of Materials Chemistry C*, 2018, 6, 12446–12467.
121. Zhao, Q.-L.; He, G.-P.; Di, J.-J.; Song, W.-L.; Hou, Z.-L.; Tan, P.-P.; Wang, D.-W.; Cao, M.-S. Flexible semitransparent energy harvester with high pressure sensitivity and power density based on laterally aligned PZT single-crystal nanowires. *ACS Applied Materials & Interfaces,* 2017, 9, 24696–24703.
122. Soman, J.; O'Neal, C.B. Fabrication and testing of a PZT strain sensor for soil applications. *IEEE Sensors Journal,* 2010, 11, 78–85.

123. Panda, P.K. Review: Environmental friendly lead-free piezoelectric materials. *Journal of Materials Science*, 2009, 44, 5049–5062.
124. Bechmann, R. Elastic, piezoelectric, and dielectric constants of polarized barium titanate ceramics and some applications of the piezoelectric equations. *Journal of the Acoustical Society of America*, 1956, 28, 347–350.
125. Koka, A.; Zhou, Z.; Tang, H.; Sodano, H.A. Controlled synthesis of ultra-long vertically aligned $BaTiO_3$ nanowire arrays for sensing and energy harvesting applications. *Nanotechnology,* 2014, 25, 375603.
126. Alluri, N.R.; Selvarajan, S.; Chandrasekhar, A.; Saravanakumar, B.; Jeong, J.H.; Kim, S.-J. Piezoelectric $BaTiO_3$/alginate spherical composite beads for energy harvesting and self-powered wearable flexion sensor. *Composites Science and Technology*, 2017, 142, 65–78.
127. Wu, Y.-R.; Singh, J. Metal piezoelectric semiconductor field effect transistors for piezoelectric strain sensors. *Applied Physics Letters*, 2004, 85, 1223–1225.
128. Wang, X.; Sun, F.; Yin, G.; Wang, Y.; Liu, B.; Dong, M. Tactile-sensing based on flexible PVDF nanofibers via electrospinning: A review. *Sensors*, 2018, 18, 330.
129. Xin, Y.; Sun, H.; Tian, H.; Guo, C.; Li, X.; Wang, S.; Wang, C. The use of polyvinylidene fluoride (PVDF) films as sensors for vibration measurement: A brief review. *Ferroelectrics*, 2016, 502, 28–42.
130. Zhou, Y.; He, J.; Wang, H.; Qi, K.; Nan, N.; You, X.; Shao, W.; Wang, L.; Ding, B.; Cui, S. Highly sensitive, self-powered and wearable electronic skin based on pressure-sensitive nanofiber woven fabric sensor. *Scientific Reports*, 2017, 7, 12949.
131. Fuh, Y.-K.; Ho, H.-C. Highly flexible self-powered sensors based on printed circuit board technology for human motion detection and gesture recognition. *Nanotechnology,* 2016, 27, 95401.
132. Lü, K.; Huang, W.; Guo, J.; Gong, T.; Wei, X.; Lu, B.-W.; Liu, S.-Y.; Yu, B. Ultrasensitive strain sensor based on flexible poly(vinylidene fluoride) piezoelectric film. *Nanoscale Research Letters*, 2018, 13, 83.
133. Nandi, A.K.; Mandelkern, L. The influence of chain structure on the equilibrium melting temperature of poly(vinylidene fluoride). *Journal of Polymer Science Part B: Polymer Physics*, 1991, 29, 1287–1297.
134. Zhang, Y.Y.; Jiang, S.L.; Yu, Y.; Xiong, G.; Zhang, Q.F.; Guang, G.Z. Phase transformation mechanisms and piezoelectric properties of poly(vinylidene fluoride)/montmorillonite composite. *Journal of Applied Polymer Science*, 2012, 123, 2595–2600.
135. Rajesh, P.S.M.; Bodkhe, S.; Kamle, S.; Verma, V. Enhancing beta-phase in PVDF through physicochemical modification of cellulose. *Electronic Materials Letters*, 2014, 10, 315–319.
136. Mazurek, B.; Rózecki, S.; Kowalczyk, D.; Janiczek, T. Influence of piezoelectric cable processing steps on PVDF beta phase content. *Journal of Electrostatics*, 2001, 51, 180–185.
137. Huang, R.; Long, Y.-Z.; Tang, C.C.; Zhang, H.-D. Fabrication of nano-branched coaxial polyaniline/polyvinylidene fluoride fibers via electrospinning for strain sensor. *Advanced Materials Research*, 2013, 853, 79–82.
138. Lou, Z.; Chen, S.; Wang, L.; Jiang, K.; Shen, G. An ultra-sensitive and rapid response speed graphene pressure sensors for electronic skin and health monitoring. *Nano Energy,* 2016, 23, 7–14.
139. Yu, G.-F.; Yan, X.; Yu, M.; Jia, M.-Y.; Pan, W.; He, X.-X.; Han, W.-P.; Zhang, Z.-M.; Yu, L.-M.; Long, Y.-Z. Patterned, highly stretchable and conductive nanofibrous PANI/PVDF strain sensors based on electrospinning and in situ polymerization. *Nanoscale,* 2016, 8, 2944–2950.

140. Zheng, J.; He, A.; Li, J.; Han, C.C. Polymorphism control of poly(vinylidene fluoride) through electrospinning. *Macromolecular Rapid Communications*, 2007, 28, 2159–2162.
141. Huang, Z.-M.; Zhang, Y.; Kotaki, M.; Ramakrishna, S. A review on polymer nanofibers by electrospinning and their applications in nanocomposites. *Composites Science and Technology*, 2003, 63, 2223–2253.
142. Sharma, T.; Je, S.-S.; Gill, B.; Zhang, J.X. Patterning piezoelectric thin film PVDF–TrFE based pressure sensor for catheter application. *Sensors and Actuators A: Physical*, 2012, 177, 87–92.
143. Hu, D.; Yao, M.; Fan, Y.; Ma, C.; Fan, M.; Liu, M. Strategies to achieve high performance piezoelectric nanogenerators. *Nano Energy*, 2019, 55, 288–304.
144. Newnham, R.; Skinner, D.; Cross, L. Connectivity and piezoelectric-pyroelectric composites. *Materials Research Bulletin*, 1978, 13, 525–536.
145. Guo, J.; Zhao, X.; De Beauvoir, T.H.; Seo, J.-H.; Berbano, S.; Baker, A.L.; Azina, C.; Randall, C.A. Recent progress in applications of the cold sintering process for ceramic-polymer composites. *Advanced Functional Materials,* 2018, 28, 1801724.
146. Patil, R.; Ashwin, A.; Radhakrishnan, S. Novel polyaniline/PVDF/$BaTiO_3$ hybrid composites with high piezo-sensitivity. *Sensors and Actuators A: Physical*, 2007, 138, 361–365.
147. Fang, F.; Yang, W.; Zhang, M.Z.; Wang, Z. Mechanical response of barium-titanate/polymer 0–3 ferroelectric nano-composite film under uniaxial tension. *Composites Science and Technology*, 2009, 69, 602–605.
148. Loh, K.J.; Chang, D. Zinc oxide nanoparticle-polymeric thin films for dynamic strain sensing. *Journal of Materials Science*, 2010, 46, 228–237.
149. Xiao, X.; Yuan, L.; Zhong, J.; Ding, T.; Liu, Y.; Cai, Z.; Rong, Y.; Han, H.; Zhou, J.; Wang, Z.L. High-strain sensors based on ZnO nanowire/polystyrene hybridized flexible films. *Advanced Materials,* 2011, 23, 5440–5444.
150. Lee, T.I.; Jang, W.S.; Lee, E.; Kim, Y.S.; Wang, Z.L.; Baik, H.K.; Myoung, J.-M. Ultrathin self-powered artificial skin. *Energy & Environmental Science,* 2014, 7, 3994–3999.
151. Alluri, N.R.; Chandrasekhar, A.; Vivekananthan, V.; Purusothaman, Y.; Selvarajan, S.; Jeong, J.H.; Kim, S.-J. Scavenging biomechanical energy using high-performance, flexible $BaTiO_3$ nanocube/PDMS composite films. *ACS Sustainable Chemistry & Engineering*, 2017, 5, 4730–4738.
152. Shin, S.; Park, D.; Jung, J.-Y.; Lee, M.H.; Nah, J.-H. Ferroelectric zinc oxide nanowire embedded flexible sensor for motion and temperature sensing. *ACS Applied Materials & Interfaces*, 2017, 9, 9233–9238.
153. Bian, J.; Wang, N.; Ma, J.; Jie, Y.; Zou, J.; Cao, X. Stretchable 3D polymer for simultaneously mechanical energy harvesting and biomimetic force sensing. *Nano Energy,* 2018, 47, 442–450.
154. Sanati, M.; Sandwell, A.; Mostaghimi, H.; Park, S.S. Development of nanocomposite-based strain sensor with piezoelectric and piezoresistive properties. *Sensors,* 2018, 18, 3789.
155. Hu, X.; Yan, X.; Gong, L.; Wang, F.; Xu, Y.; Feng, L.; Zhang, D.; Jiang, Y. Improved piezoelectric sensing performance of P(VDF–TrFE) nanofibers by utilizing BTO nanoparticles and penetrated electrodes. *ACS Applied Materials & Interfaces,* 2019, 11, 7379–7386.
156. Kim, H.; Fernando, T.; Li, M.; Lin, Y.; Tseng, T.-L.B. Fabrication and characterization of 3D printed $BaTiO_3$/PVDF nanocomposites. *Journal of Composite Materials,* 2017, 52, 197–206.

157. Ye, H.-Y.; Liao, W.-Q.; Zhou, Q.; Zhang, Y.; Wang, J.; You, Y.-M.; Wang, J.-Y.; Chen, Z.-N.; Li, P.-F.; Fu, D.-W.; et al. Dielectric and ferroelectric sensing based on molecular recognition in Cu(1,10-phenlothroline)2SeO_4·(diol) systems. *Nature Communications,* 2017, 8, 14551.
158. Liao, W.-Q.; Zhao, D.; Tang, Y.-Y.; Zhang, Y.; Li, P.-F.; Shi, P.-P.; Chen, X.-G.; You, Y.-M.; Xiong, R.-G. A molecular perovskite solid solution with piezoelectricity stronger than lead zirconate titanate. *Science,* 2019, 363, 1206–1210.
159. Ye, H.-Y.; Tang, Y.-Y.; Li, P.-F.; Liao, W.-Q.; Gao, J.-X.; Hua, X.-N.; Cai, H.; Shi, P.-P.; You, Y.-M.; Xiong, R.-G. Metal-free three-dimensional perovskite ferroelectrics. *Science,* 2018, 361, 151–155.
160. Tang, Y.-Y.; Li, P.; Liao, W.-Q.; Shi, P.-P.; You, Y.-M.; Xiong, R.-G. Multiaxial molecular ferroelectric thin films bring light to practical applications. *Journal of the American Chemical Society*, 2018, 140, 8051–8059.
161. Liao, W.-Q.; Tang, Y.-Y.; Li, P.; You, Y.-M.; Xiong, R.-G. Large piezoelectric effect in a lead-free molecular ferroelectric thin film. *Journal of the American Chemical Society*, 2017, 139, 18071–18077.
162. Khan, A.; Abas, Z.; Kim, H.S.; Oha, I.-K. Piezoelectric thin films: An integrated review of transducers and energy harvesting. *Smart Materials and Structures*, 2016, 25, 53002.
163. Guo, W.Z.; Tan, C.; Shi, K.; Li, J.-W.; Wang, X.-X.; Sun, B.; Huang, X.; Long, Y.-Z.; Jiang, P. Wireless piezoelectric devices based on electrospun PVDF/$BaTiO_3$ NW nanocomposite fibers for human motion monitoring. *Nanoscale,* 2018, 10, 17751–17760.
164. Ryu, J.; Kim, J.; Oh, J.; Lim, S.; Sim, J.Y.; Jeon, J.S.; No, K.; Park, S.; Hong, S. Intrinsically stretchable multi-functional fiber with energy harvesting and strain sensing capability. *Nano Energy,* 2019, 55, 348–353.
165. Chen, X.; Shao, J.; An, N.; Li, X.; Tian, H.; Xu, C.; Ding, Y. Self-powered flexible pressure sensors with vertically well-aligned piezoelectric nanowire arrays for monitoring vital signs. *Journal of Materials Chemistry C,* 2015, 3, 11806–11814.
166. Park, J.; Kim, M.; Lee, Y.; Lee, H.S.; Ko, H. Fingertip skin–inspired microstructured ferroelectric skins discriminate static/dynamic pressure and temperature stimuli. *Science Advances*, 2015, 1, e1500661.
167. Sohn, J.I.; Cha, S.; Song, B.G.; Lee, S.; Kim, S.M.; Ku, J.; Kim, H.J.; Park, Y.J.; Choi, B.L.; Wang, Z.L.; et al. Engineering of efficiency limiting free carriers and an interfacial energy barrier for an enhancing piezoelectric generation. *Energy & Environmental Science,* 2013, 6, 97–104.
168. Adhikary, P.; Garain, S.; Ram, S.; Mandal, D. Flexible hybrid Eu^{3+} doped P(VDF-HFP) nanocomposite film possess hypersensitive electronic transitions and piezoelectric throughput. *Journal of Polymer Science Part B: Polymer Physics*, 2016, 54, 2335–2345.
169. Garain, S.; Jana, S.; Sinha, T.K.; Mandal, D. Design of in situ poled Ce^{3+}-doped electrospun PVDF/graphene composite nanofibers for fabrication of nanopressure sensor and ultrasensitive acoustic nanogenerator. *ACS Applied Materials & Interfaces,* 2016, 8, 4532–4540.
170. Saraf, R.; Pu, L.; Maheshwari, V. A light harvesting, self-powered monolith tactile sensor based on electric field induced effects in $MAPbI_3$ perovskite. *Advanced Materials,* 2018, 30, 1705778.
171. Song, M.; Liu, Y.; Yu, A.; Zhang, Y.; Zhai, J.; Wang, Z.L. Flexible Li-doped ZnO piezotronic transistor array for in-plane strain mapping. *Nano Energy,* 2019, 55, 341–347.

172. Huang, T.; Yang, S.; He, P.; Sun, J.; Zhang, S.; Li, D.; Meng, Y.; Zhou, J.; Tang, H.; Liang, J.; et al. Phase-separation-induced PVDF/graphene coating on fabrics toward flexible piezoelectric sensors. *ACS Applied Materials & Interfaces,* 2018, 10, 30732–30740.
173. Zhang, Y.; Liu, C.; Liu, J.; Xiong, J.; Liu, J.; Zhang, K.; Liu, Y.; Peng, M.; Yu, A.; Zhang, A.-H.; et al. Lattice strain induced remarkable enhancement in piezoelectric performance of ZnO-based flexible nanogenerators. *ACS Applied Materials & Interfaces,* 2016, 8, 1381–1387.
174. Liu, C.; Yu, A.; Peng, M.; Song, M.; Liu, W.; Zhang, Y.; Zhai, J. Improvement in the piezoelectric performance of a ZnO nanogenerator by a combination of chemical doping and interfacial modification. *The Journal of Physical Chemistry C,* 2016, 120, 6971–6977.
175. Wang, X.; Song, C.; Li, D.; Geng, K.; Zeng, F.; Pan, F. The influence of different doping elements on microstructure, piezoelectric coefficient and resistivity of sputtered ZnO film. *Applied Surface Science*, 2006, 253, 1639–1643.
176. Yang, Y.C.; Song, C.; Wang, X.H.; Zeng, F.; Pan, F. Giant piezoelectric d_{33} coefficient in ferroelectric vanadium doped ZnO films. *Applied Physics Letters,* 2008, 92, 012907.
177. Yang, Y.C.; Song, C.; Wang, X.H.; Zeng, F.; Pan, F. Cr-substitution-induced ferroelectric and improved piezoelectric properties of $Zn_{1-x}Cr_xO$ films. *Journal of Applied Physics,* 2008, 103, 074107.
178. Luo, J.T.; Yang, Y.C.; Zhu, X.Y.; Chen, G.; Zeng, F.; Pan, F. Enhanced electromechanical response of Fe-doped ZnO films by modulating the chemical state and ionic size of the Fe dopant. *Physics Review B*, 2010, 82, 014116.
179. Pan, F.; Luo, J.; Yang, Y.; Wang, X.; Zeng, F. Giant piezoresponse and promising application of environmental friendly small-ion-doped ZnO. *Science China Technological Sciences*, 2011, 55, 421–436.
180. Lee, S.; Lee, J.; Ko, W.; Cha, S.; Sohn, J.; Kim, J.; Park, J.; Park, Y.; Hong, J. Solution-processed Ag-doped ZnO nanowires grown on flexible polyester for nanogenerator applications. *Nanoscale,* 2013, 5, 9609–9614.
181. Sinha, N.; Ray, G.; Bhandari, S.; Godara, S.; Kumar, B. Synthesis and enhanced properties of cerium doped ZnO nanorods. *Ceramics International*, 2014, 40, 12337–12342.
182. Zhao, T.; Fu, Y.; Zhao, Y.; Xing, L.; Xue, X. Ga-doped ZnO nanowire nanogenerator as self-powered/active humidity sensor with high sensitivity and fast response. *Journal of Alloys and Compounds*, 2015, 648, 571–576.
183. Zhu, D.; Hu, T.; Zhao, Y.; Zang, W.; Xing, L.; Xue, X. High-performance self-powered/active humidity sensing of Fe-doped ZnO nanoarray nanogenerator. *Sensors and Actuators B: Chemical*, 2015, 213, 382–389.
184. Shin, S.; Kwon, Y.H.; Lee, M.H.; Jung, J.-Y.; Seol, J.H.; Nah, J.-H. A vanadium-doped ZnO nanosheets–polymer composite for flexible piezoelectric nanogenerators. *Nanoscale,* 2016, 8, 1314–1321.
185. Yadav, H.; Sinha, N.; Goel, S.; Kumar, B. Eu-doped ZnO nanoparticles for dielectric, ferroelectric and piezoelectric applications. *Journal of Alloys and Compounds*, 2016, 689, 333–341.
186. Chen, Y.J.; Brahma, S.; Liu, C.P.; Huang, J.-L. Enhancement of the piezoelectric coefficient in hexagonal $Mg_xZn_{1-x}O$ films at lower Mg compositions. *Journal of Alloys and Compounds*, 2017, 728, 1248–1253.
187. Goel, S.; Sinha, N.; Yadav, H.; Godara, S.; Joseph, A.J.; Kumar, B. Ferroelectric Gd-doped ZnO nanostructures: Enhanced dielectric, ferroelectric and piezoelectric properties. *Materials Chemistry and Physics*, 2017, 202, 56–64.

188. Goel, S.; Sinha, N.; Yadav, H.; Joseph, A.J.; Kumar, B. Experimental investigation on the structural, dielectric, ferroelectric and piezoelectric properties of La doped ZnO nanoparticles and their application in dye-sensitized solar cells. *Physica E,* 2017, 91, 72–81.
189. Baek, C.; Wang, J.E.; Ryu, S.; Kim, J.-H.; Jeong, C.K.; Park, K.-I.; Kim, D.K. Facile hydrothermal synthesis of $BaZr_xTi_{1-x}O_3$ nanoparticles and their application to a lead-free nanocomposite generator. *RSC Advances*, 2017, 7, 2851–2856.
190. Kalyani, A.K.; Senyshyn, A.; Ranjan, R. Polymorphic phase boundaries and enhanced piezoelectric response in extended composition range in the lead free ferroelectric $BaTi_{1-x}Zr_xO_3$. *Journal of Applied Physics*, 2013, 114, 014102.
191. Zhuang, Y.; Xu, Z.; Fu, X.; Li, J.; Li, J.; Liao, Z.; Liu, W. A novel flexible tactile sensor based on Ce-doped $BaTiO_3$ nanofibers. *Semiconductor Science and Technology*, 2017, 32, 074002.
192. Bohara, B.B.; Batra, A.; Bowen, C. Lanthanum-modified lead zirconate titanate based paint for sensor and energy harvesting applications. *Journal of Materials Science: Materials in Electronics*, 2018, 29, 20931–20941.
193. Mandal, D.; Henkel, K.; Schmeißer, D. Improved performance of a polymer nanogenerator based on silver nanoparticles doped electrospun P(VDF–HFP) nanofibers. *Physical Chemistry Chemical Physics*, 2014, 16, 10403.
194. Karan, S.K.; Bera, R.; Paria, S.; Das, A.K.; Maiti, S.; Maitra, A.; Khatua, B.B. An approach to design highly durable piezoelectric nanogenerator based on self-poled PVDF/AlO-rGO flexible nanocomposite with high power density and energy conversion efficiency. *Advanced Energy Materials*, 2016, 6, 1601016.
195. Zhang, X.-J.; Wang, G.; Cao, W.-Q.; Wei, Y.-Z.; Cao, M.-S.; Guo, L. Fabrication of multi-functional PVDF/RGO composites via a simple thermal reduction process and their enhanced electromagnetic wave absorption and dielectric properties. *RSC Advances*, 2014, 4, 19594–19601.
196. Lee, Y.; Park, J.; Cho, S.; Shin, Y.-E.; Lee, H.; Kim, J.; Myoung, J.; Cho, S.; Kang, S.; Baig, C.; et al. Flexible ferroelectric sensors with ultrahigh pressure sensitivity and linear response over exceptionally broad pressure range. *ACS Nano,* 2018, 12, 4045–4054.
197. Du, C.; Hu, W.; Wang, Z.L. Recent progress on piezotronic and piezo-phototronic effects in III-group nitride devices and applications. *Advanced Engineering Materials*, 2017, 20, 1700760.
198. Hu, W.; Zhang, C.; Wang, Z.L. Recent progress in piezotronics and tribotronics. *Nanotechnology,* 2018, 30, 042001.
199. Kou, J.; Liu, Y.; Zhu, Y.; Zhai, J. Progress in piezotronics of transition-metal dichalcogenides. *Journal of Physics D: Applied Physics,* 2018, 51, 493002.
200. Wang, Z.L. Nanopiezotronics. *Advanced Materials,* 2007, 19, 889–892.
201. Wang, Z.L.; Wu, W. Piezotronics and piezo-phototronics: Fundamentals and applications. *National Science Review*, 2013, 1, 62–90.
202. Wu, W.; Pan, C.; Zhang, Y.; Wen, X.; Wang, Z.L. Piezotronics and piezo-phototronics: From single nanodevices to array of devices and then to integrated functional system. *Nano Today,* 2013, 8, 619–642.
203. Zhou, J.; Gu, Y.; Fei, P.; Maia, W.; Gao, Y.; Yang, R.; Bao, G.; Wang, Z.L. Flexible piezotronic strain sensor. *Nano Letter*, 2008, 8, 3035–3040.
204. Wei, Y.; Wu, W.; Guo, R.; Yuan, D.; Das, S.; Wang, Z.L. Wafer-scale high-throughput ordered growth of vertically aligned ZnO nanowire arrays. *Nano Letter*, 2010, 10, 3414–3419.
205. Wallentin, J.; Anttu, N.; Asoli, D.; Huffman, M.; Aberg, I.; Magnusson, M.H.; Siefer, G.; Fuss-Kailuweit, P.; Dimroth, F.; Witzigmann, B.; et al. InP nanowire array solar cells achieving 13.8% efficiency by exceeding the ray optics limit. *Science,* 2013, 339, 1057–1060.

206. Zhang, Z.; Liao, Q.; Zhang, X.; Zhang, G.; Li, P.; Lu, S.; Liu, S.; Zhang, Y. Highly efficient piezotronic strain sensors with symmetrical Schottky contacts on the monopolar surface of ZnO nanobelts. *Nanoscale,* 2015, 7, 1796–1801.
207. Li, P.; Liao, Q.; Zhang, Z.; Zhang, Y.; Huang, Y.; Ma, S. Flexible microstrain sensors based on piezoelectric ZnO microwire network structure. *Applied Physics Express,* 2012, 5, 061101.
208. Parmar, M.; Perez, E.A.L.; Ardila, G.; Saoutieff, E.; Pauliac-Vaujour, E.; Mouis, M. A demonstration of the mechanical sensing capability of individually contacted vertical piezoelectric nanowires arranged in matrices. *Nano Energy,* 2019, 56, 859–867.
209. Liao, X.; Yan, X.; Lin, P.; Lu, S.; Tian, Y.; Zhang, Y. Enhanced performance of ZnO piezotronic pressure sensor through electron-tunneling modulation of MgO nanolayer. *ACS Applied Materials & Interfaces,* 2015, 7, 1602–1607.
210. Sun, Q.; Seung, W.; Kim, B.J.; Seo, S.; Kim, S.-W.; Cho, J.H. Active matrix electronic skin strain sensor based on piezopotential-powered graphene transistors. *Advanced Materials,* 2015, 27, 3411–3417.
211. Guo, Y., et al. Thin, stretchable, universal wireless power transfer system for electric vehicle charging. *Royal Society of Chemistry Advances*, 2020, 10(58), 35426–35432.
212. Yewang, S.; Shuang, Li. A stretchable flexible electrode and its preparation method. *Institute of Mechanics, Chinese Academy of Sciences*, 2016.

3 Application and Research Trends of Flexible Piezoelectric Arrays

Han Zhang
Chinese Academy of Sciences

CONTENTS

3.1 INTRODUCTION

An ultrasonic phased array transducer [1] can be viewed as a whole piezoelectric wafer cut into many small independent crystals, which form an array, and each small crystal can be individually excited and triggered. According to the form of ultrasonic phased array arrays, phased arrays can be classified into

DOI: 10.1201/9781003182146-3

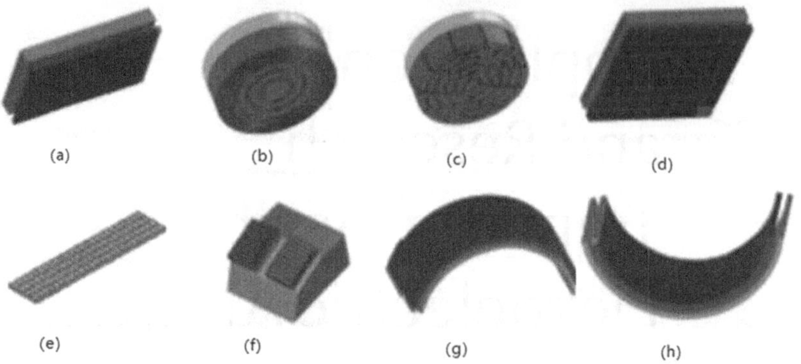

FIGURE 3.1 Different array arrangements of ultrasonic phased array transducers: (a) one-dimensional liner array, (b) circle array, (c) fan-shaped array, (d) two-dimensional matrix, (e) fractional dimension probe, (f) double liner array, (g) concave array and (h) convex array.

one-dimensional line arrays, two-dimensional surface arrays, bilinear arrays, toroidal arrays, sector arrays, convex arrays, concave arrays, etc. (Figure 3.1). One-dimensional linear array can realize the acoustic beam deflection and focus in the line array plane, and the processing technology and data processing are not too complicated, so it is the most widely used form of ultrasonic phased array transducer at present. The annular array can achieve two-dimensional focus in the sound beam profile, and the sound field energy is concentrated with strong penetrating power, but this kind of array cannot realize the sound beam deflection. Fan-shaped array and two-dimensional surface array can realize the three-dimensional space of the sound beam deflection and can focus and realize three-dimensional imaging, very development potential, but now the processing technology and data processing technology are not mature. The fractional dimensional array, as a transition between a 1D line array and a 2D surface array, has a narrow matrix shape with many dense small crystallites in the length direction and only a limited number of rows in the width direction. The bilinear array is a probe with separate ultrasonic emission and reception, which eliminates the effect of phantom echoes and provides good near-surface detection capability compared to other kinds of probes. The concave array has inherent physical focusing performance due to its unique array element arrangement and is mostly used for defect detection in pipes with a fixed curvature. Convex arrays are generally used for medical ultrasound probes for clinical medical diagnosis [2].

When an ultrasonic probe is used for non-destructive testing of a specimen, the emitting surface of the probe must fit tightly with the surface of the specimen. If the ultrasonic probe does not fit the specimen tightly, most of the ultrasonic waves will be reflected on the surface of the specimen and cannot enter the specimen for testing. For this reason, the ultrasonic testing will now be applied to the surface of the test piece with a layer of coupling agent, oil, etc., so as to make the probe emission surface and the surface of the test piece closely coupled together.

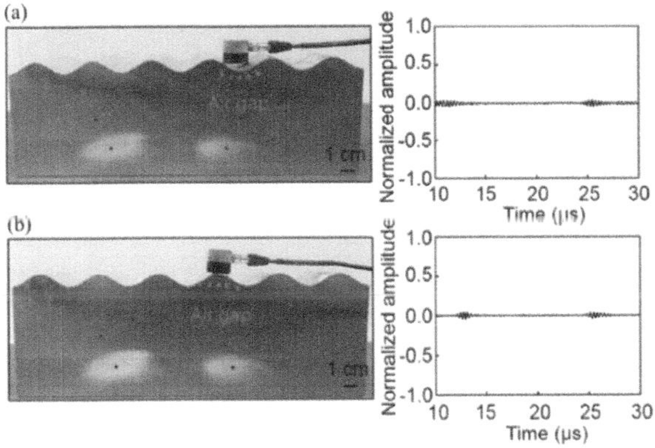

FIGURE 3.2 Fit of an ultrasonic probe to surface [3]: (a) concave surface and (b) convex surface.

However, considering that all ultrasonic probes are now rigid, for some curved test pieces, e.g., pipes and axles, a tight fit between the probe emitting surface and the surface (Figure 3.2a and b) cannot be achieved even with the use of coupling agents [3]. Therefore, researchers have invested a lot of research efforts on flexible phased array transducers.

Chatillon et al. [4] developed a contact ultrasonic phased array transducer (Figure 3.3a), which has a flexible piezoelectric array element radiation surface that can adapt to the actual surface of the inspection workpiece and fit well with the actual surface, avoiding the sensitivity variation caused by the poor contact between the probe emission surface and the workpiece surface. At the same time, the static and dynamic conditions of the defect detection results show that the contact phased array transducer has a certain degree of reliability for defect detection performance. However, the spacing between the array elements in the transducer is not fixed, which is not very accurate for each array element delay, and thus affects the accuracy of detection.

Shi Keren et al. [5] prepared a flexible phased array transducer (Figure 3.3b). In addition to the parameters used to describe the rigid phased array probe (array size, array spacing, number of arrays and array center frequency), the study also considered the array depth, array rotation angle and inter-array curve length of the flexible probe at curved surfaces for the purpose of transducer detection accuracy. Similarly, the spacing between the elements of the flexible probe cannot be fixed, which affects the accuracy of the delay algorithm and the accuracy of defect detection. In addition, the size of the probe is relatively large, which is not suitable for field work.

Zhe Wang et al. [6] prepared a flexible ultrasonic transducer array by MEMS process (Figure 3.3c) and did the pointing analysis of the transducer. However, the transducer did not have the necessary matching layer and backing, and the study of parameters, such as the array element spacing and the number of array elements, was lacking.

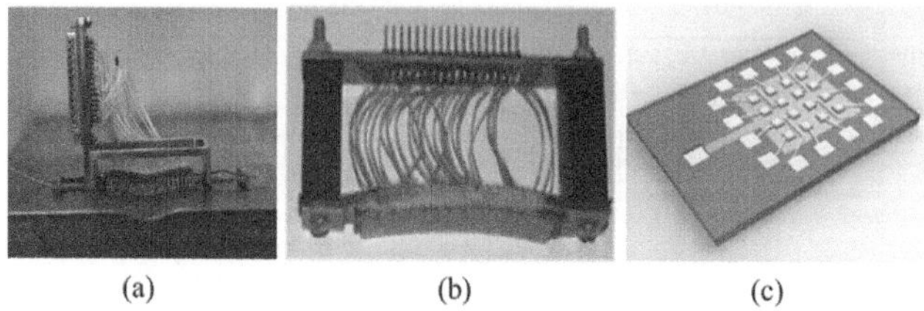

FIGURE 3.3 Flexible ultrasound probe [4–6]. (a) Contact ultrasonic phased-array transducer; (b) Flexible ultrasonic phased-array transducer; (c) Micromechanical flexible ultrasonic probe array.

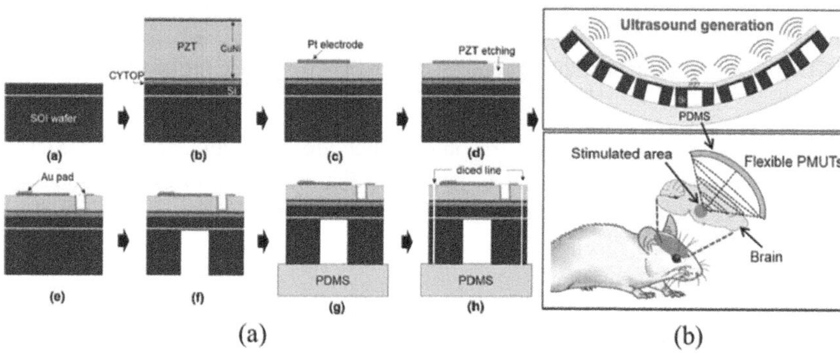

FIGURE 3.4 Flexible piezoelectric micromechanical ultrasound transducer array [8]. (a) Production process; (b) Structure and working schematic.

Jin-Hyung Lee et al. [7] prepared a flexible piezoelectric micromechanical ultrasound transducer array (pMUT) using a MEMS process (Figure 3.4a) for the study of ultrasound brain stimulation. This flexible piezoelectric micromechanical ultrasound transducer array (Figure 3.4b) can achieve a minimum bending radius of curvature of 5 mm, which can ensure a good fit to the mouse brain. At the same time, the probe can achieve a sound intensity of 44 mW/cm^2 at 4 mm from the transmitting end of the transducer array, which is sufficient to accomplish low-intensity brain stimulation. The spacing between the pMUTs of this transducer varies at different degrees of bending, a point that makes it difficult to control the acoustic beam with delayed focusing and thus cannot achieve a phased array.

Hongjie Hu et al. [3] developed a 10×10 flexible ultrasonic phased array transducer that can be both bent and stretched using flexible electronics (Figure 3.5a). To achieve the stretchability of the transducer, this work cut the negative and signal wires into snake-bent wires by laser cutting and did finite element calculations on the stretchability of the snake-bent wires to ensure the electrical and mechanical reliability of the structure when the transducer is bent or stretched. The fabrication process (Figure 3.5b): (i) Spin coat a layer of polyimide film (2 μm) on top of copper foil (20 μm thick); (ii) Cut the PI/Cu into the specified serpentine wires with a laser according to the pre-designed pattern; (iii) Transfer the first layer of serpentine wires onto the Ecoflex substrate with water soluble adhesive and treat with

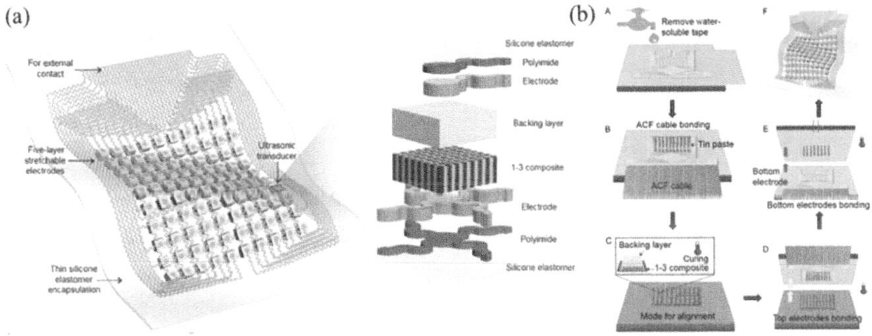

FIGURE 3.5 Stretchable ultrasound transducer array [3]. (a) Structure schematic; (b) Production process.

ultraviolet (UV) to increase the adhesive strength of PI and Ecoflex substrate; (iv) after dissolving the hydrosol with water, spin coat and cure a layer of Ecoflex film with a thickness of 35 μm on top of the snake wire, and the Ecoflex film is used as an insulating barrier layer; (v) repeat steps 2 and 3 above to make a signal line layer with four layers of snake wire; (vi) expose the pads of the signal line layer and then hot press the ACF line to the line port as the external; (vii) bond 100 pre-made 1–3 piezoelectric composites one by one with conductive silver adhesive to the 10×10 pads of the negative layer; (viii) bond the backing material on the back of each 1–3 piezoelectric composite with conductive silver adhesive; (ix) bond the signal layer to the backing material; and (x) encapsulate the whole device with Ecoflex as a protective layer. The flexible ultrasonic phased array commutator is a flexible ultrasonic commutator. The center frequency of this flexible ultrasonic phased array transducer is 3.51 MHz, and the standard deviation of the resonant frequency of 10×10 piezoelectric array elements is 56.8 kHz, which has good consistency. The crosstalk between adjacent array elements is –70 dB, which is much lower than the standard –30 dB. However, because the flexible ultrasonic phased array transducer can be stretched, the spacing between adjacent array elements changes when it is affixed to different surfaces and the spacing of the entire array elements is not uniform. Therefore, the phased array is actually difficult to realize the use of the delayed focus law and thus cannot well realize the deflection and focus of the acoustic beam.

Chonghe Wang et al. [9] prepared a flexible ultrasound array capable of detecting deep arterial and venous blood pressure using flexible electronics (Figure 3.6). Unlike conventional bulkier and more rigid ultrasound probes, this flexible ultrasound array is able to achieve up to 60% stretchability and is thin and light (240 μm thickness), which can be well coupled to the surface of human tissues to achieve continuous, non-invasive and accurate monitoring of blood pressure in deeply buried vessels, such as carotid arteries and jugular veins, which is of high clinical value for monitoring cardiovascular diseases. However, similarly, the ultrasound array cannot achieve beam deflection and focusing and lacks flexible control of the beam.

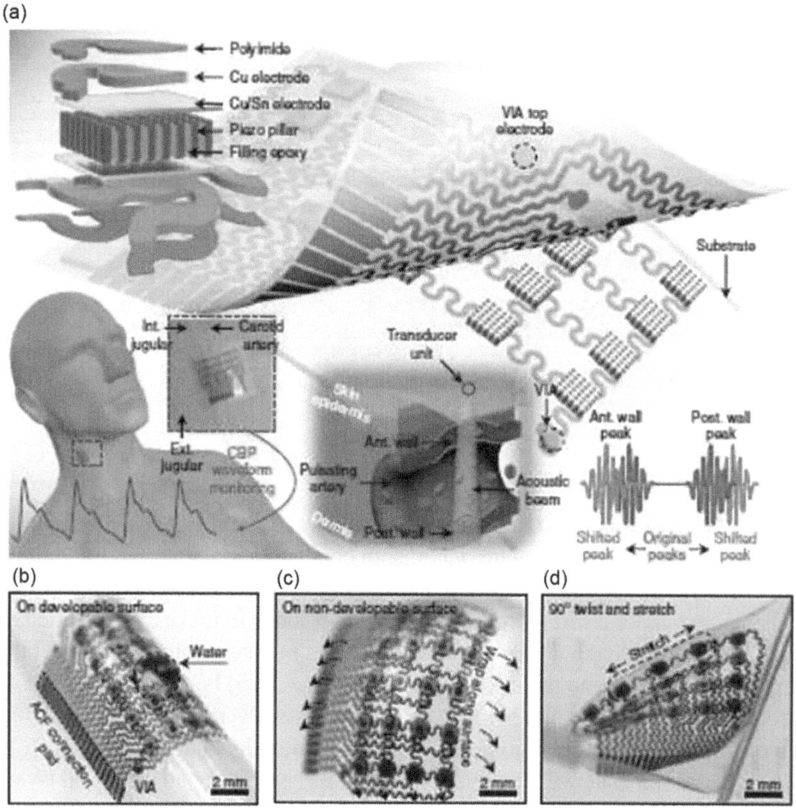

FIGURE 3.6 Design and working principle of the flexible ultrasound array [9].

Around the application of flexible piezoelectric arrays, this chapter focuses on the following seven aspects:

The first section mainly introduces the research status of ultrasonic phased array transducer.

The second section introduces the application of flexible piezoelectric array in cardiac pacemaker.

The third part introduces the application of flexible piezoelectric array in wireless power transmission system.

The fourth part introduces the application of flexible piezoelectric array in phase control transceiver.

The fifth part introduces the application of flexible piezoelectric array in large air pressure sensor.

The sixth part introduces the application of flexible piezoelectric array in acoustic tweezers for micromanipulation.

In the seventh part, the application of flexible piezoelectric array is summarized and prospected.

3.2 CARDIAC PACEMAKER WITH A FLEXIBLE PIEZOELECTRIC ARRAY

Artificial cardiac pacemakers have made a significant contribution to regulate heartbeat using electrical impulses for contracting the heart muscles of people who suffer from sick sinus syndrome or heart block, which causes abnormal heart rate, and may result in symptoms including syncope, angina, dizziness and even heart failure or heart attack [10,11]. However, due to the limited lifespan of the battery, replacement surgery for the artificial pacemaker implanted beneath chest skin should be made every 7–10 years (or even every 3–6 years for an implantable cardioverter defibrillator) [12,13]. This poses a risk to elderly persons, particularly with regard to infection or bleeding during the surgical procedure [14]. Enhancing the battery lifetime is thus a critical issue to assure longer working time of the implanted pacemakers and increase the replacement cycle. An attractive approach to address this challenge is to adopt self-powered systems, which potentially can provide low maintenance, independent operation and sustainability for implantable biomedical devices [15,16].

Energy-harvesting systems based on irregular vibrational motion and mechanical deformation are promising candidates for self-powered biomedical electronics [17–21]. Manipulating flexible piezoelectric energy harvesters [called nanogenerators (NGs)] inside the human body is of particularly medical interest, because they can scavenge inexhaustible biomechanical energy, such as cardiac motion, muscle contraction/relaxation and blood circulation, and convert it to electrical energy [22]. This could feasibly contribute to not only the operation of implantable heart rate monitoring and transmitting system but also the development of a self-powered artificial pacemaker by directly recharging the battery or stimulating the heart [23,24]. Many research teams have explored various types of flexible piezoelectric materials on thin plastic substrates including ZnO nanowires [25,26], $BaTiO_3$ thin film [27] and lead zirconatate titanate (PZT) thin films [28,29]. Although the aforementioned flexible energy harvesters can provide power for operating small electronic devices, their relatively low output current of a few μA has severely restricted the range of applications in consumer electronics as well as biomedical devices: for example, a cardiac pacemaker operates at an input of 100 μA and 3 V [26,30].

Therefore, it is highly desirable to utilize materials with a high piezoelectric charge coefficient, which represents the piezoelectric capability of converting mechanical deformation into electric charges, to increase the output current efficiency for flexible energy harvesters. One such piezoelectric material is a new generation of single crystalline $(1-x)\ Pb(Mg_{1/3}\ Nb_{2/3})O_3-xPbTiO_3$ (PMN-PT). It has exceptional piezoelectric charge constant of d_{33} up to 2500 pC/N, which is almost four times higher than that of PZT, 20 times higher than that of $BaTiO_3$ and 90 times higher than that of ZnO [31,32].

Herein, Geon-Tae Hwang et al. reported a flexible and highly efficient energy harvester enabled by single-crystalline piezoelectric PMN-PT thin film on a plastic substrate to achieve a self-powered artificial pacemaker with significantly increased electric output current. The stress-controlled exfoliating process was optimized for transferring the PMN-PT thin film from a bulk substrate onto a flexible substrate without mechanical damage by utilizing the inherent residual stress of Ni film [33]. The maximum output current and voltage of the flexible PMN-PT thin-film harvester reached 145 μA and 8.2 V, respectively, through the periodic mechanical motions of bending and unbending. Furthermore, the current signal of 223 μA, corresponding to the highest output current among piezoelectric-based flexible NGs reported to date, was obtained by finger tapping. The converted electricity was used to directly turn on 50 commercial green light-emitting diodes and charge coin batteries for driving portable electronic devices. Finally, real-time functional electrical stimulation was performed to provide the artificial heart beating for a live rat using the high-performance flexible PMN-PT energy harvester.

Figure 3.7a schematically illustrates the device fabrication process and stimulation test on a living heart. The following is a detailed explanation: (i) Single-crystalline rhombohedral 0.72 PMN-0.28 PT ingot was grown directly from the melt by a modified Bridgman method near the morphotropic phase boundary (MPB) [34,35]. Exceptionally high piezoelectric and electromechanical coupling coefficients were achieved at a composition around MPB with a single crystal structure [36]. The (001)-oriented PMN-PT crystal was cut into a square block and polished to have a mirror surface using a diamond suspension. An Au bottom electrode deposited PMN-PT plate was completely bonded on (100) silicon wafer by adhesive epoxy and then thinned to a film thickness of 8.4 μm using grinding and subsequent chemical mechanical polishing. The PMN-PT was poled along the (100) direction by applying an electric field of 1.8 kV/mm at room temperature for 1 hour followed by deposition of the top electrode layer. (ii) A mechanical exfoliating process was carried out to peel off the top piezoelectric portion of the PMN-PT metal–insulator–metal (MIM) layer (8.6 μm in thickness) from the silicon substrate using an electroplated tensile Ni stressor. Exploiting the directional stress mismatch between the top Ni film (intrinsic tensile stress) and underlying layers (against compressive stress) of the MIM/epoxy/wafer offered a simple and robust route to spontaneous and uniform exfoliation of the piezoelectric thin film from the mother substrate without mechanical damage, such as cracking, delamination or wrinkling [33]. Moreover, the thickness of the exfoliated PMN-PT can be readily controlled by modulating the residual stress of the Ni film [37], and a 20-μm-thick Ni layer was an optimized condition for safe and precise detachment of the piezoelectric MIM structure [33]. (iii) The large-area PMN-PT thin film (1.7 cm × 1.7 cm) was transferred onto UV light-cured polyurethane (PU)-coated polyethylene terephthalate (PET) substrate. The freestanding thin PMN-PT film on the thin plastic substrate provided enough flexibility to achieve conformal contact on a curved subcutaneous layer and corrugated organs in the human body, thus making it possible to serve as a self-energy generator by slight mechanical movements [38]. (iv) To stimulate a live rat's heart by the flexible PMN-PT energy harvester, Cu wires were attached to the metal pads with a silver conductive paste and then connected to stimulation electrodes. Finally, the PMN-PT stimulator was applied as an artificial cardiac pacemaker with electrical energy generated by a periodic bending/unbending cycle of the flexible device as schematically shown in Figure 3.7a–iv.

Figure 3.7b shows a cross-sectional scanning electron microscopy (SEM) image of the single-crystalline PMN-PT thin film on a flexible PET substrate after the transfer process. The PMN-PT thin film (8.4 μm in thickness) bonds on

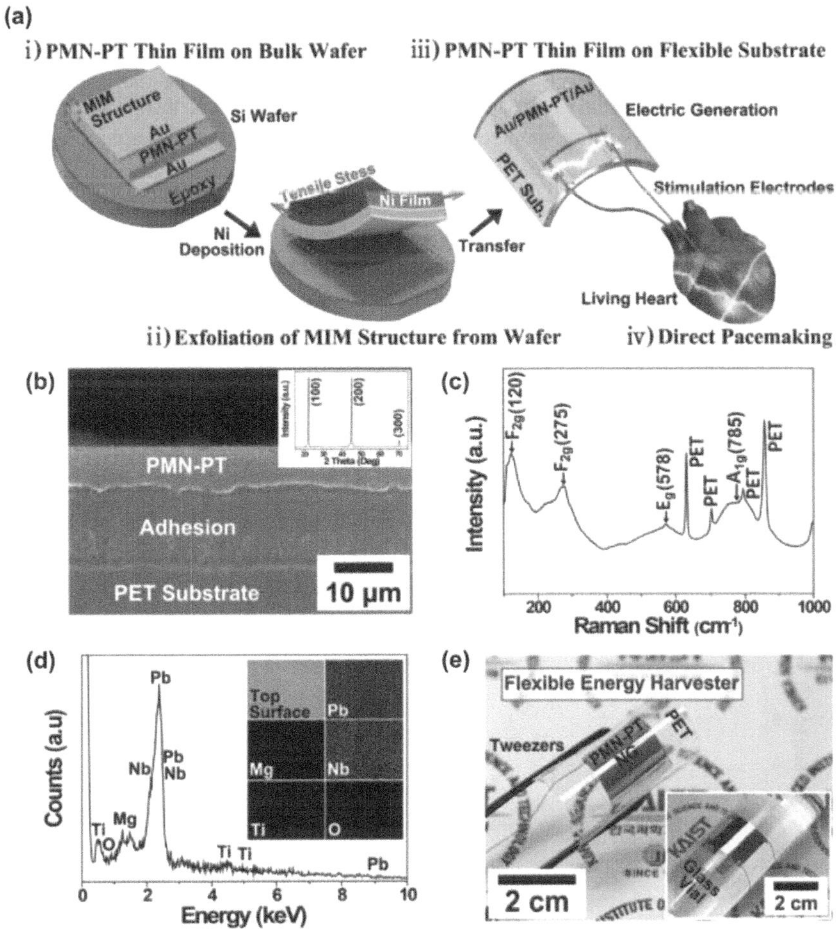

FIGURE 3.7 (a) Manufacturing process and biomedical applications of flexible PMN-PT piezoelectric energy harvesters; (b) cross-sectional SEM images; (c) Raman spectra; (d) top surface analysis of EDS curves; (e) photographs of flexible PMN-PT thin-film energy harvesters on PET substrates [39].

PU without the formation of cracks or blisters in the piezoelectric material. We conducted structural and compositional characterizations of the flexible PMN-PT thin film on plastics using X-ray diffraction (XRD), Raman spectroscopy and energy-dispersive spectroscopy (EDS). The inset of Figure 3.7b shows the results of the XRD analysis of the single crystal PMN-PT thin film with only (001)-textured peaks, confirming the formation of a pure perovskite structure [40]. Raman spectroscopy (Figure 3.7c) was carried out to analyze the phase of the PMN-PT thin film using a 514.5 nm Ar^+ laser as an excitation source at room temperature. The peaks of the Raman spectra at 120, 275, 578 and 785 cm^{-1} are ascribed to the F_{2g}, F_{2g}, E_g and A_{1g} modes, respectively, in good agreement with the typical features for perovskite relaxors [36]. Figure 3.7d and its inset show the EDS spectrum and elemental mapping results obtained from the top surface of the PMN-PT thin film. From the EDS-based compositional analysis, it is clearly proven that the single crystal PMN-PT on flexible substrate consists of all the required elements, and the Pb, Mg, Nb, Ti and O atoms are distributed evenly across the measured area. Figure 3.7e shows the flexible PMN-PT energy harvester completely bent by tweezers without notable damage, and the inset presents the device on a rounded glass vial with a curvature radius of 1.4 cm. The Ni exfoliated piezoelectric thin film on a plastic substrate could provide outstanding bendability and mechanical stability for flexible NGs [41].

PMN-PT is considered to be among the most promising piezoelectric and ferroelectric materials due to its excellent dielectric, ferroelectric and piezoelectric characteristics [42]. To evaluate the ferroelectric properties of the flexible single crystal PMN-PT thin film, we measured its dielectric constant, dielectric loss and polarization–electric field (P–E) hysteresis loop. Figure 3.8a shows the dielectric properties of the PMN-PT thin film on a PET substrate at room temperature as a function of frequency, ranging from 1 kHz to 1 MHz, with a direct current (DC) bias of 5 mV. The dielectric constant (top line) and loss tangent (lower line) of the PMN-PT thin film are 3500 and 0.045, respectively, at 1 kHz, and are stably preserved with increasing frequency. The inset of Figure 3.8a shows the dependence of the dielectric constant of MIM on the bending radius at a fixed frequency of 1 kHz. As the bending radius is varied from 60 to 10 mm

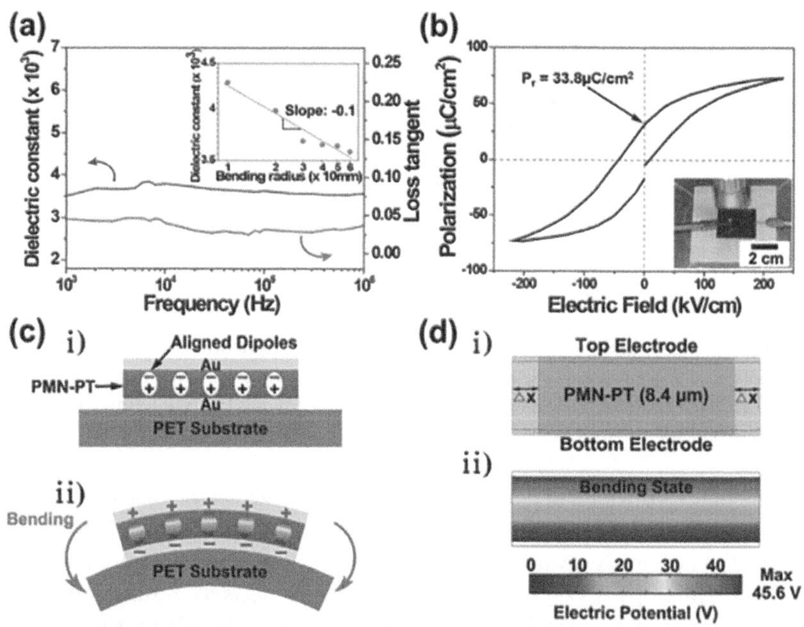

FIGURE 3.8 PMN-PT film property analysis [39]. (a) Dielectric properties of MIM structured PMN-PT thin film as a function of frequency on a plastic substrate. (b) Polarization-electric field (P-E) hysteresis loop of the ferroelectric PMN-PT thin film on a PET. (c) Schematics of the electricity generation of the flexible PMN-PT thin film NG before (i) and after (ii) bending. (d) The simulated model (i) and inside piezopotential distribution (ii) of the MIM-based PMN-PT energy harvester under tensile strain of 0.36%.

(i.e., corresponding to a gradual increase of the surface strain from 0.1% to 0.62%), the dielectric constant increases up to 4280. It seems that the lattice distortion mainly governed by tensile stress possibly enhances the dielectric constant of the PMN-PT thin film with variation of the lattice polarity and lattice dynamic property [43]. The ferroelectric materials are polarized by exposing them to a strong DC electric field below the Curie point [44]. Figure 3.8b plots the characterized P–E hysteresis curve of the PMN-PT thin film on a plastic substrate at room temperature, exhibiting good ferroelectric property with a high remnant polarization (P_r) of 33.8 μC/cm^2 at a maximum applied field of 225 kV/cm [42].

Figure 3.8c-i and c-ii schematically illustrates how electric energy is generated from the flexible energy harvester. The dipoles in the PMN-PT thin film can be aligned during the poling process under a high electric field. As depicted in the top panel of Figure 3.8c-i, the dipoles in the PMN-PT thin film are originally positioned perpendicular to the surface of the device. Once the flexible harvesting device is bent, as shown in Figure 3.8c-ii, piezopotential is generated inside the PMN-PT thin film by the tensile stress-induced deformation of the device, resulting in electrons flow in the external load to balance the electric field made by dipoles and accumulate at the top electrode. When the NG returns to the original flat state, the charges tend to move back to their original positions. Consequently, under periodic motions of bending and unbending, positive and negative electric signals are generated from the flexible device. Figure 3.8d shows the theoretical calculation of the piezopotential inside the PMN-PT MIM structure by a finite element analysis (FEA) using the COMSOL package to provide more information on the working principle. We simulated the piezopotential distribution for a PMN-PT thin film (8.4 μm in thickness) on a plastic substrate (120 μm in thickness) with a bending radius of 16.5 mm (i.e., corresponding to a tensile strain of 0.36%). A piezoelectric charge constant of $d_{31} = -1102$ pC/N, a dielectric constant of KT = 1264, a Young's modulus of $E = 133$ GPa and a mass density of $\rho = 8040$ kg/m^3 were used for the FEA [45]. The strain applied to the PMN-PT thin film nearly equals the strain at the top surface of the plastic film due to the thickness of plastic being sufficiently larger than the thickness of the piezoelectric film [46]. From the relationship ($S = \Delta L / L_0$) between the tensile strain (S) and the original width (L_0) on the X-axis of the flexible NG, the PMN-PT thin film is lengthened in the parallel direction for a displacement (ΔL) of 64 μm in the simulated model (Figure 3.8d-i). We assumed no failure modes, such as cracking, slipping or delamination of the PMN-PT thin film on plastic in the FEA [47]. As illustrated in Figure 3.8d-ii, the calculation predicted a piezopotential difference of 45.6 V inside the color-coded PMN-PT thin film. This FEA result strongly supports that mechanical deformation of the PMN-PT thin film can be immediately converted into electric energy.

The real-time functional electrical stimulation of the heart by a high-output flexible piezoelectric energy harvester has great biomedical importance for a self-powered artificial heart pacemaker [48]. Here, we used the instantaneous electric output of the harvesting device to directly stimulate a rat heart without any external power sources or circuits. Figure 3.9 provides an overall schematic illustration of artificial cardiac pacemaking using a flexible PMN-PT thin-film stimulator.

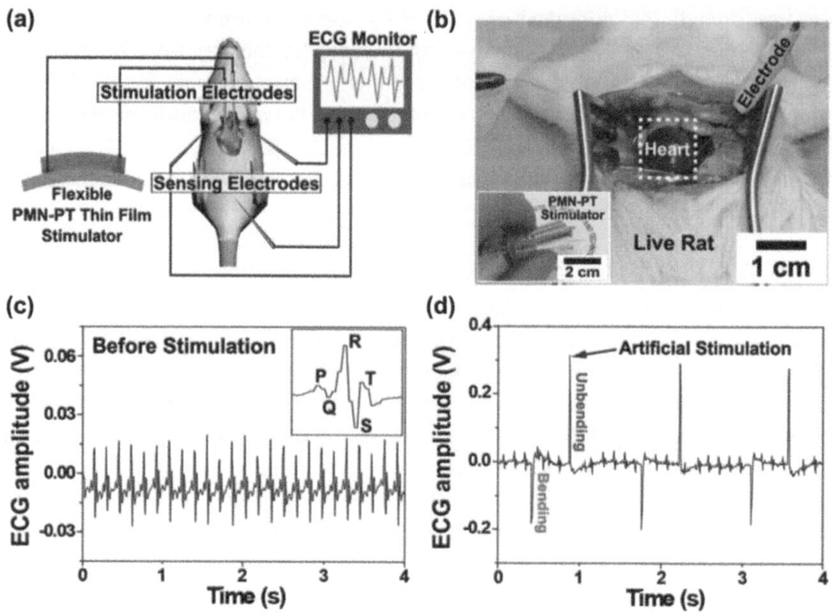

FIGURE 3.9 Experiment on artificial heart pacing using electrical energy from a flexible PMN-PT energy harvester [39]. (a) A schematic illustration of the artificial cardiac pacemaking using electrical energy from the flexible PMN-PT energy harvester. (b) A photograph of the medical experiment on a living rat for stimulating the heart. (c) The simultaneously recorded ECG in a normal rat heart before the stimulation. (d) The remarkable artificial peaks on the ECG arise from the stimuli by bending and unbending of the flexible energy harvester.

The flexible cardiac stimulator was directly linked to stimulation electrodes to provide electrical stimuli to the heart of an anesthetized rat. Three sensing terminals were pinned to the rodent, on the left posterior leg and both anterior legs, to monitor its electrocardiogram (ECG). Figure 3.9b shows the animal experiment with opening the chest of a rat for stimulation of the heart and perception of the heartbeat. The rat had a typical QRS complex (i.e., QRS is clinically acknowledged as a typical combination of waves related to the heartbeat), P wave and T wave in the ECG amplitude with a heart rate of about 6 beats per second as displayed in Figure 3.9c and its inset. In normal animals, external electric energy of 1.1 μJ is minimally needed to trigger the action potential for artificially contracting the heart [49]. When the flexible PMN-PT stimulating device was bent and unbent cyclically, the corresponding spike peaks were observed on the natural heartbeat of the rat in the ECG, as visualized in Figure 3.9d. The generated energy (2.7 μJ) from one bending motion of the flexible stimulator was larger than the threshold energy (1.1 μJ) to electrically stimulate the living heart. This result shows that the thin-film NG has potential biomedical use for the normalization of cardiac function [50].

In summary, a high-performance flexible single-crystalline PMN-PT thin-film energy harvester was successfully fabricated and applied in a practical demonstration of a self-powered cardiac pacemaker. A mechanical Ni exfoliation process was employed to transfer the entire large area of a single crystal PMN-PT thin film onto a PET substrate from a mother silicon wafer. The PMN-PT NG converted tiny biomechanical motion and mechanical deformation into electric energy with a high current signal of up to 0.223 mA and output voltage of 8.2 V. An FEA simulation theoretically shows that electrical energy can be obtained from stress-induced bending of the flexible piezoelectric harvester. More interestingly, our flexible and thin PMN-PT stimulator was readily implemented in the cardiac muscle of a live rat, enabling real-time functional electrical stimulation. This approach verified that our flexible PMN-PT harvesting device mounted on a thin plastic substrate can be further applicable as a sustainable energy source, which may be able to recharge batteries and stimulate the heart through application to artificial cardiac pacemakers. We are currently planning to investigate three-dimensional stacking of piezoelectric thin films on a single plastic substrate and energy harvesting from the movement of porcine organs.

3.3 WIRELESS POWER TRANSFER SYSTEM

For energy conservation and pollution reduction, electric vehicles have been carried out for decades [51–54]. One of the challenges of electric vehicles is the charging method of batteries, and researchers have been working on the improvement of the performance and convenience of the charging method for a long time. Meanwhile, the methods of charging are always researched to improve performance and convenience [55–59]. The wireless power transfer systems, which have demonstrated many advantages, such as wirelessness, and being waterproof, dustproof and weatherproof are gaining considerable attention for pure and plug-in hybrid electric vehicles [60–63]. The main scheme of wireless charging is usually to install the transmitting coil and module on the ground and the receiving coil on the chassis of the vehicle [64–66]. However, there are many obstacles limiting the application of wireless power transfer technology for electric vehicles charging [67–69]. Many literatures had reported several important parameters, which affect inductive transmission, such as the distance between transmitting coil and receiving coil, the frequency of transmission current, the alignment of inductive coils and the design of coils [70–73]. The distance between the vehicle's chassis and ground is long, resulting in the leakage magnetic flux and low charging efficient, and the different distances between the chassis of different vehicles and the ground cause unstable charging efficient [74–76]. In addition, it is difficult for most drivers to park at the accurate location to satisfy a good alignment of the two coils, and hence, this human factor decreases the charging efficiency [77–79]. Moreover, some debris around the charging pad, such as metal objects, cigarette packs and gum wrappers, could result in lower charging efficiency and even fire accidents due to eddy current [80–82]. Therefore, wireless charging technology still faces great challenges in efficiency, convenience, safety and universal applications for electric vehicles [83,84].

In this chapter, we develop a novel wireless power transfer system with a thin, stretchable and inductive coil. The receiving coil is integrated into the vehicle roof and has the ability for the conformal contact with various models due to the geometric characteristics of thin and stretchability. The transmitting coil is hanged over the vehicles and could be controlled individually to adjust the alignment and distance to receiving coils. This design could eliminate the perplexity of alignment for driver and shorten the distance between the two coils, which can lead to the reduction of the leakage magnetic flux and the improvement of the charging efficiency. In addition, the system is easy to clean and rarely disturbed by other sundries because it is installed on the vehicles roof, which can improve the safety. The coils can maintain stable power transfer efficiency even if subjected to bend or stretch. Moreover, this novel SWPT system demonstrates the unique advantages of outstanding charging efficient and universality and shows the potential applications for electric vehicles charging.

It is well known that the wireless power transfer system basically consists of the inductive coupling between transmitting coil and receiving coil. The transmitting coil is powered through a powder electronic converter, which provides a high-frequency current and a high-frequency magnetic field. The field coupling with the receiving coil allows the wireless transfer of electrical power. Then, a rectification stage converts the signal to DC for charging the battery of vehicle [85–87]. It is obvious that the coils, as the primary medium, play a key role in achieving the wireless power transfer, and thus, the design of coils can affect the transfer efficiency directly. Figure 3.10a exhibits a novel wireless power transfer system and its schematic diagram. The receiving coil is integrated into the car roof, and the transmitting coil is fixed by holding its four corners to cover the receiving coil and could be moved up, down, left, right, forward and backward to adjust the alignment. This design vastly shortens the transmitting distance and keeps the alignment of the two coils to avoid the decrease in inductive efficiency. Moreover, the coil is made of thin serpentine copper traces and hence can be bended and stretched, as shown in Figure 3.10b. These features of coil allow it to conform to the car roofs of different vehicle types when charging. An expended view of the structure of coil is shown in Figure 3.10c, where the coil has a multi-layer configuration of induction coil/insulation layer/induction coil, encapsulated with Ecoflex on the up and down.

Figure 3.11 outlines the fabrication process of an inductance coil, which is realized by microprocessing strategies, involving laser-cutting serpentine copper traces to forming a windings, putting the windings on the Ecoflex substrate, coating Ecoflex as insulation layer on the upside of the windings, putting another windings with opposite spiral direction on the insulation layer and encapsulating the top coil with Ecoflex.

Generally, the inductive wireless power transfer system is excited by high-frequency alternating current (AC). However, the excitation process comes with the skin effect, which leads to low current density and high resistance due to lots of current concentrating upon the surface of the wires of coil [89,90]. However, an effective method could be applied to reduce the effect greatly by dividing the wire into multiple strand wires [91,92]. In view of this, the winding of stretchable inductance coil is optimized to be multi-strand serpentine copper traces to reduce the influence of skin effect and improve the inductive performance, as shown in Figure 3.12a. The conductive coil is designed in a square, as shown in

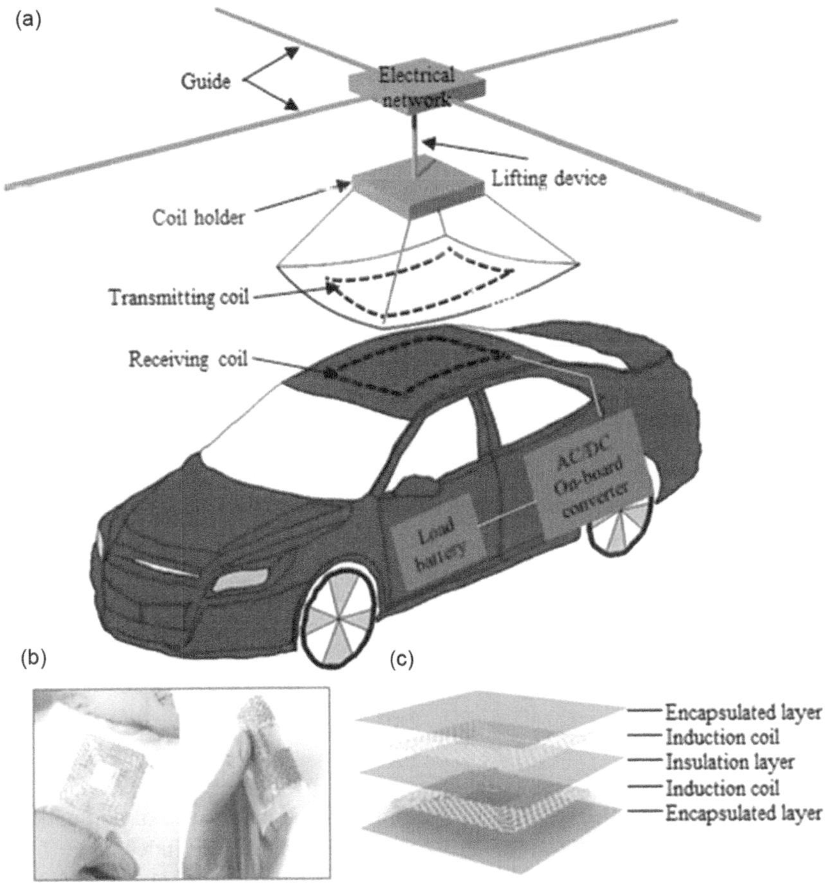

FIGURE 3.10 (a) The schematic representation of SWPT system for electric vehicles; (b) photography of the bended and stretched coil; and (c) expanded-view schematic illustration of the various constituents of the coil [88].

Figure 3.12a (left), the outside length d_{out} is 73 mm, and the inside length is d_{in}, which is obtained after determining the relevant parameters of winding as following. The windings consist of single strand, bifilar strand and treble stranded serpentine copper traces, respectively, and the total width of serpentine copper traces (w_t) in one winding keeps constant. The specific design parameters are marked in Figure 3.12a (right), d is the distance between two windings, θ is the central angle of serpentine copper traces, r is the arc radius of one winding, w is the width of a serpentine copper traces, s is the space between the two serpentine copper traces and l is the length of straight segment in serpentine copper traces.

The performance of stretchable conductive coils mainly involves two factors: one is stretchability and the other is quality factor. The quality factor

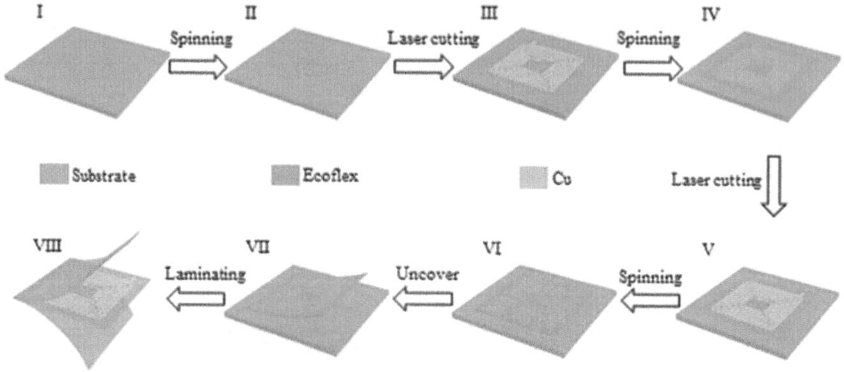

FIGURE 3.11 Fabrication of the inductance coil [88], involving (I -III) putting the wingdings on an Ecoflex encapsulation layer, (IV, V) putting another same wingdings on a new Ecoflex layer to insulate the two layers of windings, (VI, VII) encapsulating the upper wingdings with Ecoflex and peeling the complete coil off, (VIII) exploded-view schematic illustration of the reductive coil.

(Q), which reflects the electrical performance of the coil, is calculated as follows [93]:

$$Q = \frac{2\pi f L}{R} \tag{3.1}$$

where f is the resonant frequency of the coil, L is the inductance of coil and AC R is the total resistance of coil. The Q is proportional to the inductance (L) of coils and inversely proportional to the resistance (R) of winding. Hence, coil configuration is designed to reduce resistance and increase inductance for improving electrical properties.

The coil is made of two layer of the same windings, which are multiperiodic serpentine copper traces. The AC resistance (R) of coil is calculated in a representative unit cell as follows:

$$\begin{cases} R = \rho \dfrac{2\dfrac{\theta\pi}{180^\circ} r + l}{wt - (w - 2\Delta)(t - 2\Delta)} (t > 2\Delta, \quad w > 2\Delta) \\ R = \rho \dfrac{2\dfrac{\theta\pi}{180^\circ} r + l}{wt} (t \leq 2\Delta \ or \ w \leq 2\Delta) \end{cases} \tag{3.2}$$

$$\Delta = \sqrt{\frac{1}{\pi f \mu \gamma}} \tag{3.3}$$

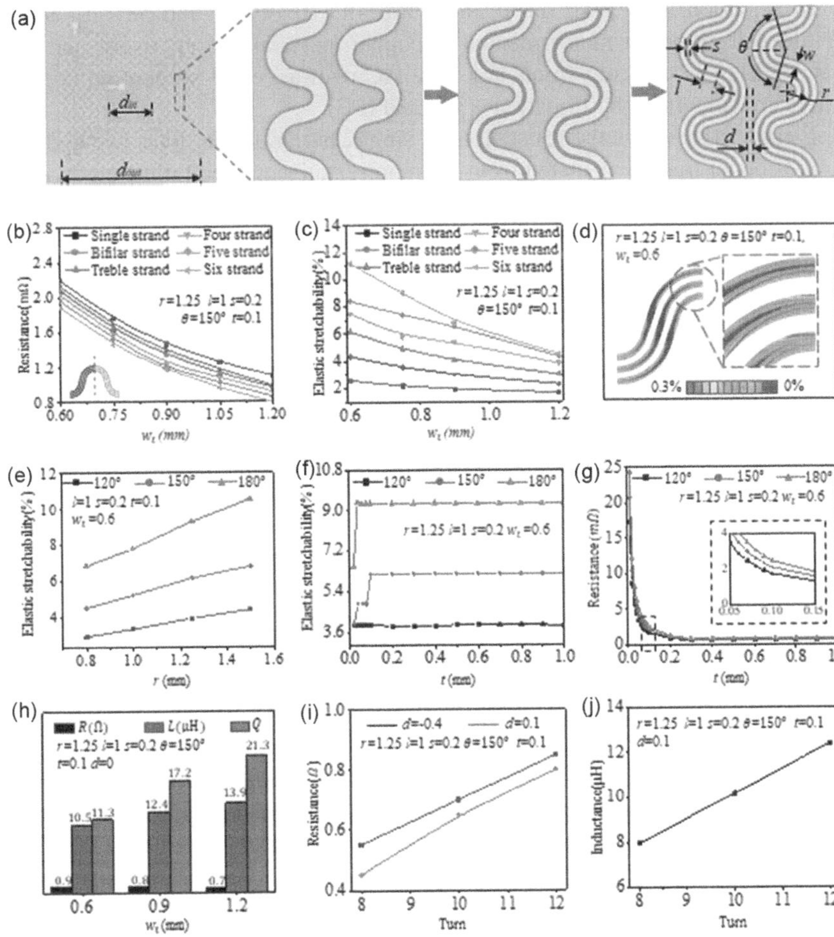

FIGURE 3.12 Design of the stretchable coil [88]. (a) The windings of the coil consisting of multi-strand serpentine copper traces; (b) the resistances of single, bifilar, treble, four, five, and six strand windings for the representative unit cell; (c) the stretchability of six kinds of windings predicated by finite element analysis (FEA); (d) FEA results of treble strand winding when is 0.6 mm; (e) the effect of r on the stretchability as is 120, 150 and 180, respectively; (f) the effect of t on the stretchability as is 120, 150 and 180, respectively; (g) the influence of t on the resistance at of 120, 150 and 180, respectively; (h) comparison among the resistance, reduction and quality factor as is 0.6, 0.9 and 1.2 mm, respectively; (i) the effect of the turn on the resistance at different, (j) the relationship between induction and turn.

where ρ is the electrical resistivity of materials, θ is the central angle of serpentine copper traces, l is the length of straight segment in serpentine copper traces, r is the radius of serpentine copper traces, w is the width of a serpentine copper traces, t is the thickness of serpentine copper traces, γ is the depth of serpentine copper traces affected by skin effect, f is the resonant frequency of the coil, μ is the permeability of vacuum and γ is the conductivity.

In Equation (3.3) [94], the μ and γ are $4\pi \times 10^{-7}$ Hm^{-1} and 58×10^6 Sm^{-1} obtained from performance manual of copper materials, respectively, and the f is 180kHz as working frequency. The AC resistances of serpentine copper traces are calculated by Equations (3.2) and (3.3).

Figure 3.12b shows the resistances of single, bifilar, treble, four, five and six strand copper traces calculated from the representative unit cell illustrated in the inset (r is 1.25mm, l is 1mm, s is 0.2mm, θ is 150° and t 0.1mm). The results demonstrate that the resistances slightly reduce with the increasing of the total width w_t of copper traces. Then, the stretchability of six kinds of copper traces are simulated by FEA, as shown in Figure 3.12c. It can be seen that w_t is an important factor to affect the stretchability of windings, and the more strand winding has obvious advantages over other. The results show the AC resistance reducing slightly, which has a slight influence on the quality factor (Q) of inductive coil. Although the stretchability increases with the increase in the number of strand, the coil consisting of treble strand serpentine copper traces is enough to accommodate car roofs with different radians. In addition, the fabrication of more strand serpentine copper traces is more complicated. Therefore, the treble strand winding as optimal design is chosen and further studied. Figure 3.12d shows the strain distribution of the treble strand winding (w_t is 0.6mm) under 6.17% tensile strain (Figure 3.12c). There is an obvious stress concentration on the arc of the serpentine copper traces. Thus, the central angle (θ) and the arc radius (r) of serpentine copper traces are the critical influence factors of stretchability. Figure 3.12e shows the stretchability improves with the increasing of r (from 0.8 to 1.5mm) when θ is 120°, 150° and 180°, respectively. Among these three central angles, coil could obtain the maximum stretchability at 180° and minimum one at 120°. Here, $r = 1.25$mm is a design parameter. Figure 3.12f demonstrates that the stretchability of treble strand serpentine copper traces keeps stable once the thickness t exceeds 0.1mm. Similarly, coil can obtain the maximum stretchability when $\theta = 180°$. The thickness of serpentine copper traces could affect the flexibility of the inductance coils directly, and the thickness 0.1mm is the optimal choice for the coil design. The resistance of copper traces is a key factor for the quality factor of coil and is inversely proportional to the quality factor according to Equation (3.1). It is necessary to calculate the resistance of treble strand serpentine copper traces again to evaluate the design parameters according to Equation (3.2), where r is 1.25mm, l is 1mm, w_t is 0.6mm, t is from 0 to 1.0mm and θ is assigned to 120°, 150° and 180°, respectively. The results illustrate that the resistance rapidly reduces with t when $0 < t < 0.2$ mm, and the resistance of copper traces obtains the maximum value when $\theta = 180°$ (inset), as shown in Figure 3.12g. In view of this, to optimize both the stretchability and the resistance of copper traces, the central angle 150° is chosen.

Some design parameters of winding have been discussed and determined, including $t = 0.1$mm, $\theta = 150°$ and $r = 1.25$mm. In addition, l and s keep 1 and 0.2mm, respectively. Meanwhile, the distance (d) between the two adjacent windings is determined as zero, which is illustrated in Figure 3.12a (right). w_t is a vital parameter, which can affect the turns and Q of coil within a set area, when the treble strand windings are fabricated as a square coil with the side length of 73 mm. 0.6, 0.9 and 1.2 mm are chosen as the total width of treble strand winding to fabricate coils according to the process described in Figure 3.11. The resistance (*R*) and inductance (*L*) of these coils are measured, and the quality factor (*Q*) is calculated by Equation (3.1). These results are exhibited in Figure 3.12h, implying the *Q* increases with the enlarging of w_t. Figure 3.12c demonstrates the relationship of w_t and stretchability of treble strand winding, illustrating the stretchability reduces with the enlarging of wt. Hence, to optimize both the quality factor and the stretchability of the coil, the w_t of 0.9 mm is chosen. The total length of the winding in a coil decreases with the increasing of *d*, leading to the reduction of resistance of coil. Two overlapping coils ($d = -0.4$ mm and $d = 0.1$ mm) are fabricated according to the fabrication process (Figure 3.11). Figure 3.12i exhibits that the actual measured resistance is lower when *d* is 0.1 mm. The square coil with the side length of 73 mm has six turns when $d = 0.1$ mm, and the receiving and transmitting coils are developed by overlapping two square coils, respectively. 0.1 mm is determined for *d*. Figure 3.12j shows the inductive performance of optimized coil, and the results illustrate that more turns can lead to a better inductive performance.

The thin, stretchable inductive coil is fitted on the vehicle roof charging for many kinds of vehicle types, where the coil size is 73 mm × 73 mm, the resonance frequency of the coil is 180 kHz. And the influence of charging efficiency coming from the deformation of the coil, such as bending and stretching, is a vital point to realize the university of SWPT system. Some experiments were carried out to verify the influence of charging efficiency produced by the deformation of coil. Figure 3.13a shows that the coils are attached on cylindrical models with different radii of curvature (30–50 mm) to measure the charging efficiency with different load resistances. The result implies that the charging efficiencies have hardly changed under bending, even compared with the state of plane, as shown in Figure 3.13b. It is well known that the coil not only is bended but also is stretched as it is used to vehicle roof. Therefore, it is necessary to test the charging efficiencies under the state of both bending and stretching. The coil was bended and stretched to conform to the cambered surface of a hemisphere with a radius of 100 mm, and the state of tensile is simulated by FEA (Figure 3.13c). The charging efficiency has no change compared with that of coil in the plane, as shown in Figure 3.13d. Moreover, to simulate the actual charging process, the coil is put on two different model car roofs for testing the charging performance (Figure 3.13e). Figure 3.13f demonstrates the charging efficiency of the coil has barely affected by the model of car and the maximum charging efficiencies are more than 50%. All the above tests prove the stretchable inductive coil is universal to the charging of any models. For the practical applications, the coils will be scaled up to match the size of real car models by just increasing the turns. Meanwhile, the processing modules and resonant circuits will have corresponding adjustment according to the relative characteristics of the enlarged transmitting and receiving coils, such as resistance and inductance, to avoid to reducing charging efficiency.

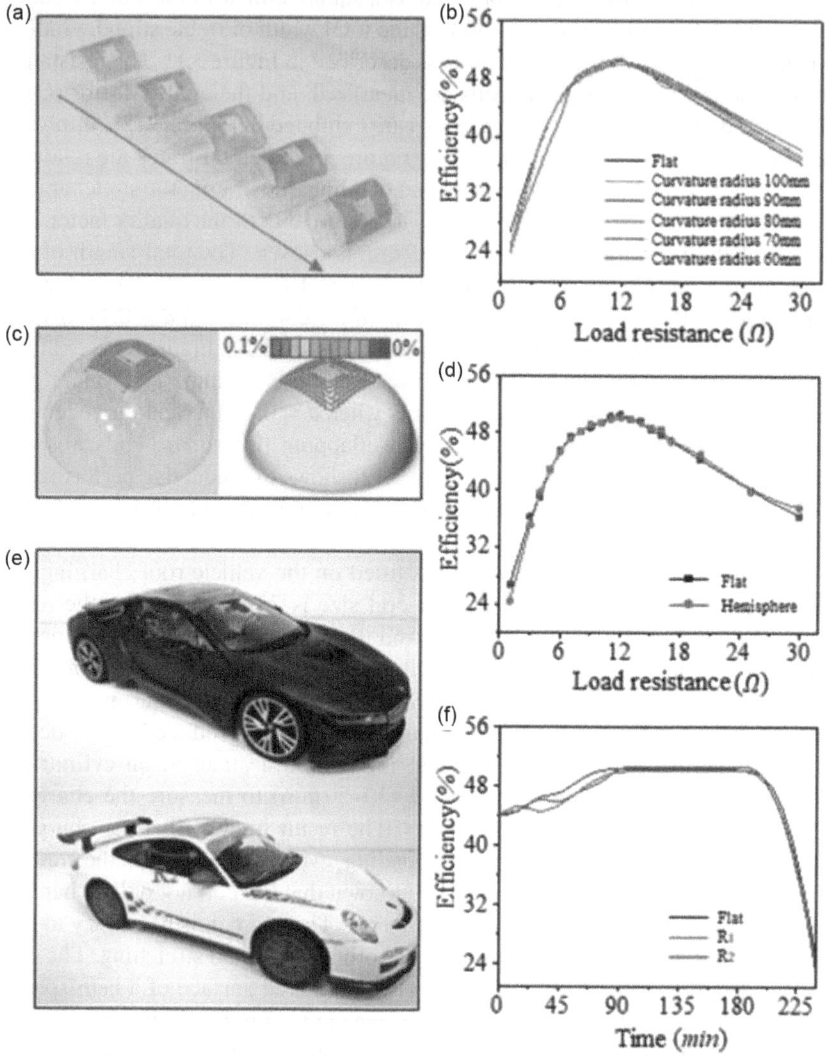

FIGURE 3.13 The charging efficiencies of the stretchable coil on (a and b) cylinders with different curvature radii, (c and d) hemisphere and (e and f) two car roofs with different radii [88].

The SWPT system different from the traditional wireless charging method was designed to improve the charging efficiency and realize universality to any vehicles. The receiving coil was integrated on the car roof for cleaning easily and enhancing safety, and the transmitting coil was hung over the car to adjust the position for shortening the distance of reductive transmission, decreasing leakage magnetic flux and improving the power transfer efficiency. The treble strand serpentine copper traces were used to fabricate the inductive coils of SWPT system for obtaining a satisfactory stretchability, which was beneficial to the conformal contact to various vehicle models. Some experiments had proved that the charging efficiency of the SWPT system was affected hardly when the coil was bended or stretched. These results reveal the significance of combining the new installation methods of the wireless charging system with structural engineering of coil to realize a novel stretchable wireless charging system for convenient and universal application to electric vehicle.

3.4 PHASE CONTROL TRANSCEIVER

Ultrasonic non-destructive testing technology is able to detect interior of structures effectively without destroying the measured objects, which promotes its broad applications in aerospace, city infrastructure, mechanical engineering and other fields that need defect detection and quality monitoring [95–101]. In general, the ultrasonic phased array realizes the acoustic beam focusing and deflection by the independent phase delay excitation of each element to achieve excellent detecting performance (Figure 3.14a) [97–101]. However, the ultrasonic phased array based on delay excitation is difficult to realize the acoustic beam controlling without prior knowledge for inhomogeneous materials [102].

Time-reversal mirror (TRM) technology has characteristics of the compensating multipath effect and adaptive focusing, which has attracted extensive research interests for the past few years. TRM technology, which is regarded as a generalized phase conjugate mirror method, is derived from the phase conjugate method in optics [104–106]. The TRM consists of transducers array, allowing the incident acoustic field to be sampled, time-reversed and re-emitted (Figure 3.14b). This process realizes the matched filtering to the inhomogeneous propagation transfer function between the array and the focal target, and all time-reversal signals are simultaneously focused on the acoustic source [104–108]. The matched filtering is equivalent to the autocorrelation processing of the transport channel function, which has the optimal signal-to-noise (SNR) output [105,109]. While multiple

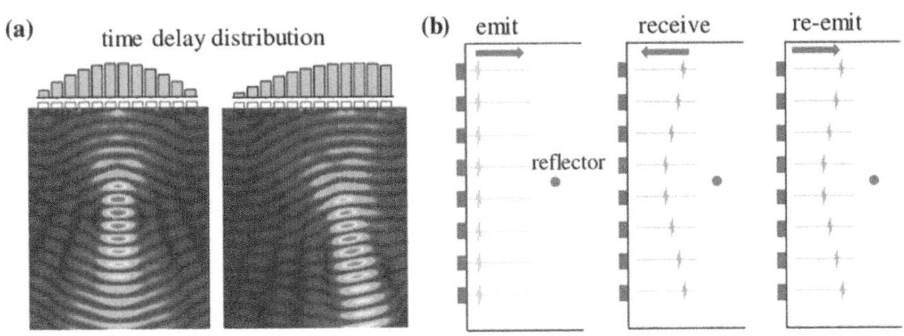

FIGURE 3.14 The detecting technology of the phased array [103]. (a) The focusing and deflection of the acoustic beam under electronic control. (b) Adaptive focusing processing of the TRM.

reflectors exist in the medium, the iterative algorithm of the time-reversal process can realize the adaptive focusing at the most reflective target [107,110]. Whatever the measured object is solid, fluid or stratified medium, the TRM always shows excellent focusing performance [111–115]. In recent years, TRM technology has showed great research significances and application values in the fields of defect detection of composite materials, medical diagnosis and treatment, underwater target recognition and so on [114–118]. However, the conventional exciting signal widely used in the TRM is a narrow pulse signal with a high bandwidth, which can obtain the high axial resolution and detect micro-size defects [107,110,119]. While the SNR and detecting range are required to be higher, the pulse width needs to be increase, which inevitably affects the axial resolution of the detecting signal.

The time bandwidth product of the conventional pulse signal is a constant, which leads to the fact that the pulse width and the bandwidth are mutually restricted. The larger pulse width represents the higher input energy while the larger bandwidth represents the higher axial resolution, which involves an inevitable contradiction. Inspired by radar signal processing technology, encoded signals based on pulse compression technology have broken the limitation of the time bandwidth product [120,121]. As the controllable and special signal with a large time bandwidth product, the linear frequency modulation (LFM) signal is initially used in the field of radar system and then gradually spreads to ultrasonic applications [119,121–124]. The application of the LFM excitation improves the SNR, detects the depth of ultrasonic detection and overcomes some defects of conventional excitations [119]. In an ultrasonic detecting system, due to the limitation of the transducer's bandwidth, the detecting echo signal obtained by LFM exciting signal is always distorted, which results in serious reduction of the matched filtering effect. In addition, distortions of the detecting signal become more serious with the increasing of the bandwidth of the exciting signal [119,125]. Testing and recording the transmitted and received signals of the determined detecting system under the impulse excitation and then the exciting signal corresponding to the desired detecting signal can be obtained by adaptive filtering deconvolution method, which can eliminate the influence of transport channel response [125].

Based on the time-reversal symmetry of the acoustic wave equation, the autocorrelation property of matched filtering for TRM technology is demonstrated and then the theoretical model of TRM technology based on the LFM signal is established in this study. The results show that the final detecting signal is in form of the dual autocorrelation function, including the autocorrelation functions of the exciting signal and the transport channel response. The transport channel response cannot be regulated, but the parameters of the LFM exciting signal can be adjusted and designed in advance. Compared with the sharp pulse exciting signal, the focusing signal's peak value of the phased array depends on the pulse width of the LFM exciting signal and the number of the array elements, which is controllable and adjustable. The adaptive filtering deconvolution method is adopted to eliminate the influence of the transport channel response, which ensures that the detecting signal of the time-reversal signal is the standard LFM waveform and improves the matched filtering effects. Combined with the adaptive filtering deconvolution method, the improved TRM technology effectively improves the SNR and the performance of ultrasonic detection.

3.4.1 Time-Reversal Symmetry and Autocorrelation

Generalized wave equations are a set of differential equations derived from Maxwell's equations to describe various wave phenomena in nature. The acoustic wave equation reflects the basic relation satisfied by the acoustic variables. Assuming that Equation (3.4) acoustic wave propagation medium is a continuous fluid medium, Equation (3.5) acoustic wave propagation medium is a non-uniform lossless medium, and Equation (3.6) acoustic wave is a small amplitude wave, the linear zigzag wave equation is determined by the continuity equation, the equation of motion and the equation of state association.

$$\nabla^2 p(r,t) - \frac{1}{c^2(r)}\frac{\partial^2 p(r,t)}{\partial t^2} = 0 \tag{3.4}$$

where $\nabla^2, p(r,t), c(r), r$ and t represent the Laplace operator, the acoustic pressure field, the acoustic velocity in the medium, spatial coordinates and time coordinates, respectively. Since only the second-order time derivative exists in the wave equation, if $p(r,t)$ is a solution to the wave equation, then $p(r,-t)$, is also a solution, which are symmetric solutions of time inversion. This property is the time-reversal symmetry of the wave equation that indicates: if the acoustic pressure field $p(r,t)$ is obtained by the reflection, refraction or scattering passing through the propagating medium from the point acoustic source, then there must be the acoustic pressure field $(r,-t)$, which reaches the initial point acoustic source along the same path. In general, time reversal in the time domain $p(r,t) \to p(r,-t)$ corresponds to phase conjugation in the frequency domain $p(r,\omega) \to p^*(r,\omega)$ for the real signal $p(r,t)$.

We analyze the autocorrelation of the TRM in time domain. If the pulse exciting signal of an element in the phased array is $\delta(t)$, then the received signal at the reflector in the propagation medium can be expressed as

$$y(t) = \delta(t) \otimes h(t) \tag{3.5}$$

where $h(t)$ is the transport channel response between the array and the reflector and $\otimes$ represents the convolution symbol. The received signal $y(t)$ is reversed in time and transmitted from the array element again as

$$y_{\text{TR}}(t) = y(-t) = \delta(-t) \otimes h(-t) \tag{3.6}$$

The second received signal at the reflector again is

$$y'_{\mathrm{TR}}(t) = \delta(-t) \otimes h(-t) \otimes h(t) \tag{3.7}$$

The process of time reversal has the characteristic of spatial matched filter, which is equivalent to autocorrelation processing only for the transport channel response. This property benefits from the time inversion symmetry of the wave (Equation 3.4). Autocorrelation processing can eliminate the multipath effect of the propagating path, extract the useful signal and improve the SNR. No matter what the channel impulse response is, the result of the convolution in Equation (3.7) is maximized at $t = 0$. The maximum value is written as

$$A_P = \int h^2(t)dt \tag{3.8}$$

which is same as the energy of the channel response signal $h(t)$.

In this study, we use the LFM signal $s(t)$ instead of the sharp pulse signal as the exciting signal, and the received signal at the reflector can be expressed as

$$y_{\mathrm{LFM}}(t) = s(t) \otimes h(t) \tag{3.9}$$

The received signal $y_{\mathrm{LFM}}(t)$ is reversed in time and transmitted from the array element again as

$$y_{\mathrm{TR\_LFM}}(t) = y_{\mathrm{LFM}}(-t) = s(-t) \otimes h(-t) \tag{3.10}$$

and then the second received signal at reflector again is

$$y'_{\mathrm{TR\_LFM}} = s(-t) \otimes h(-t) \otimes h(t) \tag{3.11}$$

The received signal at the reflector after passing through the matched filter becomes

$$y''_{\mathrm{TR\_LFM}}(t) = \left[s(-t) \otimes s(t)\right] \otimes \left[h(-t) \otimes h(t)\right] = R_{ss}(t) \otimes R_{hh}(t) \tag{3.12}$$

which is equivalent to convolving the autocorrelation function $R_{hh}(t)$ of the transport channel response with the autocorrelation function $R_{ss}(t)$ of the LFM exciting signal. The $s(t)$ and $s(-t)$ are a pair of time-reversal signals, while the signal $h(t)$ and $h(-t)$ are also a pair of time-reversal signals. Thus, we define the final detecting signal $y''_{\mathrm{TR\_LFM}}(t)$ as the dual autocorrelation function in this study. The transport channel response is invariant in a certain detecting system. The parameters of the LFM exciting signal can be designed in advance, which ensures that the peak value of the final pulse compression signal is controllable. The result of convolution in Equation (3.12) maximizes at $t = 0$ and the maximum value is written as

$$A_{\mathrm{LFM}} = \int h^2(t) \cdot s^2(t) dt \tag{3.13}$$

which depends on the transport channel response $h(t)$ and the input LFM signal $s(t)$.

3.4.2 The LFM Signal Based on Pulse Compression

As the widely used method in radar signal processing system, pulse compression technology realizes both high input energy of long pulse and high axial resolution of narrow pulse, which can resolve the contradiction between pulse width and bandwidth. In particular, as the signal with a large time bandwidth product commonly used in pulse compression technology, pulse width and bandwidth of the LFM signal are convenient for designing and adjusting. The time domain expression of the LFM signal is

$$s(t) = \mathrm{rect}\left(\frac{t}{T}\right) e^{j2\pi\left(f_0 t + \frac{K}{2}t^2\right)} \left(|t| \le \frac{T}{2}\right) \tag{3.14}$$

where f_0, B, T and $K = \frac{B}{T}$ represent the central frequency, the bandwidth, the pulse width and the slope of frequency modulation, respectively. The central frequency f_0 is equal to the central frequency f_{tr} of the transducer. $\mathrm{rect}\left(\frac{t}{T}\right)$ is the rectangular windows signal, which is expressed as

$$\mathrm{rect}\left(\frac{t}{T}\right) = \begin{cases} 1,\left(|t| \le \frac{T}{2}\right) \\ 0,(\mathrm{others}) \end{cases} \tag{3.15}$$

The instantaneous frequency of the LFM signal is obtained as

$$f(t) = f_0 + Kt \tag{3.16}$$

which indicates that the instantaneous frequency $f(t)$ is a linear function of t and the bandwidth is $B = KT$.

The LFM signal is processed of autocorrelation by matched filter at the receiver, which can extract useful signal and maximize SNR. The time domain expression of the matched filter is

$$g(t) = s^*(-t) \tag{3.17}$$

While the LFM signal $s(t)$ only contains the real part, the pulse response of the matched filter $g(t)$ is the time-reversal waveform of the LFM signal. The corresponding acoustic pressure fields of the above time-reversal signals all satisfy the same wave Equation (3.4) due to the time-reversal symmetry; thus, the convolution result of them is a function of the signal $s(t)$ and itself (i.e., autocorrelation function). Then, the waveform of the LFM signal $s(t)$ after passing through the matched filter is

$$s_0(t) = s(t) \otimes g(t) = \sqrt{B \cdot T} \cdot \frac{\sin(\pi Bt)}{\pi Bt} \cdot \text{rect}\left(\frac{t}{2T}\right) \cdot e^{j2\pi f_0 t} \tag{3.18}$$

The envelope waveform of the compressed signal $s_0(t)$ is written as

$$S_0(t) = \sqrt{B \cdot T} \cdot \text{Sa}(\pi Bt) \cdot \text{rect}\left(\frac{t}{2T}\right) \tag{3.19}$$

While solution t_1 satisfies $\pi Bt_1 = \pm\pi/2$ (the −4 dB main lobe width of the envelope signal, as shown in Figure 3.15a), the pulse width of the compressed signal is $\tau = 2t_1 = \frac{1}{B}$, which is approximate to the axial resolution of the pulse compression signal. And the compression ratio is $D = \frac{T}{\tau} = BT$, which is equivalent to the time bandwidth product.

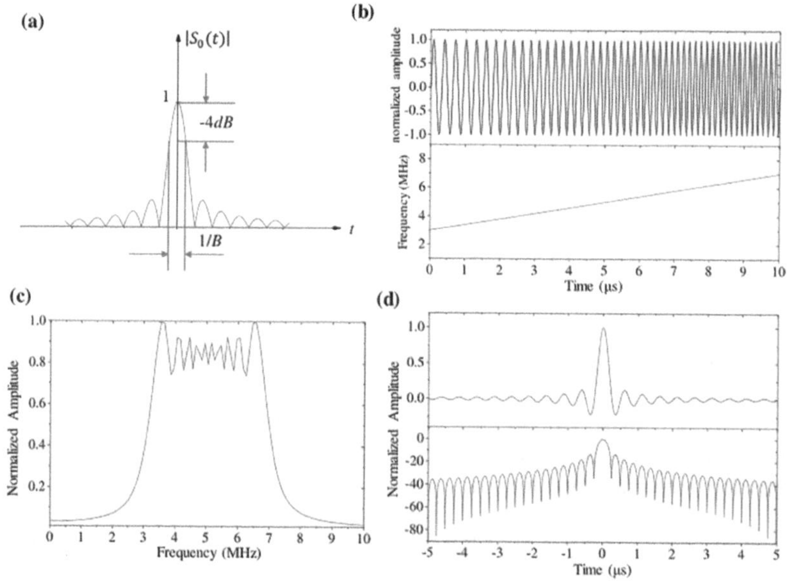

FIGURE 3.15 The LFM signal with the time bandwidth of 40 [103]. (a) The definition of pulse compression signal's pulse width. (b) Time domain waveform and instantaneous frequency of the LFM signal. (c) The amplitude-frequency characteristic curve of the LFM signal. (d) The time domain waveform of the LFM signal after pulse compression (the top part) and its corresponding signal after logarithm (the bottom part).

The LFM signal breaks the limitation of the time bandwidth product of conventional pulse exciting signals. The pulse width T and the bandwidth B of the LFM signal, which correspond to the input energy and the axial resolution, are very convenient for designing. For example, when the pulse width T, the central frequency f_0 and the bandwidth B are chosen as 10 μs, 5 MHz and 4 MHz, respectively, the time domain waveform, the instantaneous frequency (Figure 3.15b), the amplitude–frequency characteristic curve of the LFM signal (Figure 3.15c) and the time domain waveform of the pulse compression signal (Figure 3.15d) are obtained from the above theoretical equations. In the bottom part of Figure 3.15d, it is assumed that the time coordinate corresponding to first lobe value is t_{flv}, and the t_{flv} satisfies the equation $\pi B t_{\text{flv}} = \tan(\pi B t_{\text{flv}})$ because of the Sa function in Equation (3.19). Then, the first lobe value is obtained as

$$A_{\text{flv}} = 20 \cdot \log_{10}\left(\frac{S_0(t_{\text{fv}})}{S_0(0)}\right) = -13.26 \text{ dB} \tag{3.20}$$

The sidelobe level can be further reduced by weighting the window function [126–131]. The time bandwidth product D is raised to 40 here and the signal with a larger time bandwidth product is also available if need be.

3.4.3 Design of LFM Signal Parameters

As one of the LFM signal' parameters, the central frequency is generally equal to the central frequency of the transducer. Thus, the most two important parameters are pulse width T and spectrum bandwidth B. The central frequency of the LFM signal is chosen as 5 MHz here. Figure 3.16a shows that the peak value of the compressed signal positively correlated with the initial pulse width T while bandwidth B is fixed (4 MHz). While pulse width is increased from 0.25 to 10 μs (i.e., the time bandwidth product is increased from 1 to 40), the peak value of the pulse compression signal is improved by 16.2 dB $\left(20 \cdot \log_{10}\left(\sqrt{\frac{0.25 \times 4}{10 \times 4}}\right)\right)$ according to Equation (3.19). Figure 3.16b shows that the main lobe width decreases with the improvement of the bandwidth B. The smaller main lobe width represents the higher axial resolution. While the bandwidth is increased from 1 to 10 MHz (i.e., the time bandwidth product is increased from 10 to 100 with the fixed pulse width $T = 10$ μs), the axial resolution is raised ten times approximately. In general, the bandwidth of the LFM signal, which is matched with the bandwidth of the transducer, cannot be infinite in ultrasonic detection.

The finite element method (FEM) is adopted in the commercial software package COMSOL Multiphysics 5.4 and is used to simulate the transmission and reception of the LFM signals with different pulse widths. The FEM model is shown in Figure 3.16c, and the spatial dimension of the two-dimensional axisymmetric is adopted here. The transmitted transducer is at the top, the received transducer is at the bottom and the transport medium is in the middle. The transmitted

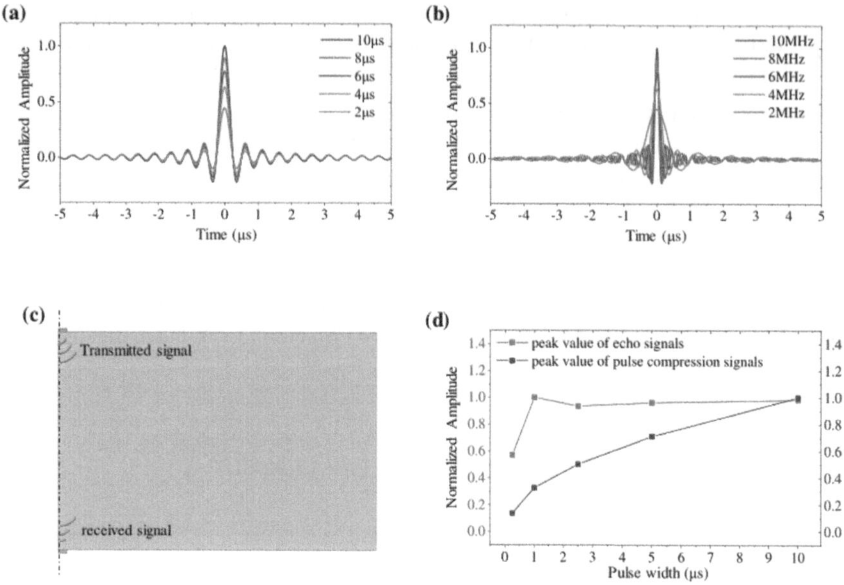

FIGURE 3.16 Parameter analysis of the LFM signal [103]. (a) Pulse compression signals with different pulse widths of LFM signals (B=4MHz). (b) Pulse compression signals with different bandwidth of LFM signals (T=10μs). (c) The finite element model for transmission and reception of the LFM signals. (d) Normalized peak values of the echo signals and pulse compression signals with different pulse widths in FEM.

transducer is excited by LFM signals, and the received transducer at the bottom is used for receiving signals by a suspension potential setting. Before the matched filtering, the peak values of received signals of LFM signals with different pulse widths are almost the equal, as shown in Figure 3.16d. The low value of the first point may be because the pulse width is too narrow to completely exciting the transducer's vibration in FEM. After passing through the matched filter, the pulse compression signal with a larger pulse width has a larger peak value, as shown in Figure 3.16d. It is verified that the parameters design of the LFM signal can effectively improve the peak value of the pulse compression signal and then improve the SNR according to the simulation results.

3.4.4 Focusing Characteristic of TRM Based on the LFM Signal

The TRM can realize the adaptive focusing of the acoustic beam in the measured medium. Conventional incentive signal to the TRM is the excitation of a narrow pulse signal to obtain the excellent axial resolution. In this study, we adopt the above LFM signal with a large time bandwidth product instead of the narrow pulse signal to improve the input energy of the transmitted signal. A linear flexible phased array with N elements is attached to a curved measured object. Each array element is regarded as a point acoustic source, and the distance between the ith element and the reflector is r_i (Figure 3.17a).

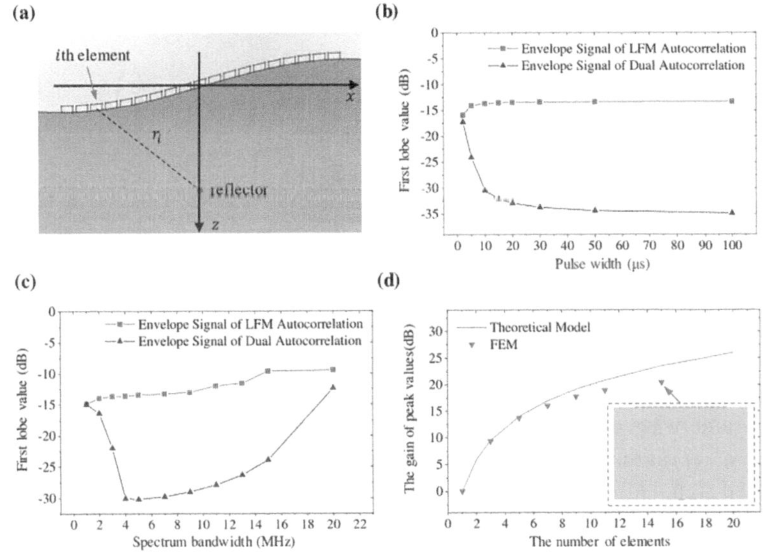

FIGURE 3.17 The TRM based on the LFM signal [103]. (a) Schematic diagram of the linear phased array on a curve surface. (b) The first lobe values of the envelope signal of the LFM autocorrelation function and the dual autocorrelation function with different pulse widths. (c) The first lobe values of the envelope signal of the LFM autocorrelation function and the dual autocorrelation function with different spectrum bandwidths. (d) The peak value gain of the phased array with the different number of elements.

The first transmitted signal of each array element is $s(t)$ and the received signal at the reflector is

$$y_i'(t) = \begin{cases} s\left(t - \frac{r_i}{v}\right) \otimes h_i\left(T_r - \frac{r_i}{v}\right)\left(\left|T_r - \frac{r_i}{v}\right| \le \frac{T}{2}\right) \\ 0\left(\left|T_r - \frac{r_i}{v}\right| > \frac{T}{2}\right) \end{cases} \tag{3.21}$$

where v and $h_i(t)$ represent the acoustic velocity in the medium and transport channel response between the ith element and the reflector. After time reversal T_r of the received signal, the signal re-emitted from the ith array element is expressed as

$$y_i^{TR}(t) = \begin{cases} s\left(t - \left(T_r - \frac{r_i}{v}\right)\right) \otimes h_i\left(t - \left(T_r - \frac{r_i}{v}\right)\right)\left(\left|t - \left(T_r - \frac{r_i}{v}\right)\right| \le \frac{T}{2}\right) \\ 0\left(\left|t - \left(T_r - \frac{r_i}{v}\right)\right| > \frac{T}{2}\right) \end{cases} \tag{3.22}$$

The signal at the reflector received again is

$$y_i^{TR'}(t) = \begin{cases} s(t-T_r) \otimes h_i(t-T_r) \otimes h_i(t) \left(|t-T_r| \le \dfrac{T}{2} \right) \\ 0 \left(|t-T_r| > \dfrac{T}{2} \right) \end{cases} \tag{3.23}$$

From Equation (3.23), it can be concluded that the signal arriving at the reflector is independent of the distance r_i, i.e., the signal secondly emitted by each array reaches the reflector at the same time. The received signal at the reflector contains the convolution of the transport channel response $h_i(t)$ and its time-reversal signal $h_i(t-T_r)$, which are two solutions satisfying the time-reversal symmetry of the same wave equation and ensure the autocorrelation property. The received signal $y_i^{TR'}(t)$ at the reflector from each array is superimposed and then the superimposed signal after the matched filter is obtained as

$$y_{\text{sum}}^{TR}(t) = s(t) \otimes \sum_{i=1}^{N} y_i^{TR'}(t) = \begin{cases} s^{-1}(t-T_r) \otimes s(t) \otimes \sum_{i=1}^{N} h_i^{-1}(t-T_r) \otimes h_i(t) (|t-T_r| \le T) \\ 0(|t-T_r| > T) \end{cases} \tag{3.24}$$

It can be found that the superimposed signals at the reflector contain the autocorrelation functions of the LFM exciting signal $s(t)$ and the transport channel response $h_i(t)$ of each array. Substitution of Equation (3.14) into Equation (3.24) gives

$$y_{\text{sum}}^{TR}(t) = \begin{cases} \left[\sqrt{B \cdot T} \cdot \text{rect}\left(\dfrac{t-T_r}{2T} \right) \cdot Sa\left(\pi B(t-T_r) \right) \right] \otimes \sum_{i=1}^{N} h_i^{-1}(t-T_r) \otimes h_i(t) (|t-T_r| \le T) \\ 0(|t-T_r| > T) \end{cases} \tag{3.25}$$

There is no distance parameter r_i in the superposition signal $y_{\text{sum}}^{TR}(t)$, and all signals reach the focal point at the same time. The superposition signal includes two parts. One part is the autocorrelation function waveform of the LFM signal. The larger the pulse width of the LFM signal represents the higher the peak value of superposition signal. The other part is the superposition of the autocorrelation function of the transport channel response of each array. As the number of array elements increases, the peak value of the superposition signal is also greatly improved.

The impulse response of the transport channel is set as the Gaussian function

$$H(f) = e^{-\alpha\left(\frac{f-f_{tr}}{B_{tr}}\right)^2} \tag{3.26}$$

where $f_{tr} = 5$ MHz and $B_{tr} = 3$MHz represent the central frequency and bandwidth of the Gaussian function and α is a constant of 4ln2. Considering a single element merely, from Equation (3.24) the final detecting signal is obtained as

$$y_{\text{sum}}^{TR}(t) = s(t - T_r) \otimes s(t) \otimes h(t - T_r) \otimes h(t) \tag{3.27}$$

The central frequency of the LFM signal $s(t)$ is chosen as 5MHz here. $y_{\text{sum}}^{TR}(t)$ in Equation (3.27) is regarded as the dual autocorrelation envelope signal and $s(t-T_r) \otimes s(t)$ is regarded as LFM autocorrelation envelope signal. While the bandwidth of the LFM signal is fixed at 4MHz and pulse width varies from 2 to 100 μs, the first lobe values of the LFM autocorrelation envelope signal and the dual autocorrelation envelope signal in Equation (3.27) are compared in Figure 3.17b. With the increase of the pulse width, the first lobe value of the envelope waveform of the LFM autocorrelation function remains about −13.3dB, which is almost unchanged and close to the theoretical value −13.26dB in Equation (3.20). And the first lobe value of the envelope waveform of the dual autocorrelation function decreases sharply and tends to be stable, while the pulse width increases. As the pulse width reaches 100 μs, the first lobe value of the envelope waveform of the dual autocorrelation function decreases to −34.85dB. The reason is that the envelope waveform of the Gaussian signal is the function with a zero sidelobe. Compared with the envelope waveform of the LFM autocorrelation function, the first lobe value of the dual autocorrelation function is greatly reduced. Similar to the window function processing, the dual autocorrelation function also increases the width of the main lobe and reduces the axial resolution to some extent. While the pulse width of the LFM signal is fixed at 10 μs and the bandwidth varies from 1 to 20MHz, the first lobe values of the LFM autocorrelation envelope waveform and the dual autocorrelation envelope waveform are shown in Figure 3.17c. With the increase in spectrum bandwidth, the first lobe value of the envelope signal of the LFM autocorrelation function rises slightly. The first lobe value of the envelope signal of the dual autocorrelation function decreases first and then increases as the bandwidth of the LFM signal increases. While the bandwidth of the LFM signal increases up to 5MHz, the first lobe value reaches the minimum of −30.19dB. The reason is that the transport channel response is the Gaussian function. As the spectral bandwidth is too large, the final signal is seriously distorted, resulting in the increase of the first lobe value.

Considering the influence of the number of the phased array elements on the detecting signal, the transport channel response between each element and the reflector is assumed to be equal to $h(t)$, and then from Equation (3.24), the final focusing signal is obtained as

$$y_{\text{sum}}^{TR}(t) = N \cdot s(t - T_r) \otimes s(t) \otimes h(t - T_r) \otimes h_i(t) \tag{3.28}$$

where N represents the number of the phased array elements. The peak value gain of the phased array compared to the single element is

$$G_{\text{peak.value}} = 20 \cdot \log_{10}\left[\frac{N \cdot s(t) \otimes s(t-T_t) \otimes h(t) \otimes h(t-T_t)}{s(t) \otimes s(t-T_t) \otimes h(t) \otimes h(t-T_t)}\right] = 20\log_{10}(N) \tag{3.29}$$

which is measured in decibels. The relationship between the number of elements and the peak value gain of the phased array is shown in Figure 3.17d. The peak value gain of the phased array is simulated in the FEM, and the results are close to the theoretical value and slightly lower than the theoretical value. The FEM model in Figure 3.17d adopts the spatial dimension of the two-dimensional plane. The theoretical model in Equation (3.29), which ignores the difference of each array's transport channel, is the optimal gain theoretically. The transport channel of the middle-transmitted transducer is selected as the reference element, and then the distance between other transmitted transducers and the received transducer is longer. So the peak value gain of the actual transducer arrays cannot reach the prediction in Equation (3.29), which is in line with our expectation.

3.4.5 Adaptive Filtering Deconvolution Method

While the LFM electrical signal is input to one side of a piezoelectric transducer, which is usually used to stimulate the mechanical vibration of the transducer, then the other side of the piezoelectric transducer generates acoustic waves into the measured object for detection. However, because of the bandwidth limitation of the transducer, the output signal of the transducer received is always deformed into a spindle-shaped waveform. In this case, the compressed effect is seriously reduced. In general, the impulse response function of the transducer is approximate to the Gaussian signal in Equation (3.26). And time domain expression of the above Gaussian signal can be written as [129]

$$s_{tr}(t) = \frac{B_{tr}}{\sqrt{\alpha}}\sqrt{\pi}e^{-\frac{B_{tr}^2}{\alpha}\pi^2 t^2 + j2\pi f_{tr} t} \tag{3.30}$$

In particular, the input LFM signal has the pulse width of 10 μs, the bandwidth of 4 MHz and the central frequency of 5 MHz and the transducer has the bandwidth of 3 MHz and the central frequency of 5 MHz. The waveform of LFM signal passing through the transducer is expressed as

$$s_{\text{out}}(t) = s(t) \otimes s_{tr}(t) \tag{3.31}$$

The time domain of the signal $s_{\text{out}}(t)$ is shown in Figure 3.17a, which deforms into a spindle-shaped signal. The corresponding amplitude–frequency characteristic curve is also provided in Figure 3.18a. Defining the expression of ratio c as

$$c = \frac{A_s}{A_l} \tag{3.32}$$

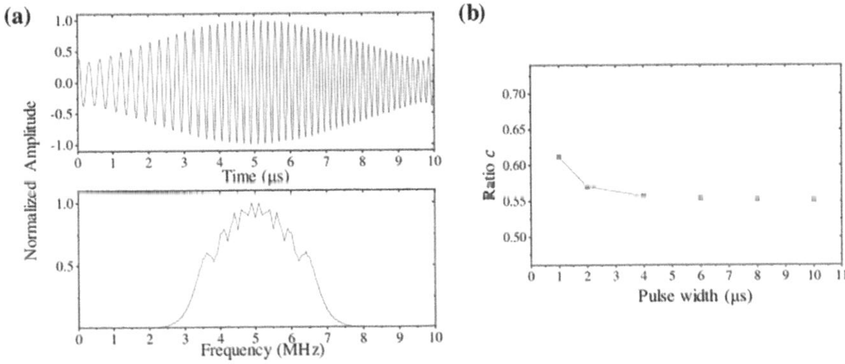

FIGURE 3.18 The LFM signal and its corresponding spindle signal [103]. (a) The time domain waveform of the spindle signal and its amplitude-frequency characteristic curve. (b) The relationship between the ratio c of the peak values and pulse width T.

where A_s and A_l represent the peak value of spindle-sharped waveform and LFM waveform after pulse compression, respectively. The ratio c is much lower than 1, and the ratio c decreases slightly as the pulse width increases, as shown in Figure 3.18b. The peak value of the spindle-sharped signal after pulse compression is obviously smaller than the peak value of the initial LFM signal. Compared with the LFM signal, the input energy and the bandwidth of the spindle-shaped signal are reduced.

In this study, the adaptive filtering method is used to realize the controllable and adjustable detecting signal in ultrasonic detecting. First, according to the actual detecting demand, the waveform of the expected detecting signal is designed in advance. Then, according to the response function of the selected detecting system, the exciting signal corresponding to the detecting signal designed in advance is solved by means of adaptive filtering deconvolution method. Finally, the exciting signal obtained is input to the detecting system by the arbitrary signal generator to get the desired detecting signal.

The detecting system consists of the transmitted transducer, the acoustic medium and the received transducer here. The input exciting signal, the impulse response of the detecting system and the received detecting signal are expressed as $x_i(n)$, $h_i(n)$ and $y_i(n)$, respectively. The subscript i represents the ith element of the phased array. The received detecting signal is the convolution of the input exciting signal and the impulse response of the detecting system. The time domain model is expressed by sampling discrete sequence as

$$y_i(n) = x_i(n) \otimes h_i(n) \tag{3.33}$$

If the expected detecting signal $y_{i\_ecpected}(n)$ of the ith element is determined in advance, then the exciting signal and the impulse response of the detecting system need to be known to obtain the detecting signal $y_{i\_ecpected}(n)$. Inputting a sharp pulse exciting signal $x_{i0}(n)$ to the certain detecting system and the received signal at the reflector is $y_{i0}(n)$. The received signal $y_{i0}(n)$ is re-emitted after time reversal as

$$y_{i0\_TR}(n) = y_{i0}(-n) = x_{i0}(-n) \otimes h_i(-n) \tag{3.34}$$

The signal received secondly at the reflector is

$$y'_{i0}(n) = x_{i0}(-n) \otimes h_i(-n) \otimes h_i(n) \tag{3.35}$$

where the detecting system response function $h_i(n)$ and its time-reversal signal $h_i(-n)$ are a pair of time-reversal solutions satisfying the same wave equation and the convolution result of them has the autocorrelation property. Collecting this pair of time series data $x_{i0}(-n)$ and $y'_{i0}(n)$ and then the autocorrelation function of the detecting system's impulse response are obtained by means of adaptive deconvolution as

$$R_{hh}(n) = h_i(-n) \otimes h_i(n) = y'_{i0}(n) \otimes^{-1} x_{i0}(-n) \tag{3.36}$$

where $\otimes^{-1}$ represents the deconvolution operation. $y'_{i0}(n)$ is regarded as the reference output of the adaptive filter and $x_{i0}(-n)$ is regarded as the input signal of the adaptive filter. The actual output $y_f(n)$ of the adaptive filter is the convolution of $x_{i0}(-n)$ and the weight sequence $w_i(n)$ of the adaptive filter. According to the error value $e_i(n)$ between the actual output $y_f(n)$ and the reference output $y'_{i0}(n)$, the adaptive filter regulates the weight sequence by a certain algorithm to control the approximation of $y_f(n)$ to $y'_{i0}(n)$, so as to obtain the optimal weight sequence of the adaptive filter $w_i^*(n) = h_i(-n) \otimes h_i(n)$. The operation process is shown in Figure 3.19a.

The expected detecting signal $y_{i\_ecpected}(n)$ of the ith element is determined in advance according to the detecting requirements and then the corresponding exciting signal is expressed as

$$x_{i\_ecpected}(n) = y_{i\_ecpected}(-n) \otimes^{-1} \left[h_i(-n) \otimes h_i(n)\right] \tag{3.37}$$

The adaptive filtering is used for deconvolution again to control the output signal $y_{f1}(n)$ of the adaptive filter to approximate $y_{i\_ecpected}(-n)$, which obtains the optimal weight sequence of the adaptive filter $w_i^*(n) = x_{i\_ecpected}(n)$. The operation process is shown in Figure 3.19b.

Simultaneously, Equations (3.36) and (3.37) give

$$x_{i\_ecpected}(n) = y_{i\_ecpected}(-n) \otimes^{-1} \left[h_i(-n) \otimes h_i(n)\right] \tag{3.38}$$

While the error e_i is small enough, the optimal weight sequence of the adaptive filter is $w_i^*(n) = x_{i\_ecpected}(n)$. After merging the previous two steps, only one time operation of the adaptive filter is needed to obtain the exciting waveform corresponding to the designed detecting signal. The operation process is shown in Figure 3.19c. First, the adaptive filter needs to select a certain algorithm, and then,

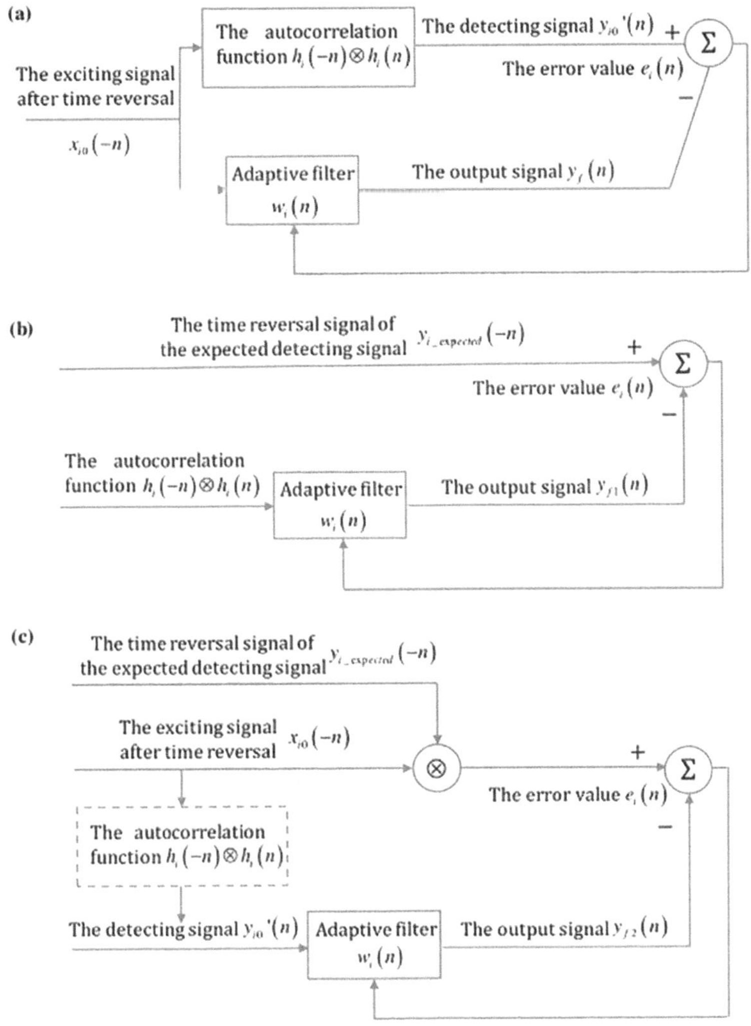

FIGURE 3.19 The adaptive filtering deconvolution operation [103]. (a) The flow diagram to calculate the autocorrelation function of the detecting system's impulse response. (b) The flow diagram to calculate the exciting signal corresponding to the expected detecting signal. (c) The process diagram after the combination.

the corresponding exciting signal waveform can be calculated according to the pair of time series data of the exciting signal and the detecting signal collected in practice. In this way, the detecting signal at the reflector of a single element can be designed in advance. The focusing result with the standard LFM signal can be obtained by applying the adaptive filtering method for each element in turn, and then, the superposition signal of Equation (3.25) becomes

$$y_{\mathrm{sum}_A}^{TR}(t) = \begin{cases} N \cdot \sqrt{B \cdot T} \cdot \mathrm{rect}\left(\frac{t - T_r}{2T}\right) Sa\left(\pi B(t - T_r)\right)\left(|t - T_r| \le T\right) \\ 0\left(|t - T_r| > T\right) \end{cases} \tag{3.39}$$

The superposition signal obtained by the adaptive filtering method is the autocorrelation function of the standard LFM signal, which is able to further improve the controllability and SNR of the focusing signal. The adaptive filtering deconvolution method, which is also applicable well to the transport channel response of other waveforms, is a universal method. When the ultrasonic phased array has the self-transmitted and self-received detecting mode, the adaptive filtering deconvolution method is also capable of regulating the waveform of the detecting signal. An array element of the phased array is excited first, and each array element receives its corresponding echo signal of the reflector. Then, the received signal is excited by each array element in turn after time reversal and later receives its corresponding second echo signal of the reflector again. Collecting this pair of time series data of the first excited signal and the second received signal. By imitating the solving method in Equation (3.38), the excited signal waveform $x_{i\_\mathrm{ecpected}}(n)$ corresponding to the expected detecting signal $y_{i\_\mathrm{ecpected}}(n)$ can be obtained. The adaptive filtering deconvolution method has good generality in controlling the waveform of the detecting signal.

In summary, we demonstrate autocorrelation property of the TRM from the time-reversal symmetry of the wave equation and establish the theoretical model of TRM method based on the standard LFM signal. Applying the LFM signal as the exciting signal to the TRM, the received signal is the dual autocorrelation function of the exciting signal and the transport channel response. With the increase of the pulse width and bandwidth of the LFM signal, the peak value and the axial resolution of the compressed pulse signal can be improved significantly. The peak value of the final superimposed signal in TRM depends on the pulse width of the LFM exciting signal and the number of the array elements actually. The theoretical model predicts that the peak value optimal gain of the TRM's final focusing signal is 26.02 dB as the number of the array elements increases to 20. According to the impulse response of the detecting system, we adopt the adaptive filtering deconvolution method to regulate the input signal, which can ensure that the detecting signal is the standard LFM signal designed in advance and further enhance the matched filtering effects. It is expected that the LFM signal would be applied to the TRM technology, and then, the adaptive filtering deconvolution method is used to regulate the input signal, so as to promote the further applications of the TRM technology.

3.5 AIR PRESSURE SENSOR FOR LARGE SCALE

The speed increase of high-speed trains has been taken as a major strategy for national development, but with the speed increase of high-speed trains, the gas dynamics problems of high-speed trains, such as air drag, crosswind stability, train intersection pressure waves and tunnel micro-pressure waves, have become increasingly prominent.

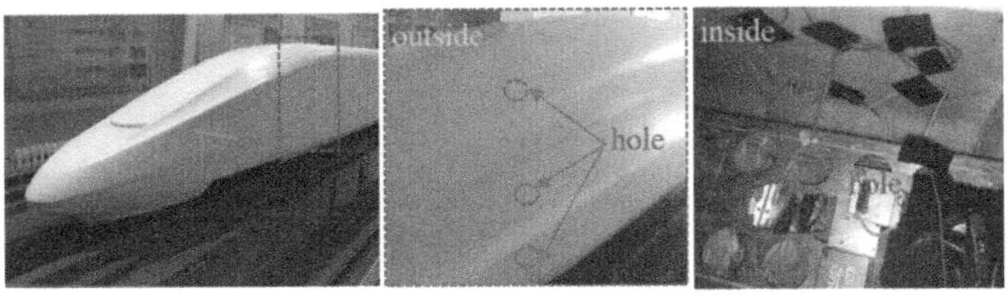

FIGURE 3.20 Schematic diagram of the pressure hole technique [1]. (a) High-speed rail model; (b) Pressure measuring hole (surface); (c) Pressure measuring hole (interior).

According to previous studies, the aerodynamic shape of HSR has a great influence on the gas dynamics performance of HSR, so many researchers use a combination of numerical simulation and experiments to optimize the design of the aerodynamic shape of HSR (especially the shape of HSR front end). The experiments need to collect the pressure on the surface of the HSR so that the dynamics can be analyzed. At present, the scaling experiments of HSR are usually performed in wind tunnels or on top of the HSR dynamic model platform, and the surface pressure is collected by the pressure tap technique (Figure 3.20). However, the pressure tap technique requires drilling holes on the model of the HSR, which will cause damage to the model; the pressure tap technique is used for single-point pressure monitoring at certain characteristic points of the area, which cannot realize the pressure field monitoring in a certain area (especially considering the complex surface of the car head); it is quite tedious to install and replace the pressure measuring system.

In view of these, and considering the rapid development of flexible electronics nowadays, a large-area stretchable air pressure sensor array is fabricated in this section using flexible electronics. The large-area stretchable barometric pressure sensor array adopts a curved bridge interconnected island–bridge structure, which can be stretched in both directions and then can be fitted to the complex non-expandable surface of the head, and used for monitoring the pressure field on the head surface in the high-speed rail dynamic model experiments.

3.5.1 Introduction of Large-Area Stretchable Air Pressure Sensor Array

Figure 3.21a shows a display of this large-area stretchable barometric sensor array. The structure as a whole is a curved bridge interconnected stretchable island–bridge structure with a total of 30 (6×5) islands that are mechanically and electrically connected to each other by serpentine wires, in which 16 (4×4) islands hold barometric pressure sensors for monitoring the surface pressure of the HSR. Considering that this air pressure sensor array is used for pressure field monitoring on the head of the experimental model of the HSR, the size of the whole island–bridge structure is 200 mm × 195 mm, a scale much larger than the size of today's flexible electronics. Figure 3.21b shows the large-area stretchable air pressure sensor array on the experimental model of HSR, and it can be seen in the figure that the air pressure sensor array can fit well to the HSR head with

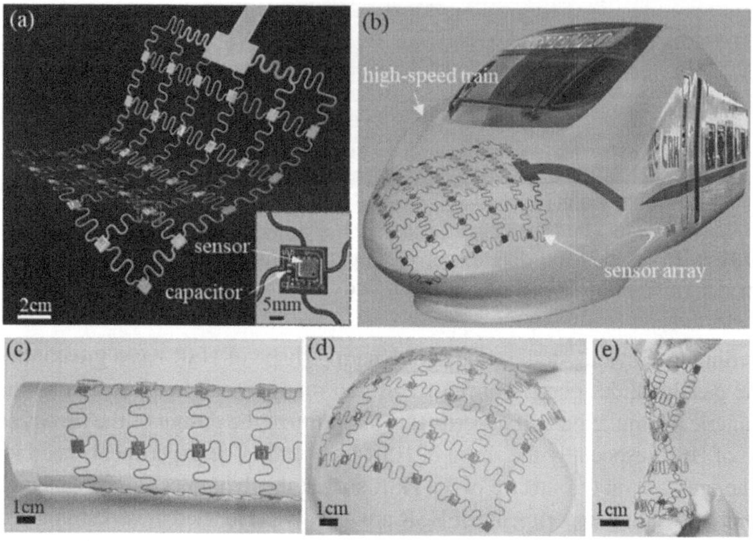

FIGURE 3.21 Large-area stretchable air pressure sensor display [1]. (a) device; (b) device and high speed rail experimental model paste fit.

complex curved surfaces. Figure 3.21c–e shows the air pressure sensor array wrapped around a cylinder, fitted to a spherical surface and subjected to torsion, respectively. These demonstrations show the flexible and stretchable properties of the large-area stretchable air pressure sensor array.

The large-area stretchable barometric pressure sensor array is composed of a barometric pressure sensor at the top, a capacitor and a multilayer flexible stretchable circuit board (FPCB) at the bottom. The barometric pressure sensor is the HP303B (Figure 3.22), a small, high-precision intelligent digital sensor from Shenzhen Huiyi Huapu Electronics Co. The HP303B is a high-resolution pressure sensor with an I2C interface that includes a silicon piezoresistive pressure element and a high-resolution 24-bit ΔΣADC. The HP303B provides a high-precision 24-bit pressure digital output, and all built-in calculations are performed using high-speed 4MHz floating-point arithmetic with minimal computational error, and data compensation is integrated internally. The specific parameters of the HP303B can be found in Table 3.1.

The data acquisition of the sensors in the array is generally achieved by using the scanning circuit, but considering that the operation time of the high-speed rail model in the subsequent experiments is only 4 seconds, and the operation is divided into acceleration, uniform speed and deceleration sections, if the scanning circuit is used, on the one hand, the microcontroller is collecting data from each sensor in turn during the scanning process, so the data collected by each sensor are not the same moment. This is not the same moment, especially for the acceleration and deceleration stages of the high-speed rail model, each sensor is not collected to the same speed pressure value, which is meaningless for the analysis of the pressure field of the head of the high-speed rail; on the other hand, 16 sensors scanned once down

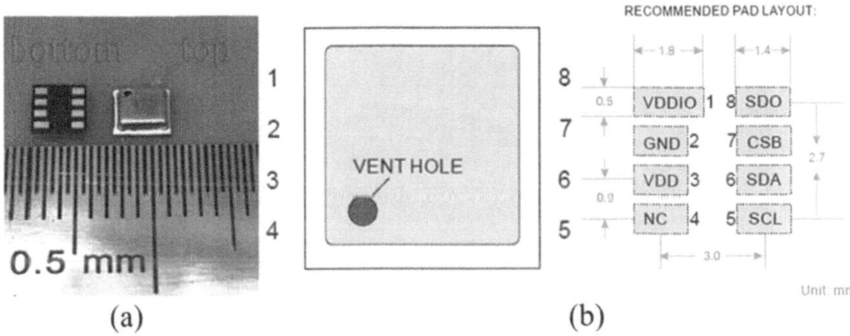

FIGURE 3.22 HP303B air pressure sensor [1]. (a) physical drawing; (b) pin configuration.

TABLE 3.1
HP303B Air Pressure Sensor Parameters [1]

Pressure Range (hPa)	Precision of Pressure (Pa, High-Precision Mode)	Temperature Sensitivity Coefficient (Pa/K)	Service Voltage (V)	Communication Mode	Packaging	Size
300–1200	0.5	0.5	1.7–3.7	I^2C and SPI	8 pin	3.7 × 3.9 × 1.2

time is nearly 1 second (equivalent to a sampling frequency of 1 Hz), and then, the entire experimental process can only collect four sets of data, which is not enough for a completed experimental. This is far from enough for the data analysis. Therefore, each barometric pressure sensor in this study is individually controlled by a micro-controller, so the positive power line, signal line, clock line and control circuit of each sensor are unique, except for the negative power line, which is shared by all sensors. Figure 3.23 shows the circuit connection of the barometric sensor array in normal operation, in which the control board (green part) has 16 independent control circuits (including microcontroller and SD data memory card), 16 independent control circuits correspond to 16 barometric sensors and the other 16 independent control circuits are through the same power switch to realize the power on and off, so as to ensure that all barometric sensors are starting and ending data at the same time. The other 16 independent control circuits are powered on and off by the same power switch, thus ensuring that all sensors start and end data acquisition at the same time. In this way, the 16 sensors are sampled at 16 Hz and stored at the same time.

Because the power positive, signal and clock wires are unique to each sensor, this resulted in a total of 49 wires for the entire sensor array that needed to be brought together at one end to connect from to the signal processing module. In addition, the island bridge structure has very limited space for wiring, so wiring and routing on the island bridge structure is a challenge. To solve this problem, this research work uses a multilayer flexible circuit board, the layered structure of the flexible circuit board schematic diagram in Figure 3.24a, the flexible circuit board

FIGURE 3.23 Circuit connection of large-area stretchable pressure sensor array [1].

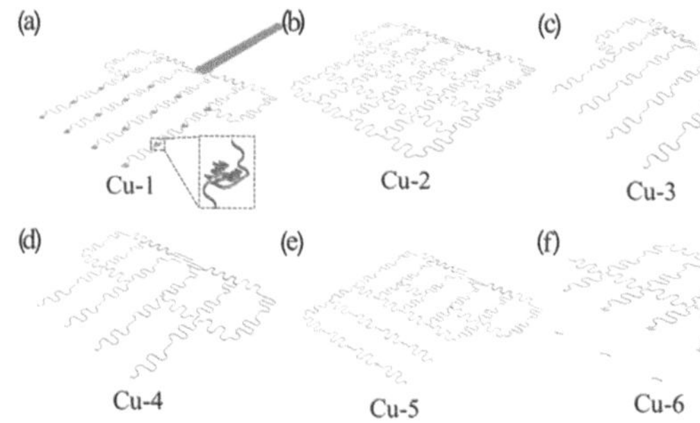

FIGURE 3.24 Schematic representation of each Cu layer [1].

has a total of six layers of Cu layer (wire layer, as shown in Figure 3.24) and seven PI (polyimide) layer, Cu layer (thickness of 35 μm) and PI layer (thickness of 25 μm) in the PI/Cu/…/Cu/PI layers are stacked together in the order PI/Cu/.../PI. The Cu layers are connected to each other by through-holes (Figure 3.25b and c) at certain locations according to the connection requirements, thus realizing the circuit connection of the whole sensor array. The topmost Cu layer is made with pads on the islands (as in Figure 3.25d), and these pads are used for the soldering of the air pressure sensors and capacitors. The width of the curved bridge (PI width) between two adjacent islands is 700 μm and the width of the Cu conductors is 250 μm.

3.5.2 Preparation of Large-Area Stretchable Air Pressure Sensor Arrays

This subsection describes the preparation of a large-area stretchable barometric sensor array. Figure 3.26 shows the preparation process of the device:

a. Cutting and chemical cleaning (Figure 3.26a). Cut the double-sided copper clad board to the right size and then chemically clean the board so that the surface of the copper foil reaches a certain degree of roughening, while washing off the oxide layer, oil, dust and other dirt on it to ensure that the surface of the board can be firmly bonded with the photoresist of the later process.

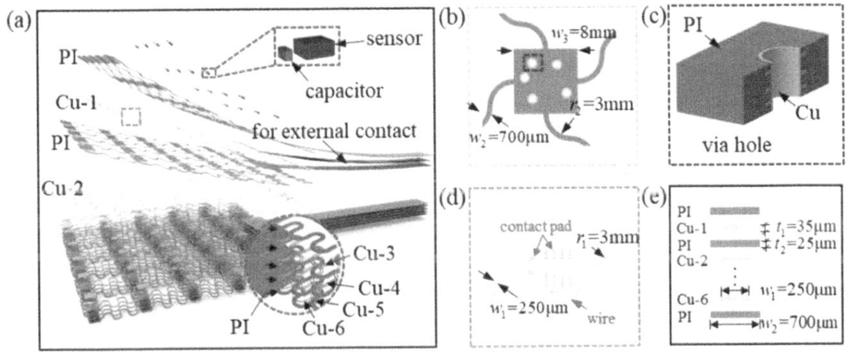

FIGURE 3.25 Large-area stretchable barometric sensor array structure [1]. (a) Schematic diagram of the layered structure; (b) island and through-hole above the island; (c) semi-sectional view of the through-hole; (d) pad; (e) cross-sectional dimensional parameters of the curved bridge.

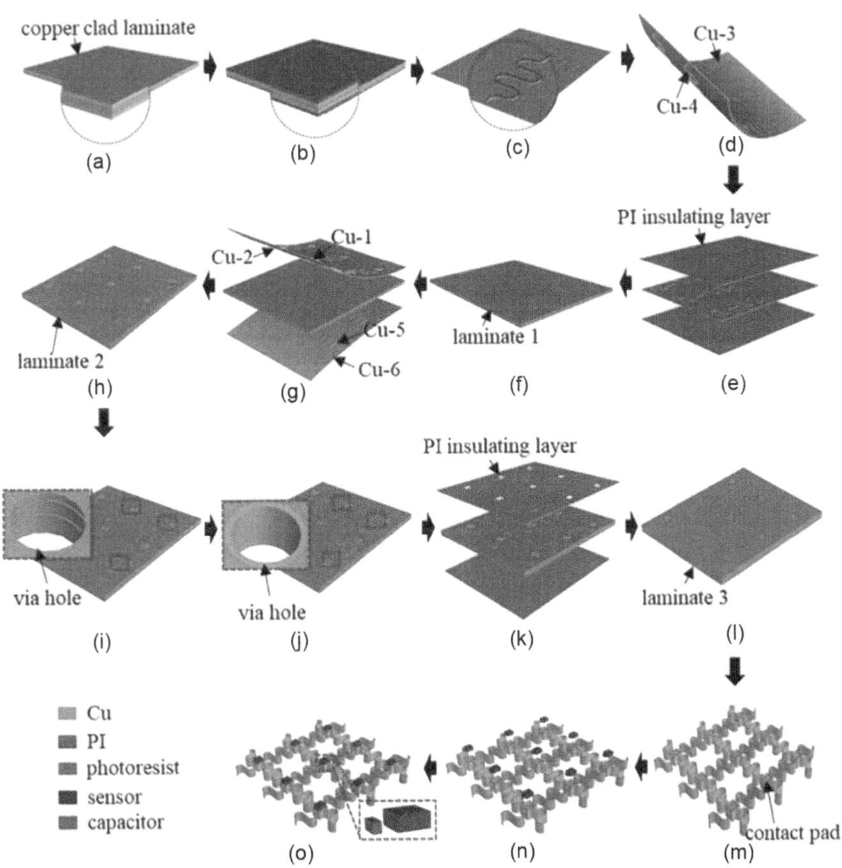

FIGURE 3.26 Preparation process of large-area stretchable barometric sensor array [1].

b. Apply photoresist (Figure 3.26b). A dry film (photoresist, photoresist) is applied to the copper laminate. The dry film is composed of three parts: polyester film, photoresist film and polyethylene protective film. When applying the film, the polyethylene protective film is peeled off from the top of the dry film, and then the dry film is applied to the copper surface under heating and pressure.
c. Exposure, development and etching (Figure 3.26c). Exposure: under the irradiation of ultraviolet light, the photoinitiator absorbs the light energy decomposed into free radicals, and then the free radicals trigger the polymerization of monomer polymerization cross-linking reaction, the reaction formed in the insoluble in dilute alkali solution of polymer structure. The polymerization reaction will continue for a short period of time. To ensure the stability of the process, it should be left for more than 15 minutes after exposure and the polyester film should not be removed immediately. Development: the reactive groups in the unexposed part of the photographic film react with the dilute alkali solution to generate soluble substances and dissolve down, leaving the sensitized cross-linked cured part of the graphics. Etching: the copper foil is removed by chemical reaction to form the desired circuit pattern, and the copper below the photoresist is retained unaffected by the etching.
d. Removal of film (Figure 3.26d). The purpose of the de-filming is to clear the resist layer left on the plate surface after etching, so that the copper underneath is exposed. By this point, the third and fourth Cu layers in the curved bridge composite structure, Cu-3/Cu-4, are obtained.
e. For the insulating layer (Figure 3.26e and f). A PI layer is laminated on the outside of each Cu-3 and Cu-4 with a laminating machine to serve as an insulating layer, which results in composite layer 1.
f. Stacked plate (Figure 3.26g and h). Repeat (a) to (d) to obtain Cu-1/Cu-2 and Cu-5/Cu-6. Cover the top and bottom sides of composite layer 1 with Cu-1/Cu-2 and Cu-5/Cu-6, respectively and then press them together in a press to obtain composite layer 2.
g. Drilling (Figure 3.26i). Drill through-holes at the specified locations on the CNC machine.
h. Plating through-holes (Figure 3.26j). Copper is plated in the hole so that the through-hole can conduct between the layers (metalizing the non-conductor part of the hole wall).
i. Make the insulating layer (Figure 3.26k and l). Repeat step (e) to obtain composite layer 3.
j. Laser cutting (Figure 3.26m). The composite layer 3 is cut with a laser in the specified pattern to form a curved bridge interconnected island–bridge structure.
k. Solder the barometric pressure sensor and capacitor (Figure 3.26n and o). Solder the sensors and capacitors to complete the large-area stretchable air pressure sensor array.

3.5.3 Study of Mechanical Properties and Deformation of Curved Bridges

Many researchers have pointed out that when Cu and PI are compounded together, the ductility of Cu is much better compared to Cu that is not compounded with PI. The curved bridge structure in this large-area stretchable barometric sensor array is a composite layer structure consisting of six layers of Cu and seven layers of PI alternately, and the ductility of Cu in the curved bridge structure is mainly studied in this subsection.

As a comparison, we conducted tensile experiments on Cu specimens without composite and composite layer specimens compounded in a curved bridge manner, respectively, where the thickness of Cu tensile specimens is 35 μm, the thickness of each Cu layer in the composite layer is 35 μm and the thickness of each PI layer is 25 μm (which is the same as the thickness of Cu and PI in the barometric sensor array). The experimental results are shown in Figure 3.27, where the horizontal axis is the tensile strain and the vertical axis is the tensile force.

The experimental results show that when the tensile strain is >1.5%, the uncompounded Cu is already pulled off, while the tensile strain of the composite layer structure compounded by the curved bridge structure needs to reach about 15% before the composite layer structure is pulled off. This is because the PI substrate can inhibit the formation of cracks in the Cu layer locally in the transverse width direction and instead produce uniform microcracks in the Cu layer, thus delaying the fracture of the Cu layer and improving the ductility of Cu [132–135].

In this subsection, finite element calculations and tensile experiments are performed to study the tensile performance and deformation of the curved bridge structure, a representative structure of the curved bridge interconnected island bridge structure. The purpose is to ensure that the tensile performance of the large-area stretchable air pressure sensor array can make the device fit well to the complex curved surface of the high-speed rail head, and to ensure that the circuit connection of the device can remain normal after stretching (Cu conductors do not break and break), so that the device can collect pressure data on the surface of the head properly. The tensile performance and deformation

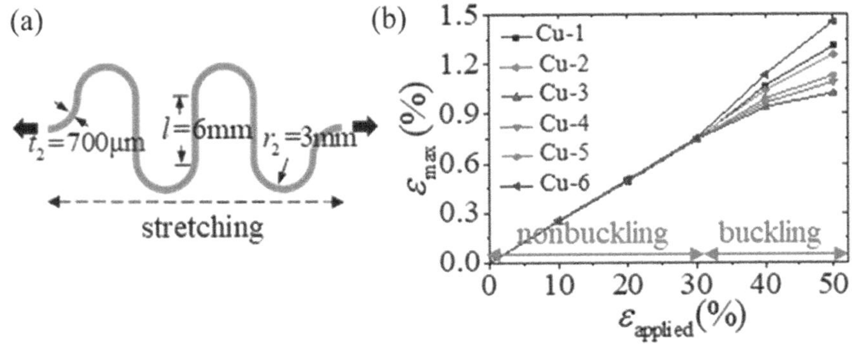

FIGURE 3.27 Finite element results for tension [1]. (a) Finite element model of the curved bridge; (b) maximum strain of each Cu layer at different tensile strains.

of the curved bridge structure are studied by finite element calculations and tensile experiments.

Figure 3.28a shows the finite element model of the curved bridge structure using conventional shell cells with displacement loads applied at both ends of the curved bridge. To determine the mechanical state of Cu in the composite layer structure of the curved bridge, we observe the maximum strain of each Cu layer at different tensile strains, and the results are shown in Figure 3.28b.

From the strain results we can get: (i) even if the tensile strain reaches 50%, the maximum strain of Cu in all layers does not exceed 1.5%, while according to the ductile experiment of PI/Cu multilayer composite strip structure. We know that the tensile strain of Cu in PI/Cu multilayer composite strip structure can reach nearly 15% under the constraint of PI, so the maximum strain 1.5% is far within the safe range, and Cu will not fracture and damage; (ii) when the tensile strain is below 30%, the maximum strains of the six layers of Cu are almost equal, and when the tensile strain exceeds 30%, the maximum strains of the six layers of Cu appear to be different, and this difference is the largest when the tensile strain is 50%. The reason for this phenomenon is that when the tensile strain is below 30%, the curved bridge only undergoes non-flexural deformation, the deformation mode is bending, and the deformation of each Cu layer is basically similar; when the tensile strain is higher than 30%, the curved bridge undergoes out-of-plane flexural deformation, and the deformation mode is a coupling of bending and torsion, and the presence of torsion causes the deformation of the outer Cu (e.g., Cu-1 and Cu-6) to be larger than that of the inner Cu (e.g., Cu -3 and Cu-4), and thus the difference in maximum strain.

The unidirectional tensile experiments were also performed on the curved bridge structure. As shown in Figure 3.29a, the left column from top to bottom shows the tensile results at 0%, 10%, 20%, 30%, 40% and 50% of tensile strain, and the right column shows the corresponding finite element results. From the results, it can be seen that the curved bridge structure gradually expands as the tensile strain increases; the finite element results and the tensile experimental results are in good agreement. In addition, the U_z displacement cloud of the curved bridge at each tensile strain shows that when the tensile strain is <30%, the displacement of the curved bridge in the z-direction is small and cannot be seen obviously, while when the tensile strain is >30%, for 40% and 50%, it can be seen that there is an obvious displacement in the z-direction between the two circular arc segments in the middle of the curved bridge. Changing the viewing angle (Figure 3.29b), when the tensile strain is 40% and 50%, the middle part of the curved bridge obviously

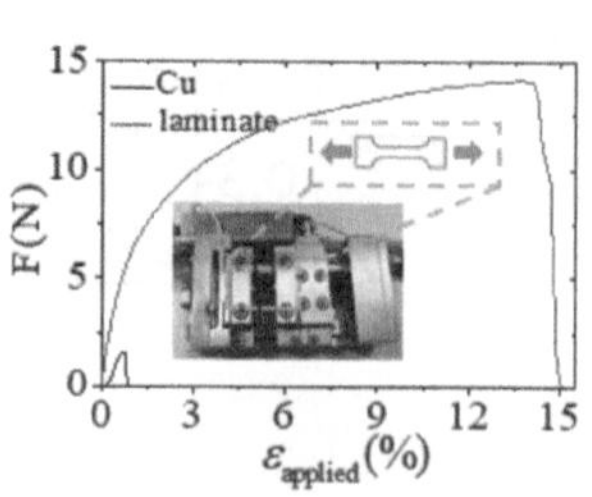

FIGURE 3.28 Ductility experiments [1].

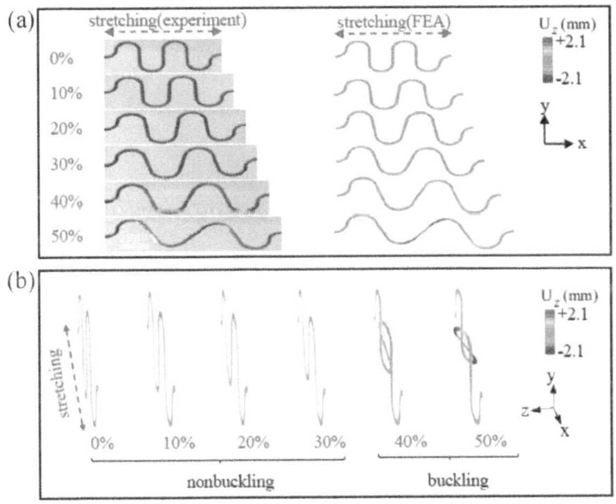

FIGURE 3.29 Results of tensile experiments and finite elements [1]. (a) Comparison of experimental and finite element; (b) Finite element results: non-flexural to flexural deformation.

leaves the XY plane. This indicates that when the tensile strain is less than 30%, the curved bridge has no or little out-of-plane buckling, while when the tensile strain is >30%, the curved bridge undergoes significant out-of-plane buckling. Therefore, the curved bridge structure undergoes a transition from non-flexural to flexural deformation during the process from smaller tensile strains to larger tensile strains; when the tensile strain is small, the curved bridge undergoes non-flexural deformation, and when the tensile strain is larger, the curved bridge structure undergoes flexural deformation. This also directly explains the difference in the relationship between the maximum strain and the tensile strain for each layer of Cu in Figure 3.27b when the tensile strain exceeds 30%.

3.5.4 Experiments with Large-Area Stretchable Barometric Sensor Arrays

The starting point for the large-area stretchable barometric sensor array is for high-speed rail head pressure field monitoring. First, we must ensure that the response of each sensor in the array to the outside world is the same, i.e., that the same outside pressure change measured for all sensors in the array is equal (consistency of the barometric sensor array). To do this, we first need to check the consistency of the barometric sensor array.

We use the vacuum drying chamber DZF-6050 (Figure 3.30a) for the consistency experiments of the air pressure sensor array. The vacuum drying oven can be operated by the external air pump to evacuate the working chamber when it is working, and the vacuum level in the working chamber can be controlled by the air pump running time and can be displayed directly. We first start the air pressure sensor array (Figure 3.32b) and place it in the working chamber of the vacuum

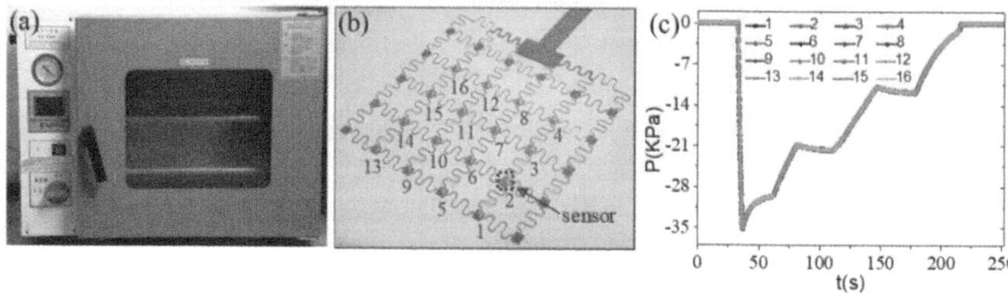

FIGURE 3.30 Consistency experiments [1]. (a) vacuum drying chamber; (b) air pressure sensor array; (c) experimental results (vertical axis: relative pressure; horizontal axis: time).

drying oven, and after the experiment starts, let the working chamber be kept as indoor atmospheric pressure (about 101 kPa) for a period of time, then pump the working chamber air pressure to 66 kPa for a short period of time, stabilize it for a period of time and then gradually open the air valve of the vacuum drying oven to let the working chamber air pressure rise until the working chamber air pressure returns to the indoor atmospheric pressure. The experimental results are shown in Figure 3.30c (the horizontal axis is the time, and the vertical axis is the change value relative to the room air pressure measured by the air pressure sensor). From the experimental results, it can be seen that: on the one hand, the pressure change measured by the air pressure sensor and the experimental operation correspond to each other very well, which shows the accuracy of the pressure change measured by the air pressure sensor; on the other hand, the pressure change curves measured by the 16 sensors in the air pressure sensor array completely overlap together, which shows the consistency of the response of the 16 sensors to the external pressure change.

The high-speed train dynamic model experimental system (Figure 3.31) is an experimental platform dedicated to the research of mechanical problems related to high-speed trains at the Institute of Mechanics, Chinese Academy of Sciences. The system is mainly composed of these parts, such as high-speed rail scale model, 364 meters long upper track (running high speed rail scale model), 111 meters long lower track (running trailer), trailer, tow rope, trailer reduction plate, gas storage tank, piston and air outlet. When the system is working, the compressed air stored in the storage tank is shot into the slide of the piston through the air gun, and the piston moves to the right side with a great acceleration under the huge pressure difference between the two sides (until it passes the air outlet and the acceleration ends) and drives the scale model of high-speed rail in the upper track to do acceleration movement through the tow rope. In this way, the scale model can reach a high speed in a very short time and distance and then pass the test section of 103.5 m with this speed. When the piston passes through the air outlet, the air pressure difference between the two sides rapidly decreases to zero, and the traction force on the tow rope also rapidly decreases, while the trailer enters the deceleration plate and starts to decelerate, giving a reverse traction force to the high iron scale model through the tow rope, so that the traction force of the high iron scale model rapidly decreases and the reverse traction force rapidly increases, thus entering the braking stage until it comes to a complete stop [99]. In this way, the experimental platform

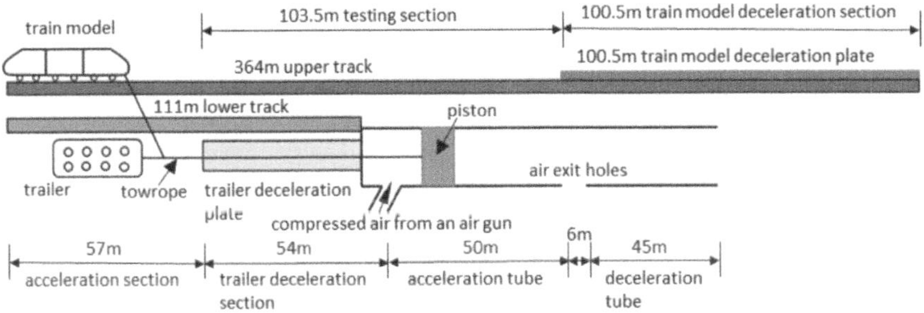

FIGURE 3.31 Schematic diagram of the high-speed train dynamic model experimental system [136].

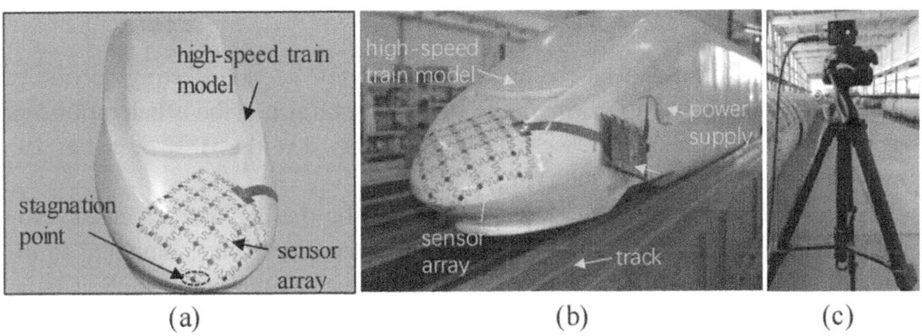

FIGURE 3.32 Experiments on the dynamic model of HSR [1]. (a) Front view; (b) Side view; (c) Photoelectric door.

can accomplish the acceleration of high-speed trains to a high-speed running state within a short distance, which in turn provides the possibility to study the mechanics of high-speed trains and related problems in a high-speed running state. A physical photograph of the high-speed train dynamic model experimental system can be seen in Figure 3.30b. This high-speed train dynamic model experimental platform can realize the stress measurement on the body surface of high-speed trains under various complex operating conditions, such as open-line operation, converging cars and crossing tunnels, and then realize the research on the aerodynamics of high-speed trains and other related mechanical problems.

In our experiments, we fit a large-area stretchable barometric sensor array to the head of the HSR scale model (Figure 3.32a and b). The position of each barometric sensor is shown in Figure 3.32a, and the barometric sensor 1 (Figure 3.30b) is affixed to the stationary position (zero air velocity at the stationary point) of the HSR scale model. The large size stretchable air pressure sensor array has the following advantages: (i) the large-area stretchable air pressure sensor array does not need to cut pressure measurement holes on top of the model; (ii) it can monitor the pressure at multiple points in the head area at the same time, thus realizing the monitoring of the head pressure field; (iii) it does not need bulky and complicated signal processing equipment; and (iv) it is easy to install and ready to use, avoiding the tedious wiring of the traditional pressure measurement hole technology.

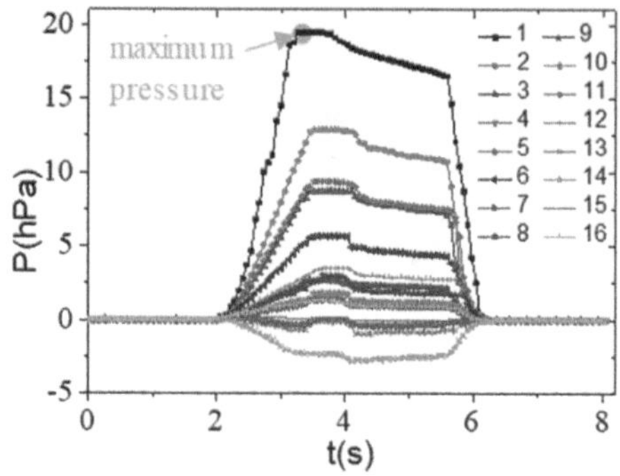

FIGURE 3.33 Relative pressure measured during the experiment [1].

During the experiment, at the entrance of the test section (testing section in Figure 3.33), i.e., the position where the HSR model finished the acceleration and reached the maximum speed, we measured the speed of the HSR model as 193 km/h with the photoelectric gate (Figure 3.32c).

Figure 3.33 shows the pressure data measured by the barometric sensor array during the experiment [the vertical axis is the relative pressure (compared to the room atmospheric pressure) and the horizontal axis is the time], and the curve numbers in the equation correspond to the barometric sensor numbers (Figure 3.30b). From the experimental results, it can be seen that:

1. The sensors in the barometric sensor array are somewhat in the positive pressure region (positive relative pressure) and somewhat in the negative pressure region (negative relative pressure), where a rough indication of the distribution of head pressure can be measured by the barometric sensor array.
2. All the pressure data measured by the barometric pressure sensors are in 0–2 seconds, and the relative pressure is basically zero; in 2–6 seconds, the absolute value of the relative pressure (considering that some barometric pressure sensors are in the negative pressure region) now increases for a short time, then maintains for a short time, then slowly decreases and finally decreases rapidly in a short interval. The change in relative pressure coincides with the experimental process, in 0–2 seconds, the HSR model is at rest; in 2–6 seconds, the HSR model first experiences acceleration and then uniform speed, then starts to decelerate and finally decelerates sharply; in the 6th second, the HSR decelerates to 0 and then is at rest.

Furthermore, the results measured by the barometric sensor array are compared here with those obtained using the conventional orifice technique. Figure 3.34a shows the Endevco Model 8530C-15 pressure transducer used in the conventional orifice technique, the parameters of which can be found in Table 3.2. Figure 3.34b

(a) (b) (c)

FIGURE 3.34 Endevco Model 8530C-15 pressure sensor results [1]. (a) Sensor; (b) Hole h1; (c) Experimental results.

TABLE 3.2
Endevco Model 8530C-15 Pressure Sensor Parameters [1]

Range (psi)	Sensitivity (mV/psi)	Resonance Frequency (Hz)	Non-Linearity Independent (%FSO typ)	Non-Repeatability (%FSO typ)	Acceleration Sensitivity (Equiv psi/g)	Package Dimensions (mm)
0–15	15.3	180,000	0.15	0.1	0.00015	$11.13 \times \varphi 7.92$

shows the location of the orifice h1. Figure 3.34c shows the pressure trend measured by the Endevco Model 8530C-15 pressure transducer during the experiment, where there is a (in the red box). This pressure bump is caused by the airflow at the outlet, which is filtered out by the large-area stretchable barometric sensor array using a filtering algorithm. The pressure trend obtained by the conventional orifice technique is essentially the same as that obtained by the large-area stretchable barometric sensor array. Therefore, the monitoring results of the barometric sensor array have a certain accuracy.

Curve 1 in Figure 3.35 shows the relative pressure values measured during the experiment for air pressure sensor 1 attached to the head station (Figure 3.32a). The absolute value of the relative pressure measured by barometric sensor 1 during the operation of the HSR model is larger than that measured by other barometric sensors, which is consistent with the characteristics of the standing point. The standing point is the position of the moving object facing the fluid, the fluid is completely blocked at this position, the fluid velocity is zero and all its kinetic energy is converted into pressure energy, so the standing point is the point where the pressure value on the surface of the moving object is the largest. The pressure value at the standing point, $P_{\text{stagnation}}$ is:

$$P_{\text{stagnation}} = \frac{1}{2}\rho v^2 \tag{3.40}$$

where ρ is the fluid density and v is the relative velocity of motion of the object and the fluid.

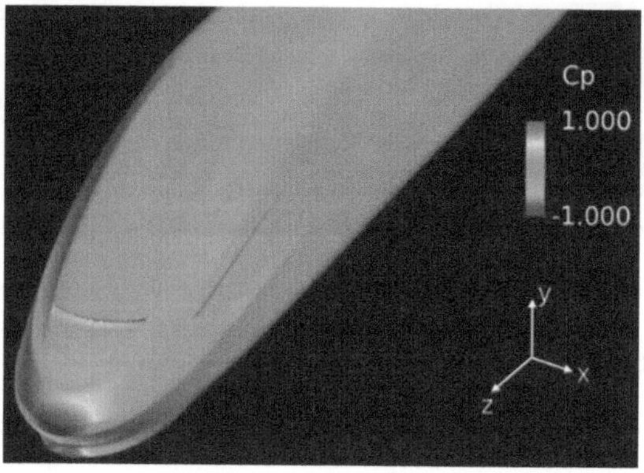

FIGURE 3.35 Cloud plot of pressure coefficient distribution [1].

In addition, the maximum relative pressure measured by air pressure sensor 1 during the experiment (Figure 3.33 maximum pressure) is 1944.25 Pa, that is, the maximum relative pressure at the stationing point during the experiment is 1944.25 Pa. According to Equation (3.40), we can get the theoretical value of the stationing point pressure at this speed is 1858.14 Pa and then the deviation of the pressure. The deviation of the pressure value at the standing point measured by the sensor from the theoretical value is 4.63%.

On the other hand, a numerical simulation of the pressure coefficient field on the surface of a high-speed railway at 193 km/h was completed using Fluent, and the simulation results are shown in Figure 3.35. Figure 3.35 shows a cloud plot of the pressure coefficient on the surface of a high-speed railway, which is a dimensionless number used to describe the relative pressure spread throughout the flow field in fluid dynamics, and its expression is

$$C_p = \frac{P - P_\infty}{1/2\rho_\infty v_\infty^2} = \frac{P - P_\infty}{P_0 - P_\infty} \tag{3.41}$$

where P is the static pressure at the point of the pressure coefficient sought, P_∞ is the independent static pressure away from any perturbation, P_0 is the independent standing pressure away from any disturbance, ρ_∞ is the independent fluid density and v_∞ is the independent fluid density or the velocity of the object through the fluid. From the simulation results, it can be learned that the pressure at the very front of the head of the high-speed train (standing point region) is the largest, and there is a negative pressure at the shoulder of the car (the transition region between the head and the body).

The numerically simulated pressure coefficient field results are compared with the experimental results to test the performance of the large-area stretchable barometric sensor array to monitor the pressure field. To achieve the comparison, the pressure coefficients at the locations of the barometric sensors are all projected onto the XY plane.

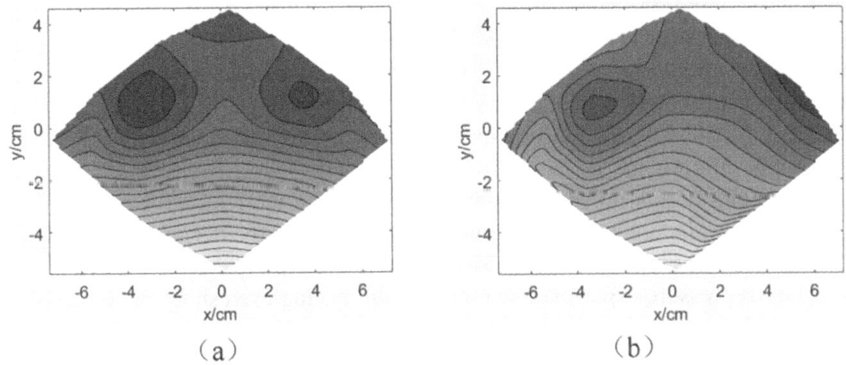

FIGURE 3.36 Discrete pressure coefficient point interpolation results [1]. (a) Numerical simulation; (b) Experiment.

The interpolation results of numerical simulation and experiments are shown in Figure 3.36a and b. Both interpolation results show that: (i) the high-pressure region is near the stationary point (0.1, −5.5) and the low-pressure region is far from the stationary point, which is consistent with the results in Figure 3.35, and (ii) there are two more obvious low-pressure areas. The above two points indicate that the large-area stretchable air pressure sensor array can basically monitor the pressure coefficient field in the head of the high-speed rail. However, the interpolation results of the numerical simulation are basically symmetric about the line $x = 0$ (not completely symmetric because the barometric sensor array paste position is not symmetric about the surface $x = 0$); in this regard, there is some difference between the experimental interpolation results and the numerical simulation interpolation results. The reason for the difference is that the air pressure sensor has a certain thickness, and when the paste position of the air pressure sensor array on the head of the high-speed train is not symmetric about the surface $x = 0$, the thickness of the air pressure sensor on both sides of the surface $x = 0$ has an asymmetric effect on the flow field.

Currently, to study aerodynamic problems of high-speed rail in the laboratory, such as aerodynamic shape optimization, researchers are generally using the pressure measurement hole technique to measure the pressure on the surface of the model. However, the pressure hole technique requires chiseling and destroying the model and can only perform regional single-point pressure monitoring, and the installation and commissioning of the pressure measurement system is tedious. In this chapter, a large-area stretchable air pressure sensor array is designed and fabricated using flexible electronics for pressure field monitoring on the surface of a high-speed rail model.

Current flexible electronic devices are miniaturized, while the size of this barometric pressure sensor array is 20 cm × 19.5 cm, which is one of the few large-area flexible electronic devices. The barometric sensor array adopts a curved bridge interconnected island–bridge structure, in which the design of the multilayer composite strip structure in the curved bridge can increase the Cu fracture strain from about 1% to 15%, which greatly improves the ductility of Cu. In addition, the stretchable air pressure sensor array can achieve a large tensile strain of nearly 50%, which allows it to fit well to vehicle heads with complex surfaces. Meanwhile, the tensile experiments of the curved bridge show that when the tensile strain is <30%, the curved bridge undergoes non-bending deformation and the deformation mode is mainly bending; when the tensile strain is >30%, the curved bridge undergoes bending deformation and the deformation mode is mainly coupled with bending and torsion.

The lift pressure experiments in the vacuum drying oven show a good consistency of the sensors in this barometric sensor array. The experimental results of the barometric sensor array for the dynamic model of the high-speed railway are consistent with the experimental procedure and the results of the conventional pressure measurement hole technique, which shows the accuracy of the barometric sensor array for measuring the surface pressure of the high-speed railway model. In addition, the comparison between the experimental results of the air pressure sensor array and the finite element simulation results shows that the air pressure sensor array can monitor the head pressure distribution well. Therefore, the large-area stretchable barometric pressure sensor array can be used as a method for the surface pressure field of high-speed rail.

The lift pressure experiments in the vacuum drying oven show a good consistency of the sensors in this barometric sensor array. The experimental results of the barometric sensor array for the dynamic model of the high-speed railway are consistent with the experimental procedure and the results of the conventional pressure measurement hole technique, which shows the accuracy of the barometric sensor array for measuring the surface pressure of the high-speed railway model. In addition, the comparison between the experimental results of the air pressure sensor array and the finite element simulation results shows that the air pressure sensor array can monitor the head pressure distribution well. Therefore, the large-area stretchable barometric pressure sensor array can be used as a method for the surface pressure field of high-speed rail.

Compared with the traditional pressure hole technology, the large-area stretchable air pressure sensor array not only does not damage the model, is easy to use and greatly reduces the cost of monitoring but also enables the monitoring of the pressure field on the model surface.

3.6 ACOUSTIC TWEEZERS FOR MICROMANIPULATION

Acoustic waves with screw wavefront dislocation are known as acoustic vortex beams, which are able to carry orbital angular momentum (OAM). In the wavefront dislocation of a screw monochromatic continuous traveling acoustic wave beam, there is a spiral phase $\exp(i\ell\theta)$ that is linearly proportional to the azimuthal angle θ, where the integer ℓ is called "topological charge" or the order of the screw beam. The sign and the value of ℓ determine the orientation of the spiral and the magnitude of the OAM, respectively [137–139].

In recent years, acoustic vortex beams have attracted considerable attention, showing great potential for applications of particles trapping, acoustic levitation, contact-less manipulation and high-speed acoustic communication [140–149]. Acoustic vortex beams have unprecedented combination of selectivity and trapping force. Non-contact transfer of the angular momentum enables them to be used as acoustic wrenches [147]. The circular distribution of acoustic power enables them to trap particles as acoustic tweezers [141,148,149]. The topological charges of OAM provide intrinsically orthogonal channels, which indicates a great potential of channel multiplexing in acoustic communication [140,145].

There are two basic methods to produce desired acoustic field. The first is designing specific transducers or transducer arrays to transmit the required acoustic beams [150–152], such as annual focus transducers [152]. The second is designing specific passive structures to modulate the amplitude, phase or path of the acoustic wave propagating in them [153,154], such as acoustic grating [154]. Therefore, acoustic vortex beams are usually generated by active methods with phase-array transducers electrically or passive structures physically [137]. The phase-spiral source consists of an array of individually addressed transducers, which are excited with appropriate screw phases to produce the expected phase profile in acoustic vortex beams [140,155,156]. Its disadvantage is the need of sophisticated electronic control, which causes substantial difficulty in integration and is expensive [157]. The physical spiral source produces acoustic vortex beams using a passive structure with screw dislocated profiles [137,139,158–161]. It avoids sophisticated electronic control, but it needs elaborate spiral profiles and may also have a bulky size. For example, Jiang et al. [161] used a disk with multi-arm coiling slits, which is a passive structure similar to optical grating. A transmitting transducer produced plane waves, and then, the disk fixed in front of the transducer was used to modulate the amplitude and phase of the waves in the diffraction field to transform them into vortex beams. In addition to these methods, Jiang et al. [162] used acoustic resonances in a planar layer with simple structure to twist wave vectors of an in-coming plane wave into a spiral phase dislocation of an outgoing vortex beam, while it still needs to select alternative materials and carry out further experiments for underwater propagation. Therefore, a more effective and generally applicable method with more compact transducer system is needed.

In this Letter, an innovative method is proposed to satisfy the demand, based on piezoelectric transducer coated with multi-arm coiled spiral electrodes, which is more compact than other phase-spiral sources [159,160,162] and easier to prepare than other passive structures [137,139,159–161]. In this method, the phases of the acoustic beams are controlled by the design of the electrodes, and then a focused acoustic vortex is generated with beam synthesis principle. That is to say, the piezoelectric transducer can produce vortex beams directly, indicating higher accuracy compared with the passive methods [137,139,158,159], where the acoustic vortex beams are formed by the delay phase difference of acoustic waves propagating in the structure. Moreover, the two separate components [161], the transmitting transducer and the disk with multi-arm coiling slits, are synthesized together into a piezoelectric disk in this work, indicating that our acoustic source is more compact and easier to install. Using the finite element simulation software, the acoustic field excitation characteristics of the piezoelectric transducer are studied and analyzed. The results show that our proposed method is able to generate an acoustic vortex with the advantages of simple operation, general applicability, great compactness with low cost. This study provides theoretical basis for promoting the practical application of acoustic vortex beams in manipulation of particles, micro-organisms, cells and acoustic communication in the future.

Figure 3.37 shows a schematic diagram of the transducer coated with multi-arm coiled spiral electrodes and its working principle of radiating acoustic vortices field. Different points of the electrodes on transducer disk radiate the same frequency acoustic wave. The amplitude and phase spatial distribution of the wavefront are affected by the difference of wave path; thus, a stable focused acoustic vortex can be synthesized in the small radiation space around the acoustic axis.

For a piezoelectric disk, one of the surfaces is coated with the electrode for total area and connected to ground in the actual drive. We coat the other surface with the multi-arm spiral electrodes and apply sinusoidal voltage with the same frequency and phase to drive the piezoelectric disk. Its operation is much simpler than the using transducer arrays to generate vortex. Logarithmic helix, $r = ae^{b\theta}$ in polar coordinates, is chosen as the helix function of the spiral arm electrodes, where r is the radial coordinate, θ is the angle coordinate, a is the initial radius and b is the azimuth coefficient, which indicates the growth rate of radial distance with the increase of angle. The helix expressions of the inner and outer edges lines of the electrode bars are

$$r_1 = a_1 e^{b\theta}, \quad r_2 = a_2 e^{b\theta}, \quad \theta \in [\theta_1, \theta_2] \tag{3.42}$$

which have different initial radii a_1 and a_2, and θ_1 and θ_2 are termination angle and origin angle, respectively. Without loss of generality and for the sake of simplicity, we first introduce the structure and sound field of transducer coated with

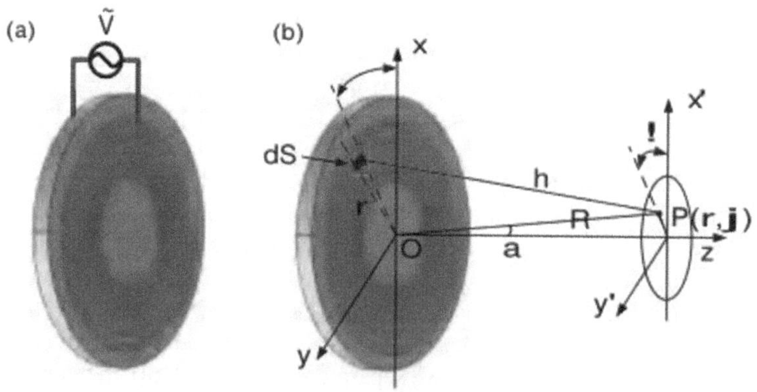

FIGURE 3.37 (a) Schematic diagram of transducer based on multi-arm coiled spiral electrodes and (b) working principle of radiating acoustic vortex field [162].

single-arm electrode which is capable of generating vortex beams with a topological charge $\ell = 1$. The origin is the central point of the circular area on the disk surface. The initial radii a of inner and outer edges are 13.8 and 14.85 mm, respectively. The azimuth coefficient b is designed to be 0.0225. The angle θ rotates from 0 to 12π. In the structure shown in Figure 3.39, the piezoelectric material is PZT-4 and the transducer is 1 mm thick. The material of coated electrode is set to silver. The width of the electrode bar d decides the available frequency bandwidth of the transducer. This can produce acoustic vortex beams with the wavelength in the range of $[2d_{\min}, 2d_{\max}]$. For the one-armed coiling electrode bar, $d_{\min} = 1$ mm and $d_{\max} = 2.47$ mm, which determine that the corresponding low and high critical frequencies are 305 and 750 kHz. The operating frequency is 425 kHz in this study. According to the relationship between electrode widths and critical frequencies,

$$d_{\max} = \frac{c}{2f_{\min}}, \quad d_{\min} = \frac{c}{2f_{\max}} \tag{3.43}$$

A wider frequency bandwidth $[f_{\min}, f_{\max}]$ is obtained by enlarging $d_{\max}$ or reducing $d_{\min}$. In this work, when the excitation voltage is applied to drive the spiral electrodes, the coated area generates longitudinal vibration through piezoelectric effect, and then the sound waves are radiated into space and the focused acoustic vortex beam is synthesized near the acoustic axis.

In the theoretical model, as shown in Figure 3.37b, the spiral electrode can be regarded as a potential energy line. Therefore, the amplitude and phase of the vibration generated by the electric excitation are the same at every point on the surface of spiral electrode. It is assumed that the piezoelectric disk is embedded in an infinite baffle plate. The coordinate origin O is set at the center of the acoustic radiation surface of the piezoelectric disk, and the polar coordinates of the plane are represented by r and θ, while the observation plane at the distance z is represented by ρ and φ. A certain observation point P is set in the observation plane. The angle between the position vector of P and the z axis is represented by α. The Rayleigh–Sommerfeld diffraction integral is used to calculate the acoustic pressure at the observation point P. The normalized pressure at P is calculated by

$$p(\rho,\varphi,z) = v_a \rho_0 c_0 \int_{\theta_1}^{\theta_2} \int_{a_1 e^{b\theta}}^{a_2 e^{b\theta}} \frac{i\omega\rho_0}{2\pi h} e^{i(\omega t - kh)} r dr d\theta \tag{3.44}$$

where ρ_0 is the density of acoustic transmission medium, c is sound velocity in the medium, ω is the angular frequency of vibration of the particle on the surface of the transducer, $k = 2\pi/\lambda$ is the wave number and $\lambda = c/f$ is the wavelength of acoustic wave in acoustic transmission medium. There is no limitation on acoustic transmission medium, that is to say, this transducer system is generally applicable. Here, h is the distance between the observation point P and the surface element $dS = r dr d\theta$ in the spiral electrode,

$$h = \sqrt{\rho^2 + r^2 + z^2 - 2rR\cos\varphi\sin\alpha} \tag{3.45}$$

where R is the straight-line distance between the observation point P and the origin point O,

$$R = \sqrt{\rho^2 + z^2} \tag{3.46}$$

Similarly, transducers coated with multi-arm coiled spiral electrodes are obtained. The expression of inner and outer edge lines of spiral electrodes is $r = ae^{b\theta}$, which is the same as the single-arm electrode. The parameters of the spiral electrodes coated on four transducers are listed in Table 3.3, which can produce acoustic vortices with topological charges of 1, 2, 3 and 4, respectively. According to the largest and smallest widths of the electrode bars, the low and high critical frequencies are 305 and 750 kHz for the four cases.

The synthetic acoustic field excited in space with $\ell = 1$ can be calculated by Equation (3.44), which can be extended to the case of two-arm electrodes. The helix expression of inner and outer edges for one of the electrodes is $r = ae^{b\theta}$. The origin angle difference between the helices of two electrodes is π. Thus, the helix expression for the other electrode is $r = ae^{b(\theta+\pi)}$. The acoustic field caused by each electrode is similar to the case of single-arm electrode. According to Equation (3.44), the synthesis acoustic field is obtained as

$$p_{\ell=2}(\rho,\varphi,z) = v_a \rho_0 c_0 \int_{\theta_1}^{\theta_2} \int_{a_1 e^{b\theta}}^{a_2 e^{b\theta}} \frac{i\omega\rho_0}{2\pi h} e^{i(\omega t - kh)} r dr d\theta + \int_{\theta_1}^{\theta_2} \int_{a_1 e^{b(\theta+\pi)}}^{a_2 e^{b(\theta+\pi)}} \frac{i\omega\rho_0}{2\pi h} e^{i(\omega t - kh)} r dr d\theta \tag{3.47}$$

Then, Equation (3.47) is extended to the case of circular piezoelectric transducer coated with multiple-arm coiled spiral electrodes as follows:

TABLE 3.3
Parameters of the Spiral Electrodes [162]

Number of Spiral Arms	Initial Radius of Inner Edge a_1 (mm)	Initial Radius of Outer Edge a_2 (mm)	Azimuth Coefficient b	Origin Angle θ_1	Termination Angle θ_2
1	13.8	14.85	0.0225	0	12π
2	13.8	14.85	0.0451	0	6π
3	13.8	14.85	0.0676	0	4π
4	13.8	14.85	0.0902	0	3π

$$p_{\ell}(\rho,\varphi,z) = v_a \rho_0 c_0 \sum_{n=1}^{\ell} \int_{\theta_1}^{\theta_2} \int_{a_1 e^{b\left[\theta+\frac{2\pi(n-1)}{\ell}\right]}}^{a_2 e^{b\left[\theta+\frac{2\pi(n-1)}{\ell}\right]}} \frac{i\omega\rho_0}{2\pi h} e^{i(\omega t-kh)} r dr d\theta \tag{3.48}$$

To verify this method, we use finite element simulation to obtain the solution of Equation (3.48) and analyze the amplitude and phase distribution of the acoustic pressure field on the observation plane. The FEM has been widely used in acoustic field calculation to solve multiphysics coupling problems [163–165]. In this work, the acoustic radiation of a piezoelectric ceramic disk with one to four coiled arm electrodes in water is simulated in the coupling multiphysics fields. The sound velocity in water is 1500 m/s, and its density is 1000 kg/m^3. In the physical field of solid mechanics, the piezoelectric material is PZT-4. The diameter and the thickness of the ceramic disk are 38 and 1 mm, respectively, indicating the compactness of this system without the necessity of extra integration cost. In the electrostatic field, the surface of the piezoelectric circular plate with spiral electrodes is set near the water and its electric potential is set 100 V. The surface with fully coated electrode is further from the water and set grounded. The excited acoustic pressure field at frequency 425 kHz is simulated and observed in the frequency domain. The outer surface of the water area radiates spherical wave corresponding to the thickness vibration mode of the piezoelectric ceramic disk. The models of piezoelectric ceramic disks with four kinds of electrode-coating structures are shown in Figure 3.38. The blue areas are the coated spiral electrodes. In the simulation, the spiral strip electrode is modeled as is potential.

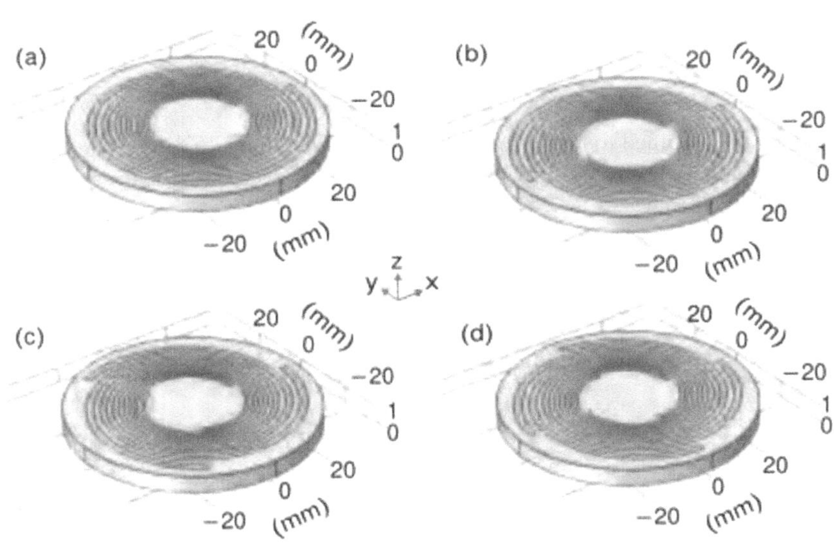

FIGURE 3.38 Simulation models of transducers coated with (a) single-arm, (b) two-arm, (c) three-arm and (d) four-arm coiled spiral electrodes [162].

With the increase in the numbers of spiral electrodes, we can obtain acoustic vortex beams with larger topological charge l, while there is a limitation of the largest topological charger ℓ_{max}, which is mainly decided by the radius of piezoelectric disk, and meanwhile related to frequency bandwidth.

The following derivation uses Equation (3.42). The smallest and largest widths of the electrode bars, d_{min} and d_{max}, which are decided by the frequency bandwidth according to Equation (3.43), are

$$d_{min} = a_2 - a_1, d_{max} = (a_2 - a_1)e^{b\theta_2} = d_{min}e^{b\theta_2} \quad (3.49)$$

The spiral electrodes and the piezoelectric disk satisfy the following geometrical constraint conditions, from which the largest topological charge ℓ_{max} is deduced. First, the radius of piezoelectric disk R should be greater than the largest radius of outer edge of the electrode bar $r_{2,max}$, and meanwhile the edge of electrode should be apart from the edge of disk for a certain distance d',

$$R \geq r_{2,max} + d' \quad (3.50)$$

where the parameters are marked in Figure 3.39, and $r_{2,max}$ is

$$r_{2,max} = a_2 e^{b\theta_2} = (a_1 + d_{min})e^{b\theta_2} \quad (3.51)$$

From Equations (3.50) and (3.51), the limitation of the initial radius of inner edge a_1 is obtained as

$$a_1 \leq \frac{(R - d')d_{min}}{d_{max}} - d_{min} \quad (3.52)$$

where d' is assumed to be d_{min} and $R = 38\,\text{mm}$. Equation (3.52) shows a positive correlation between R and the maximum value of a_1; accordingly, the limitation of a_1 is $a_1 \leq 14\,\text{mm}$.

$$r' - r_2 \geq H \quad (3.53)$$

where r_1' and r_2 are marked in Figure 3.39, and the expression of r_1' is

$$r_1' = a_1 e^{b(\theta + 2\pi/\ell)} \quad (3.54)$$

where ℓ is the topological charger; i.e., the number of electrodes. The thickness of piezoelectric disk H should be not more than half of the wavelength,

$$H \leq \lambda_{min}/2 \quad (3.55)$$

Assume that H is set to be its maximum value $\lambda_{min}/2$, i.e., d_{min}, derived from Equations (3.43) and (3.55). From Equations (3.42), (3.49), (3.53) and (3.54), we can obtain

$$\ell \leq \frac{2\pi}{\theta_2} \frac{\ln\left(d_{max}/d_{min}\right)}{\ln\left(2d_{min}/a_1+1\right)} \tag{3.56}$$

which indicates that the max topological charger λ_{min} increases with the growth of the radius of piezoelectric disk R. In this study, we have $\ell_{max} = 6$ (Figure 3.39) according to the disk radius and frequency bandwidth. It is expected that the largest topological charger is obtained by increasing the radius of piezoelectric disk.

Figures 3.40 and 3.41 show the simulation results of the radiated acoustic pressure fields excited by the transducers in Figure 3.38, where the observation plane is located at 2.5 wavelengths perpendicular to the piezoelectric transducers in water. Figure 3.40 is the normalized amplitude distributions of the acoustic pressure for transducers with single-arm to four-arm electrodes, and Figure 3.41 is the phase distribution. It can be seen that the structures in Figure 3.38 successfully produce acoustic vortices on the observation plane, focusing on the acoustic axis with the topological charge $\ell = 1,2,3$ and 4. In the ring of diffraction phase field generated by the transducers with single-arm electrode to four-arm electrode, the phase has changed by 2π, 4π, 6π and 8π, respectively, which proves the topological structure of the corresponding acoustic vortex field.

In summary, a method to generate focused acoustic vortex has been proposed based on a piezoelectric transducer coated with multi-arm coiled electrodes. Finite element simulation results for single-arm to four-arm systems show that the method successfully modulates the amplitude and phase spatial distribution of acoustic wave near the acoustic axis, by the principle of acoustic field synthesis, making the acoustic wave rotate spirally in space and form a stable focused vortex beam. This method has the advantages of compact size, simple operation and generally applicable, breaking through the scale limitation of the traditional active generation of vortex beams with

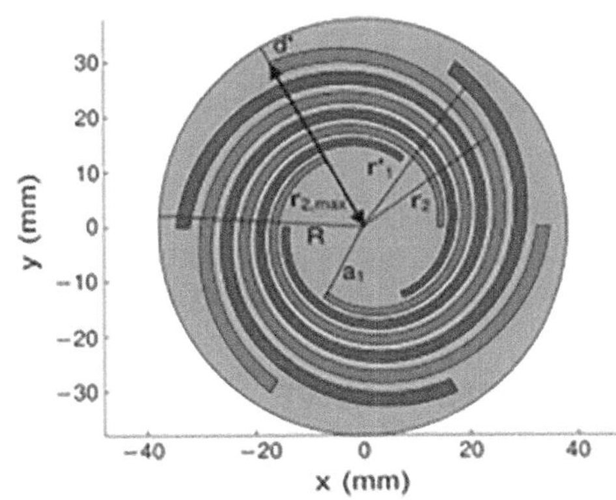

FIGURE 3.39 Simulation models of transducers coated with six-arm coiled spiral electrodes (most electrodes) [162].

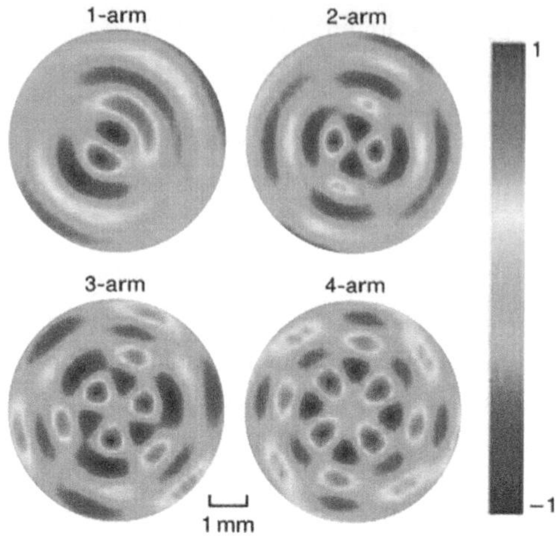

FIGURE 3.40 Normalized amplitude distributions of simulated acoustic pressure field [162].

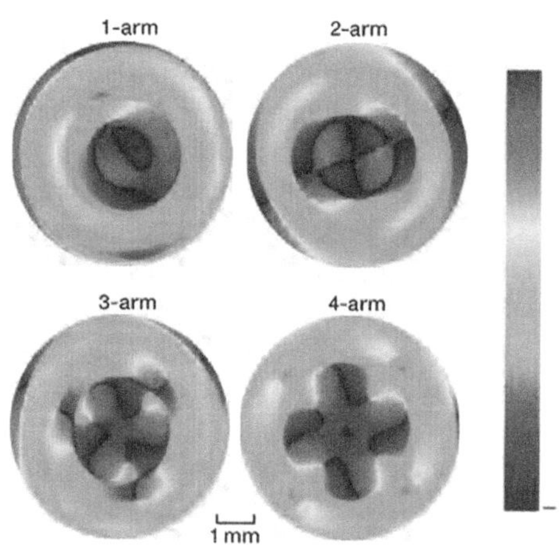

FIGURE 3.41 Phase distributions of simulated acoustic pressure field [162].

arrays of transducers and making it possible to miniaturize the active mode. It can be utilized in the manipulation and capture of micro-particles, acoustic detection, imaging and so on. This method has particularly good prospects in medicine.

3.7 SUMMARY AND PERSPECTIVE

This chapter mainly introduces some representative applications of flexible piezoelectric array, such as pacemaker, wireless power transmission system, odor control transceiver, large surface retractable pneumatic sensor array and acoustic tweezers. These technologies break through the technical barriers of traditional piezoelectric arrays facing complex surfaces and multifunctional integration, especially suitable for large-scale high-efficiency or high-precision microoperation. These results indicate that flexible piezoelectric arrays have broad application prospects in biomedical, micromechanical and sensor fields.

REFERENCES

1. 尹世珍,柔性电子多层岛桥结构在高铁结构监测和检测中的理论与应用研究，中国科学院大学.
2. 陈虎， 黄辉， 陈定岳,钱盛杰. 无损检测 2018, 40(03), 66–72.
3. H. Hu, Z. Xuan, & C. Wang, et al. *Sci. Adv.* 2018, 4(3), eaar3979.
4. S. Chatillon, G. Cattiaux, M. Serre, et al., *Ultrasonics*, 2000, 38(1–8), 131–134.
5. 施克仁， 阙开良， 郭大勇. 清华大学学报自然科学版(英文版, 2004, 9(5), 574–577.
6. Z. Wang, et al. *Sensors (Basel, Switzerland)* 2015, 15, 2538–2547.
7. J.H. Lee, I.J. Cho, K. Ko, et al. *Microsyst. Technol.* 2016, 23, 2321–2328.
8. Y. Duan, Y. Huang, Z. Yin, N. Bu, & W. Dong, *Nanoscale* 201, 6(6), 3289–3295.
9. C. Wang, X. Li, H. Hu, et al. *Nat. Biomed. Eng.* 2018, 2(9), 687–695.
10. H. G. Mond, J. G. Sloman, & R. H. Edwards, *Pacing Clin. Electrophysiol.* 1982, 5, 278.
11. G. D. Nelson, *Texas Heart Inst. J.* 1993, 20, 12.
12. V. S. Mallela, V. Ilankumaran, & N. S. Rao, *Indian Pacing Electrophysiol. J.* 2004, 4, 201.
13. F. W. Horlbeck, F. Mellert, J. Kreuz, G. Nickenig, & J. O. Schwab, *J. Cardiovasc. Electrophysiol.* 2012, 23, 1336.
14. M. J. Wilhelm, C. Schmid, D. Hammel, S. Kerber, H. M. Loick, M. Herrmann, & H. H. Scheld, *Ann. Thorac. Surg.* 1997, 64, 1707.
15. Z. L. Wang, *Adv. Mater.* 2012, 24, 280.
16. K. I. Park, M. Lee, Y. Liu, S. Moon, G. T. Hwang, G. Zhu, J. E. Kim, S. O. Kim, D. K. Kim, Z. L. Wang, & K. J. Lee, *Adv. Mater.* 2012, 24, 2999.
17. S. Xu, Y. Qin, C. Xu, Y. G. Wei, R. S. Yang, & Z. L. Wang, *Nat. Nanotechnol.* 2010, 5, 366.
18. X. D. Wang, *Am. Ceram. Soc. Bull.* 2013, 92, 18.
19. C. Dagdeviren, S. W. Hwang, Y. W. Su, S. Kim, H. Y. Cheng, O. Gur, R. Haney, F. G. Omenetto, Y. G. Huang, & J. A. Rogers, *Small* 2013, 9, 3398.
20. K. I. Park, C. K. Jeong, J. Ryu, G. T. Hwang, & K. J. Lee, *Adv. Energy Mater.* 2013, 3, 1539.
21. C. K. Jeong, I. Kim, K. I. Park, M. H. Oh, H. Paik, G. T. Hwang, K. No, Y. S. Nam, & K. J. Lee, *ACS Nano* 2013, 7, 11016.
22. Z. L. Wang, & J. H. Song, *Science* 2006, 312, 242.
23. Y. F. Hu, Y. Zhang, C. Xu, L. Lin, R. L. Snyder, & Z. L. Wang, *Nano Lett.* 2011, 11, 2572.
24. C. K. Jeong, K.-I. Park, J. Ryu, G.-T. Hwang, & K. J. Lee, *Adv. Funct. Mater.* 2014. DOI: 10.1002/adfm.201303484.
25. M. Y. Choi, D. Choi, M. J. Jin, I. Kim, S. H. Kim, J. Y. Choi, S. Y. Lee, J. M. Kim, & S. W. Kim, *Adv. Mater.* 2009, 21, 2185.
26. Y. F. Hu, L. Lin, Y. Zhang, & Z. L. Wang, *Adv. Mater.* 2012, 24, 110.
27. K. I. Park, S. Xu, Y. Liu, G. T. Hwang, S. J. L. Kang, Z. L. Wang, & K. J. Lee, *Nano Lett.* 2010, 10, 4939.
28. J. Kwon, W. Seung, B. K. Sharma, S. W. Kim, & J. H. Ahn, *Energ. Environ. Sci.* 2012, 5, 8970.
29. K.-I. Park, J. H. Son, G.-T. Hwang, C. K. Jeong, J. Ryu, M. Koo, I. Choi, S. H. Lee, M. Byun, Z. L. Wang, & K. J. Lee, *Adv. Mater.* 2014, 26, 2154.

30. M. Southcott, K. MacVittie, J. Halamek, L. Halamkova, W. D. Jemison, R. Lobel, & E. Katz, *Phys. Chem. Chem. Phys.* 2013, 15, 6278.
31. S. Y. Xu, Y. W. Yeh, G. Poirier, M. C. McAlpine, R. A. Register, & N. Yao, *Nano Lett.* 2013, 13, 2393.
32. M. Lucas, W. J. Mai, R. S. Yang, Z. L. Wang, & E. Riedo, *Nano Lett.* 2007, 7, 1314.
33. S. W. Bedell, K. Fogel, P. Lauro, D. Shahrjerdi, J. A. Ott, & D. Sadana, *J. Phys. D Appl. Phys.* 2013, 46.
34. S. G. Lee, R. G. Monteiro, R. S. Feigelson, H. S. Lee, M. Lee, & S. E. Park, *Appl. Phys. Lett.* 1999, 74, 1030.
35. H. S. Luo, G. S. Xu, H. Q. Xu, P. C. Wang, & Z. W. Yin, *Jap. J. Appl. Phys.* 2000, 39, 5581.
36. S. Kamba, E. Buixaderas, J. Petzelt, J. Fousek, J. Nosek, & P. Bridenbaugh, *J. Appl. Phys.* 2003, 93, 933.
37. Y. J. Zhai, L. Mathew, R. Rao, D. W. Xu, & S. K. Banerjee, *Nano Lett.* 2012, 12, 5609.
38. D. H. Kim, J. Viventi, J. J. Amsden, J. L. Xiao, L. Vigeland, Y. S. Kim, J. A. Blanco, B. Panilaitis, E. S. Frechette, D. Contreras, D. L. Kaplan, F. G. Omenetto, Y. G. Huang, K. C. Hwang, M. R. Zakin, B. Litt, & J. A. Rogers, *Nat. Mater.* 2010, 9, 511.
39. G.-T. Hwang, et al., *Adv. Mater.* 2014, 26, 4880–4887.
40. L. P. Chen, & J. Gao, *Epl-Europhys. Lett.* 2011, 93, 57003.
41. G. T. Hwang, D. Im, S. E. Lee, J. Lee, M. Koo, S. Y. Park, S. Kim, K. Yang, S. J. Kim, K. Lee, & K. J. Lee, *ACS Nano* 2013, 7, 4545.
42. Z. Y. Feng, D. Q. Shi, & S. X. Dou, *Solid State Commun.* 2011, 151, 123.
43. J. Kim, Y. Kim, Y. S. Kim, J. Lee, L. Kim, & D. Jung, *Appl. Phys. Lett.* 2002, 80, 3581.
44. I. R. Henderson, *Piezoelectric Ceramics: Principles and Applications.* APC International Ltd., Mackeyville, PA, 2006.
45. A. Nechibvute, A. Chawanda, & P. Luhanga, *ISRN Mater. Sci.* 2012, 2012, 1–11.
46. R. S. Yang, Y. Qin, L. M. Dai, & Z. L. Wang, *Nat. Nanotechnol.* 2009, 4, 34.
47. S. I. Park, J. H. Ahn, X. Feng, S. D. Wang, Y. G. Huang, & J. A. Rogers, *Adv. Funct. Mater.* 2008, 18, 2673.
48. G. Zhu, A. C. Wang, Y. Liu, Y. S. Zhou, & Z. L. Wang, *Nano Lett.* 2012, 12, 3086.
49. T. E. Starzl, R. A. Gaertner, & R. C. Webb Jr., *Circulation* 1955, 11, 952.
50. L. Gu, N. Y. Cui, L. Cheng, Q. Xu, S. Bai, M. M. Yuan, W. W. Wu, J. M. Liu, Y. Zhao, F. Ma, Y. Qin, & Z. L. Wang, *Nano Lett.* 2013, 13, 91.
51. Z. Dai, J. Wang, M. Long, & H. Huang, *Energies* 2017, 10, 323.
52. M. Liu, Y. Choi, L. Yang, K. Blinn, W. Qin, P. Liu, & M. Liu, *Nano Energy* 2012, 1, 448–455.
53. L.-H. Bjornsson, & S. Karlsson, *Appl. Energy* 2015, 143, 336–347.
54. E. R. Grijalva, & J. M. Lopez Martinez, *Energies* 2019, 12, 1–31.
55. M. Taiebat, & M. Xu, *J. Cleaner Prod.* 2019, 230, 794–797.
56. Z. Bi, R. De Kleine, & G. Keoleian, *J. Ind. Ecol.* 2016, 21, 344–355.
57. Z. Li, K. Song, G. Wei, J. Jiang, & C. Zhu, *IEEE Trans. Power Electron.* 2016, 32, 1.
58. X. Lu, P. Wang, D. Niyato, D. Kim, & Z. Han, *IEEE Commun. Surv. Tutor.* 2016, 18, 1413–1452.
59. T. Imura, T. Yasuda, K. Oshima, T. Nayuki, M. Sato, & A. Oshima, *IEEJ Trans. Electr. Electron. Eng.* 2016, 11, S91–S99.
60. A. Kurs, A. Karalis, R. Moffatt, J. D. Joannopoulos, P. Fisher, & M. Soljacic, *Science* 2007, 317, 83–86.
61. Z. Khan, S. Khan, M. Chowdhury, I. Safro, & H. UshijimaMwesigwa, *Comput. Aided Civ. Infrastruct. Eng.* 2019, 34, 547–568.
62. N. Mohamed, F. Aymen, M. Ben Hamed, & S. Lassaad, *Energy Storage* 2020, 2, e117.

63. L. Sun, D. Ma, & H. Tang, *Renewable Sustainable Energy Rev.* 2018, 91, 490–503.
64. Y. Hsieh, Z. Lin, M. Chen, H. Hsieh, Y. Liu, & H. Chiu, *IEEE Trans. Circuits Syst. II Express Briefs* 2017, 64, 942–946.
65. W. Dghais, & M. Alam, *Mobile Network Appl.* 2016, 23, 1151–1164.
66. P. Joseph, & E. Devaraj, *J. Energy Storage* 2018, 16, 145–155.
67. J. He, H.-J. Huang, H. Yang, & T.-Q. Tang, *Phys. A* 2017, 481, 119–126.
68. Y. D. Chung, E. Park, W. Lee, & J. Lee, *IEEE Trans. Appl. Supercond.* 2018, 28(3), 1–5.
69. S. Li, & C. Mi, *IEEE J. Emerging Sel. Top. Power Electron.* 2015, 3, 4–17.
70. Z. Luo, & X. Wei, *IEEE Trans. Ind. Electron.* 2017, 65(1), 331–341.
71. J. Xu, Y. Xu, & Q. Zhang, *Comput. Electr. Eng.* 2019, 80, 106470.
72. K. Aditya, *Electr. Eng.* 2018, 100, 1819–1826.
73. M. Z. Chaari, *Design and Production of a System for Wireless Charging the Battery.* Scholars' Press, Atlanta, GA, 2018.
74. C. Zheng, J.-S. Lai, & L. Zhang, *IEEE Trans. Power Electron.* 2015, 30, 1.
75. J. Acero, C. Carretero, I. Lope, R. Alonso, O. Lucia, & J. Burdio, *IEEE Trans. Ind. Electron.* 2013, 60, 410–420.
76. D. Kim, H. Kim, A. Huang, Q. He, H. Zhang, S. Ahn, Y. Zhu, & J. Fan, *Energies* 2019, 12, 4797–4808.
77. K. A. Kalwar, M. Aamir, & S. Mekhilef, *Measurement* 2018, 118, 237–245.
78. D. Wu, Q. Sun, X. Wang, & F. Yang, *IET Power Electron.* 2018, 11, 781–786.
79. J. Y. Seong, & S. Lee, *Energies* 2019, 12, 4689.
80. S. Y. Jeong, H. G. Kwak, G. C. Jang, S. Y. Choi, & C. T. Rim, *IEEE Trans. Power Electron.* 2018, 33, 7387–7397.
81. X. Ren, R. Corcolle, & L. Daniel, *Eur. Phys. J. Appl. Phys.* 2016, 73, 20902.
82. L. Xiao, M. Li, W. J. Cheng, H. W. Fan, G. Liu, & Y. H. Sun, *Int. J. Appl. Electromagn. Mech.* 2016, 52, 1433–1441.
83. P. Machura, & Q. Li, *Renewable Sustainable Energy Rev.* 2019, 104, 209–234.
84. B. Zhang, R. Carlson, J. Smart, E. Dufek, & B. Y. Liaw, *eTransportation* 2019, 2, 100012.
85. M. Kiani, B. Lee, P. Yeon, & M. Ghovanloo, *IEEE J. SolidState Circuits* 2015, 50, 1–10.
86. C. Panchal, & S. Stegen, *Eng. Sci. Technol. Int. J.* 2018, 21, 922–937.
87. V. Cirimele, M. Diana, F. Freschi, & M. Mitolo, *IEEE Trans. Ind. Appl.* 2018, 54, 4069–4079.
88. Y. Guo, et al., *R. Soc. Chem. Adv.* 2020, 10(58), 35426–35432.
89. J. Liu, Q. Deng, D. Czarkowski, M. K. Kazimierczuk, H. Zhou, & W. Hu, *IEEE Trans. Power Electron.* 2019, 34, 2355–2363.
90. H. Igarashi, *IEEE Trans. Magn.* 2017, 53, 1–7.
91. M. Kiani, U.-M. Jow, & M. Ghovanloo, *IEEE Trans. Biomed. Circuits Syst.* 2012, 5, 579–591.
92. C. Sullivan, *IEEE Trans. Power Electron.* 1999, 14, 283–291.
93. A. Karalis, J. D. Joannopoulos, & M. Soljacic, *Ann. Phys.* 2008, 323, 34–48.
94. O. M. O. Gatous, & J. Pissolato, *IEE Proc.-H Microwaves, Opt. Antennas* 2004, 151, 212–216.
95. C. Boller, *Int. J. Syst. Sci.* 2000, 31, 1333–1349.
96. A. Nair, & C. Cai, *Eng. Struct.* 2010, 32, 1704–1714.
97. H. Hu, et al., *Sci. Adv.* 2018, 4, eaar3979.
98. T.E. Michaels, & J.E. Michaels, *AIP Conf. Proc.* 2004, 700, 1468–1475.
99. D.A. Sack, & L.D. Olson, *NDT&E Int.* 1995, 28, 349–357.
100. M.J.S. Lowe, D.N. Alleyne, & P. Cawley, *Ultrasonics* 1998, 36, 147–154.
101. A. Raghavan, & C.E.S. Cesnik, *Smart Mater. Struct.* 2007, 16, 355–366.

102. X. Gao, et al., *Chin. J. Acoust.* 36, 385–399(2017).
103. Y. Wang, H. Zhang, & H. Li, et al. *Sci. Rep.* 11, 194 (2021).
104. M. Fink, C. Prada, F. Wu, & D. Cassereau, *Proc.-IEEE Ultrason. Symp.* 1989, 1989, 681–686.
105. M. Fink, *J. Phys. D Appl. Phys.* 1993, 26, 1333–1350.
106. M. Fink, *IEEE Trans. Ultrason. Eng.* 1992, 39, 555–566.
107. C. Dorme, & M. Fink, *J. Acoust. Soc. Am.* 1995, 98, 1155–1162.
108. D. Cassereau, F. Wu, & M. Fink, *Proc. IEEE Ultrason. Symp.* 1990, 1990, 1613–1618.
109. M. Tanter, J.L. Thomas, & M. Fink, *J. Acoust. Soc. Am.* 2000, 108, 223–234.
110. C. Prada, F. Wu, & M. Fink, *J. Acoust. Soc. Am.* 1991, 90, 1119–1129.
111. C. Draeger, D. Cassereau, & M. Fink, *J. Acoust. Soc. Am.* 1997, 102, 1289–1295.
112. N. Chakroun, M. Fink, & F. Wu, *Proc. IEEE Ultrason. Symp.* 1992, 2, 809–814.
113. W. Wei, C. Liu, & C. Wang, *Chin. J. Acoust.* 2000, 19, 89–95.
114. M. Lu, B. Zhang, & C. Wang, *Chin. J. Acoust.* 2004, 23, 340–349.
115. W. Wei, & C. Wang, *Chin. J. Acoust.* 2000, 19, 83–88.
116. J.L. Thomas, & M. Fink, *IEEE Trans. Ultrason. Eng.* 1996, 43, 1122–1129.
117. J.L. Thomas, F. Wu, & M. Fink, *Ultrason. Imaging* 1996, 18, 106–121.
118. M. Fink, G. Montaldo, & M. Tanter, *Annu. Rev. Biomed. Eng.* 2003, 5, 465–497.
119. P. Kosmas, & C.M. Rappaport, *IEEE Trans. Microwave Theory Tech.* 2005, 53, 2317–2323.
120. G.F. Edelmann, et al., *IEEE J. Oceanic Eng.* 2002, 27, 602–609.
121. S. Cong, T. Gang, J. Zhang, & C. Sheng, *J. Nondestr. Eval.* 2014, 33, 684–693.
122. R.Y. Chiao, & X. Hao, *IEEE Trans. Sonics Ultrason.* 2005, 52, 160–170.
123. M. O'Donnell, *IEEE Trans. Sonics Ultrason.* 1992, 39, 341–351.
124. M. Roberton, & E.R. Brown, *Int. Microwave Symp. Digest. IEEE* 2003, 2003, 611–614.
125. R. Tao, N. Zhang, & Y. Wang, *IET Radar Sonar Navig.* 2011, 5, 12–22.
126. B. Lamboul, M.J. Bennett, T. Anderson, & N.W. McDicken, *Proc. IEEE Ultrason. Symp.* 2007, 2007, 1251–1254.
127. H. Zhang, M. Li, J. Mao, & X. Wang, *Acta Acust.* 2009, 34, 539–547.
128. J.E. Wilhjelm, & P.C. Pedersen, *IEEE Trans. Ultrason. Eng.* 1993, 40, 366–372.
129. M. Pollakowski, & H. Ermert, *IEEE Trans. Ultrason. Eng.* 1994, 41, 655–659.
130. T. Misaridis, & J.A. Jensen, *IEEE Trans. Ultrason. Eng.* 2005, 52, 192–207.
131. Q. Wang, J. Mao, X. Ding, G. Lian, & Y. Hu, *Acta Acust.* 2020, 45, 227–234.
132. T. Li, Z. Huang, Z. Suo, et al. *Appl. Phys. Lett.* 2004, 85(16), 3435–3437.
133. Y. Xiang, T. Li, Z. Suo, et al. *Appl. Phys. Lett.* 2005, 87(16), 161910.
134. S.P. Lacour, D. Chan, S. Wagner, et al. *Appl. Phys. Lett.* 2006, 88(20), 204103.
135. N. Lu, X. Wang, Z. Suo, et al. *Appl. Phys. Lett.* 2007, 91(22), 221909.
136. Q.S. Yang, J.H. Song, & G. Yang, *Lect. Notes Electr. Eng.* 2012, 148, 81–86.
137. X. Jiang, Y. Li, B. Liang, et al., *Phys. Rev. Lett.* 2016, 117, 034301.
138. Z.Y. Hong, J. Zhang, & B.W. Drinkwater, *Phys. Rev. Lett.* 2015, 114, 214301.
139. R. Wunenburger, J.I.V. Lozano, & E. Brasselet, *New J. Phys.* 2015, 17, 103022.
140. C. Shi, M. Dubois, M. Wang, et al., *Proc. Natl. Acad. Sci. U. S. A.* 2017, 114, 7250.
141. A. Marzo, M. Caleap, & B.W. Drinkwater, *Phys. Rev. Lett.* 2018, 120, 044301.
142. K.D. Skeldon, C. Wilson, M. Edgar, et al., *New J. Phys.* 2008, 10, 013018.
143. A.M. Yao, & M.J. Padgett, *Adv. Opt. Photon.* 2011, 3, 161.
144. D. Baresch, J.L. Thomas & R. Marchiano, *Phys. Rev. Lett.* 2016, 116, 024301.
145. J. Lekner, *J. Acoust. Soc. Am.* 2006, 120, 3475.
146. Y. Ren, L. Li, Z. Wang et al., *Sci. Rep.* 2016, 6, 33306.
147. A. Anhäuser, R. Wunenburger, & E. Brasselet, *Phys. Rev. Lett.* 2012, 109, 034301.
148. C.R.P. Courtney, C.E.M. Demore, H. Wu, & A. Grinenko, et al., *Appl. Phys. Lett.* 2014, 104, 154103.

149. S.B.Q. Tran, P. Marmottant, & P. Thibault, *Appl. Phys. Lett.* 2012, 101, 114103.
150. M. Qi, J. Liu, Y. Mao, et al., *Chin. Phys. B* 2018, 27, 14301.
151. W. Sun, G. Liu, H. Xia, et al., *Chin. Phys. B* 2018, 27, 84301.
152. S. Wang, & S. Lin, *Acta Phys. Sin.* 2019, 68 024303 (in Chinese).
153. Z. Xu, M. Qian, Q. Cheng, & X. Liu, *Chin. Phys. Lett.* 2016, 33, 114302.
154. S. Yu, X. Ni, Y. Xu, C. He, et al., *Chin. Phys. Lett.* 2016, 33, 044302.
155. B.T. Hefner, & P.L. Marston, *J. Acoust. Soc. Am.* 1999, 1063313.
156. R. Marchiano, & J.L. Thomas, *Phys. Rev. E* 2005, 71, 066616.
157. M.J. Padgett, *Opt. Express* 2017, 25, 11265.
158. J.L. Ealo, J.C. Prieto, & F. Seco, *IEEE Trans. Ultrason. Ferroelectr. Freq. Control* 2011, 58, 1651.
159. S. Gspan, A. Meyer, S. Bernet, & M. Ritsch-Marte, *J. Acoust. Soc. Am.* 2004, 115, 1142.
160. X. Jiang, B. Liang, J. Cheng, & C. Qiu, *Adv. Mater.* 2018, 301800257.
161. X. Jiang, J. Zhao, S. Liu, et al., *Appl. Phys. Lett.* 2016, 108203501.
162. 张晗, 高阳, 中国物理快报：英文版 2019, 11, 35–39.
163. Z. Qian, D. Shang, Q. Sun, et al., *Acta Phys. Sin.* 2019, 68024301 (in Chinese).
164. J. Tang, S. Piao, & H. Zhang, *Chin. Phys. B* 2017, 26, 114301.
165. D. Yang, R. Zhang, & S. Shi, *Acta Phys. Sin.* 2018, 67, 244301 (in Chinese).

4 Advanced Manufacturing of Carbon Fiber Material

Xinzhu Wang and Linji Xu
Chongqing University

CONTENTS

DOI: 10.1201/9781003182146-4

4.1 INTRODUCTION TO CARBON FIBERS

4.1.1 Crystal Structure of Carbon Fibers

Carbon fibers (CFs) are composed of carbon atoms that are bonded together to form a long chain and have at least 92 wt.% carbon in composition.[1] They are extremely stiff, strong, and light. They can be short or continuous. Their structure can be crystalline, amorphous, or partly crystalline, and their crystalline form has the crystal structure of graphite.[2] As shown in Figure 4.1, graphite is composed of layers of carbon atoms that are arranged in 6-membered, hexagonal rings.[3] These rings are attached to one another on their edges. Layers of fused rings can be modeled as an infinite series of fused benzene rings (without the hydrogen atoms). Carbon atoms in these ring arrays are in the sp2-hybridized state. In the sp2 molecular orbital model, each carbon atom is attached to three other species, three other carbon atoms in the case of graphite. In this bonding mode, the bond angle between adjacent carbon atoms is 120. These "ring arrays" are arranged in large sheets of carbon atoms, and individual sheets are known as graphene layers. The carbon-carbon bond length in a layer plane is 1.418. Graphene layers are stacked one on top of another parallel with the "C" crystallographic axis of the hexagonal 4-axis system in which graphite crystallizes.

Graphene layers are precisely "indexed" so that an exact three-dimensional (3D) repeating order is developed and maintained throughout the crystal.[4] Each graphene layer is indexed to the layer above in this arrangement and the layer below in such a way as to form a repeating alignment every second layer. In other words, the carbon atoms in the first layer indexed are located in precise alignment with the carbon atoms of the third layer indexed. Correspondingly, the carbon atoms of the second graphene layer are in alignment with the atoms of the fourth layer. This configuration is known as an "ABABAB" structure. All the "A" layers are aligned with other "A" layers, and all the "B" layers are aligned with other "B" layers. A simplified view of this model is to imagine the "A" layer fixed in space. Then align the "B" layer in the "ABABAB" structure so that directly below the center of each "A" layer hexagon there is located, dead center a carbon atom in the "B" layer.

Since graphite is a crystalline solid it can be defined by a basic unit cell. Graphite is hexagonal and therefore is defined by a hexagonal unit cell. The hexagonal unit cell of graphite is a rhombic prism (not a hexagonal prism). The prism is defined by six surfaces: four sides and a top and bottom.[5] This cell encloses or partially encloses the carbon atoms from three graphene layers, so one could say the unit cell starts at layer "A1" and finishes at layer "A2." Layer "B1" would then be located within the unit cell. Each unit cell contains a total of four carbon atoms ($z=4$). One and one-sixth (1 1/6) atoms each come from the first and third layer of the unit cell (top and bottom) and one and two-third (1 2/3) atoms come from the second layer (middle layer). Since each unit cell contains "centered" and "edge" carbon atoms, unit cell is considered a "centered" unit cell as opposed to a primitive unit cell. The unit cell dimensions are the lengths of the imaginary line connecting in-plane carbon atoms (a) along cell edges that are parallel with the graphite layers, and the length of the imaginary line connecting between-plane carbon atoms (c) between the first and third layers (the edge perpendicular to the graphene layer). These dimensions are: $a=2.456$, $c=6.694$. Since the distance between two layers (c) is 6.694, the distance between any two adjacent layers is 6.692/2=3.346 ($c/2$).

Carbon atoms in the graphite crystal are in the sp2-hybridized state. As discussed in the section on organic chemistry, this means that carbon atoms support two bonding components a sigma (σ) component, and pi (π) component.[6] The sigma bonds are "hard" bonds and are similar to the pure sp3 type sigma bonds, which form the 3D giant molecular structure of diamond. The sigma bonding system acting within a single graphene layer is exceedingly vital. In fact, it is probably the strongest 2D network structure known and is the structural backbone of all carbon/graphite fibers and carbon nano-tubes. The high hardness of diamond also illustrated the strength of the sigma bond in carbon. Each carbon atom within the graphene layer has three of these sigma atomic orbitals, which combine with similar orbitals of adjacent carbon atoms forming the molecular bonds that hold the layer together (see sp2 bonding in the organic chemistry section).

In contrast, the carbon atoms of the diamond structure each have four such bonds. From an atomic perspective, one could "walk" from one carbon atom, to any carbon atom within a graphene layer using the "hard" sigma bonds as bridges to traverse the pathway between carbon atoms. However, your stroll would be limited only to the two dimensions covered by the graphene layer you were on. In diamond, the entire diamond molecule can be traversed, in a similar fashion, in any direction, up, down, or side-to-side so that any carbon atom can be reached from any other carbon atom using "sigma bond bridges" to traverse the path.

The stiff sigma bonds that give rigidity and high tensile strength to a graphene layer do not provide any support in the perpendicular direction.[4] They do nothing to hold the graphite crystal together in the "c" direction. Graphene layers are natural structures and require some force to maintain sheet alignment as well as the proper equilibrium separation distance. The function of crystal support in the third dimension, the "c" dimension, is supplied by the second bonding component of the sp2 model known as the pi component or pi bond. The pi component results from weak, secondary electrical bonds formed by the overlapping pi (p) orbitals of the sp2 carbon network within each graphene sheet. Each carbon atom has one pi-electron. This electron is highly likely to be found in a region just above or below the plane formed by the carbon atoms in a graphene layer.[7] Since each graphene layer can be viewed as a system of fused six-carbon rings, one can also imagine that a ring or donut-shaped region exists above and below each ring, containing six pi electrons (one from each carbon atom). These pi electrons are not stationary but are "delocalized" within the confines of the donut-shaped region they occupy. This delocalization of pi electrons results in what is known as resonance energy or resonance stability.[8] Here, the stability here is chemical stability, which is spread throughout the entire graphite structure. In non-quantum terms, the effect is one of "spreading" out the electrical charge across the entire array of carbon atoms. In addition to the delocalization of the pi orbital electrons within the region located immediately above given a ring, this delocalization is "conjugated" throughout the contiguous graphene layer. This means that there is some mechanism for pi electrons to be delocalized not only in their immediate orbital but also to move back and forth between and within adjacent orbitals.[9]

Carbon atoms within a layer are bonded by (i) covalent bonds provided by the overlap of the sp2 hybridized orbitals and (ii) metallic bonding provided by the delocalization of the pz orbitals, Le, the 7r electrons.[10] This delocalization makes graphite a good electrical conductor and a good thermal conductor in the *x–y* plane. The bonding between the layers is van der Waals bonding, so the carbon layers can easily slide with respect to one another; graphite is an electrical insulator and a thermal insulator perpendicular to the layers. Due to the difference between the in-plane and out-of-plane bonding, graphite has a high modulus (HM) of elasticity parallel to the plane and a low modulus perpendicular to the plane. Thus, graphite is highly anisotropic.

The HM of a CF stems from the fact that the carbon layers, though not necessarily flat, tend to be parallel to the fiber axis. This crystallographic preferred orientation is known as a fiber texture. As a result, a CF has a higher modulus parallel to the fiber axis than perpendicular to the fiber axis. Similarly, the electrical and thermal conductivities are higher along the fiber axis, and the coefficient of thermal expansion is lower along the fiber axis.[11] The greater the degree of alignment of the carbon layers parallel to the fiber axis, i.e., the stronger the fiber texture, the greater the *c*-axis crystallite size (L_c), the density, the carbon content, and the fiber's tensile modulus, electrical conductivity, and thermal conductivity parallel to the fiber axis; the smaller the fiber's coefficient of thermal expansion and internal shear strength.

The carbon layers in graphite are stacked in an AB sequence, such that half of the carbon atoms have atoms directly above and below them in adjacent layers (Figure 4.1). Note that this AB sequence differs from that in a hexagonal close-packed crystal structure. In a CF, there can be graphite regions of size L, perpendicular to the layers and size L_c parallel to the layers. There can also be crystalline regions in which the carbon layers, though well developed and parallel to one another, are not stacked in any particular sequence; the carbon in these regions is said to be turbostratic carbon. Yet another type of carbon that can exist in CFs is amorphous carbon, in which the carbon layers, though well developed, are not even parallel to one another.[12]

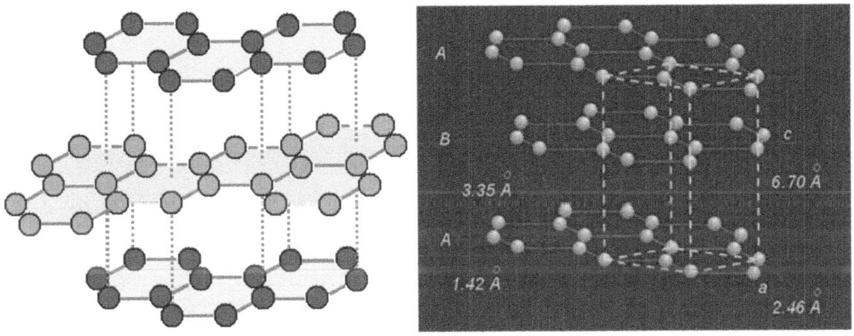

FIGURE 4.1 Half of the carbon atoms have atoms directly above and below them in adjacent layers.[3]

4.1.2 Properties of Various Fibers and Whiskers

The proportion of graphite in a CF can range from 0% to 100%. When the proportion is high, the fiber is said to be graphitic, and it is called a graphite fiber. However, a graphite fiber is polycrystalline, whereas a graphite whisker is a single crystal with the carbon layer rolled up like a scroll. Because of their single crystal nature, graphite whiskers are virtually flaw-free and have exceptionally high strength (HS). However, the production yield of graphite whiskers is too low for them to be commercially significant. This book only deals with fibers and not whiskers.[12]

Table 4.1 compares the mechanical properties, melting temperature, and density of CFs with other types of fibers.[13] There are numerous grades of CFs; Table 4.1 only shows the two high-performance (HP) grades, which are labeled "HS" and "HM." Among the fibers (not counting the whiskers), high-strength CFs exhibit the highest strength while HM CFs exhibit the highest modulus of elasticity.

Moreover, the density of CFs is quite low, making the specific modulus (modulus/density ratio) of high-modulus CFs exceptionally high. The polymer fibers, such as polyethylene and Kevlar fibers, have densities even lower than CFs, but their melting temperatures are low. The ceramic fibers, such as glass, SO_2, $A1_2O_3$, and SiC fibers, have densities higher than CFs. Most of them (except glass fibers) suffer from high prices or are not readily available in a continuous fiber form. The tensile stress–strain curves of the fibers are straight lines all the way to fracture, so the strength divided by the modulus gives the ductility (strain at break) of each fiber, as shown in Table 4.1. The main drawback of CFs' mechanical properties is the low ductility, which is lower than those of glass, SiO_2, and Kevlar fibers. The ductility of high-modulus CFs is even lower than that of high-strength CFs.

CFs that are commercially available are divided into three categories, namely general purpose (GP), HP, and activated carbon fibers (ACFs). The GP type is characterized by an amorphous and isotropic structure, low tensile strength, low tensile modulus, and low cost. The HP type is characterized by relatively HS

TABLE 4.1
Properties of Various Fibers and Whiskers

Material	Density[a] (g/cm³)	Tensile Strength[a] (GPa)	Modulus of Elasticity[a] (GPa)	Ultimate Strain[a] (%)	Melting Temperature[a] (°C)	Specific Modulus[a] (10^6 m)	Specific Strength[a] (10^4 m)
E – glass	2.55	3.4	74	2.5–4.0	<1725	2.90	14
S – glass	2.50	4.5	86.9	5.2	<1725	3.56	18
SiO_2	2.19	5.9	72.4	8.1	1728	3.38	27.4
Al_2O_3	4.0	2.4	380	0.55	2015	9.86	5.3
ZrO_2	4.84	2.1	340	0.62	2677	7.26	4.3
Carbon (High – Strength)	1.80	7.15	324	2.0	3700	180	18
Carbon (High – Modulus)	1.91	3.8	588	0.7	3700	308	19
BN	1.90	1.4	90	1.6	2730	4.74	7.4
Boron	2.5	3.5	400	0.9	2030	16.4	12
B_4C	2.36	2.3	480	0.48	2450	20.9	9.9
SiC	4.09	2.1	480	0.44	2700	12.0	5.1
TiB_2	4.48	0.10	510	0.02	2980	11.6	0.3
Be	1.83	1.28	300	0.4	1277	19.7	7.1
W	19.4	4.0	410	0.98	3410	2.2	2
Polyethylene	0.97	2.59	120	2.2	147	12.4	27.4
Kevlar	1.44	3.22	64.8	1.44	500	4.5	22.4

(Continued)

TABLE 4.1 (*Continued*)
Properties of Various Fibers and Whiskers

Material	Density[a] (g/cm^3)	Tensile Strength[a] (GPa)	Modulus of Elasticity[a] (GPa)	Ultimate Strain[a] (%)	Melting Temperature[a] (°C)	Specific Modulus[a] (10^6 m)	Specific Strength[a] (10^4 m)
Al_2O_3 whiskers	3.96	21	430	4.9	1982	11.0	53.3
BeO whiskers	2.85	13	340	3.8	2550	12.3	47.0
B_4C whiskers	2.52	14	480	2.9	2450	19.5	56.1
SiC whiskers	3.18	21	480	4.4	2700	15.4	66.5
Si_3N_4 whiskers	3.18	14	380	3.7	—	12.1	44.5
Graphite whiskers	1.66	21	703	3.0	3700	43	128
Cr whiskers	7.2	8.90	240	3.7	1 890	3.40	12

Data from Ref. [12].

and HM. A higher modulus is associated with a higher proportion of graphite and more anisotropy among the HP CFs. ACFs are characterized by the presence of a large number of open micropores, which act as adsorption sites. ACFs are characterized by the presence of a large number of open micropores, which act as adsorption sites. The adsorption capacity of ACFs is comparable to that of activated carbons, but the fiber shape of ACFs allows the adsorbate to get to the adsorption site faster, thus accelerating the adsorption and desorption processes.[14] As a result, the amount adsorbed increases with the severity of activation. Severe activation may be achieved by treating commercial ACF with sulfuric acid followed by heating at up to 500°C.[15]

4.1.3 Precursor-Based Carbon Fibers

Commercial CFs are fabricated by using pitch or polyacrylonitrile (PAN) as the precursor. The processes for both precursors are shown in Figure 4.2.[16] Precursor fibers are fabricated by conventional spinning techniques, such as wet spinning for PAN and melt spinning for pitch. They must be converted to a form which is flameproof and stable at the high temperatures (>700°C) involved in carbonization. Therefore, before carbonization (pyrolysis), they are stabilized for the case of the PAN precursor, or infusiblized for the case of the pitch precursor. Both stabilization and infusiblization are carried out in an oxidizing atmosphere. After that, GP and HP fibers are obtained by carbonization in an inert atmosphere, followed by graphitization at >2500°C in an inert atmosphere if an HM is desired, whereas ACFs are obtained by activating in a reactive atmosphere, such as steam at elevated temperatures. To enhance the preferred orientation in the HP CFs,

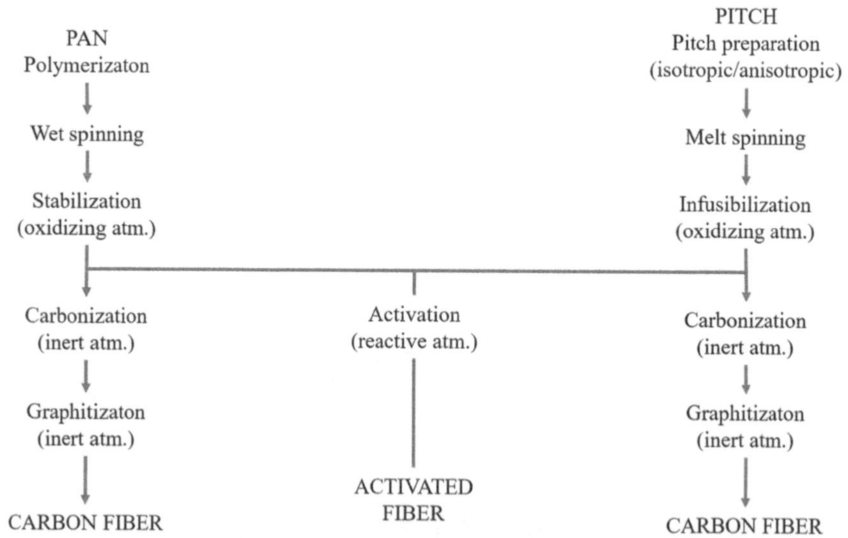

FIGURE 4.2 The processes for making carbon fibers from PAN and pith precursors.

graphitization can be performed while the fibers are under tension. The higher the graphitization temperature, the greater the preferred orientation.

For the case of pitch as the precursor, isotropic pitch gives an isotropic CF, which belongs to the category of GP CFs, whereas anisotropic pitch (such as mesophase pitch) gives HP CFs which have the carbon layers preferentially parallel to the fiber axis.

Table 4.2 shows the tensile properties of various CFs on the market. Among the HP CFs, those based on pitch can attain a higher modulus than those based on PAN, because pitch is more graphitizable than PAN. In particular, the HP fiber designated E-130 by du Pont exhibits a modulus of 894 GPa, which is over 80% of the theoretical value of a single graphite crystal (1000 GPa). A higher modulus is associated with a lower elongation at break, as shown by comparing the group of Amoco HP fibers in the order P-25, P-75S, and P-12OS, and the group of du Pont HP fibers in the order E-35, E-75, and E-130. The du Pont HP fibers exhibit higher tensile strengths and greater elongations than the Amoco HP fibers of similar moduli. Among the HP fibers, those based on PAN can attain a higher tensile strength and greater elongation than those based on pitch, because (i) shear is easier between the carbon layers in a graphitized fiber, (ii) pitch is more graphitizable than PAN, and (iii) the oriented graphitic structure causes the fibers to be more sensitive to surface defects and structural flaws. In particular, the HP PAN-based fiber designated T-1000 by Toray exhibits a tensile strength of 7060 MPa and an elongation of 2.4%. The GP fibers tend to be low in strength and modulus but high in elongation at break. Table 4.2 also shows the diameters of various commercial CFs.

Pitch-based CFs (GP and HP) represent only about 10% of the total CFs produced worldwide around 1990,[17] but this percentage is increasing due to the lower cost and higher carbon content of pitch compared to PAN.

Mesophase pitch-based CFs are currently the most expensive due to the processing cost. Isotropic pitch-based CFs are the least expensive. PAN-based CFs are intermediate in cost. CFs' prices and tensile strengths, aramid fibers, and glass fibers. Although CFs are more expensive than aramid fibers or glass fibers, they mostly provide higher tensile strengths.[18] Among the different grades of CFs, the prices differ greatly. In general, the greater the tensile strength, the higher the price. The price of CFs has been decreasing, while the consumption has been increasing.[18] The decreasing price is broadening the applications of CFs from military to civil applications, from aerospace to automobile applications, and from biomedical devices to concrete structures.

Under rapid development are short CFs grown from the vapor of low-molecular-weight hydrocarbon compounds, such as acetylene. This process involves catalytic growth using solid catalyst particles (e.g., Fe) to form carbon filaments, which can be as small as 0.1 pm in diameter. Subsequent chemical vapor deposition from the carbonaceous gas in the same chamber causes the filaments to grow in diameter, thus resulting in vapor grown CFs (VGCFs) or gas-phase grown CFs.

CFs can alternatively be classified on the basis of their tensile strength and modulus. The nomenclature given below was formulated by IUPAC.

TABLE 4.2
Tensile Properties and Diameters of Commercial CFs

Type	Fiber Designation	Tensile Strength (MPa)	Tensile Modulus of Elasticity (GPa)	Elongation at break (%)	Diameter (μm)	Manufacturer
GP	TS-101S	720	32	2.2	14.5	Kureha Chem
	TS-201S	690	30	2.1	14.5	Kureha Chem
	S-210	784	39	2.0	13	Donac
	P-400	690	48	1.4	10	Ashland Petroleum
	GF-20	980	98	1.0	7–11	Nippon Carbon
HP (PAN)	T-300	3530	230	1.5	7.0	Toray
	T-400HB-3000	4410	250	1.8	7.0	Toray
	T-800HB	5490	294	1.9	5.2	Toray
	T-1000GC	7000	324	2.0	5.3	Toray
	MR 50	5490	294	1.9	5	Mitsubishi Rayon
	MRE 50	5490	323	2.0	6	Mitsubishi Rayon
	HMS-40	3430	392	0.87	6.2	Toho Rayon
	HMS-40X	4700	392	1.20	4.7	Toho Rayon
	HMS-60X	3820	588	0.65	4.0	Toho Rayon
	AS-1	3105	228	1.32	8	Hercules
	AS-2	2760	228	1.2	8	Hercules
	AS-4	3795	235	1.53	8	Hercules
	AS-6	4140	242	1.65	5	Hercules
	IM-6	4382	276	1.50	5	Hercules
	HMS4	2484	338	0.7	8	Hercules
	HMU	2760	380	0.70	8	Hercules

(Continued)

TABLE 4.2 (*Continued*)
Tensile Properties and Diameters of Commercial CFs

Type	Fiber Designation	Tensile Strength (MPa)	Tensile Modulus of Elasticity (GPa)	Elongation at break (%)	Diameter (μm)	Manufacturer
HP (pitch)	P-25	1400	160	0.9	11	Amoco
	P-75S	2100	520	0.4	10	Amoco
	P-120S	2200	827	0.27	10	Amoco
	E-35	2800	241	1.03	9.6	du Pont
	E-75	3100	516	0.56	9.4	du Pont
	E-130	3900	894	0.55	9.2	du Pont
	F-140	1800	140	1.3	10	Donac
	F-600	3000	600	0.52	9	Donac
ACF	FX-100	—	500	18	15	Toho Rayon
	FX-600	—	1000	50	7	Toho Rayon
	A-10	245	1000	20	14	Donac
	A-20	98	2000	45	11	Donac

- *UHM (ultra-high-modulus) type*: CFs with modulus >500 GPa.
- *HM-type*: CFs with modulus >300 GPa and strength-to-modulus ratio <1%.
- *IM (intermediate-modulus) type*: CFs with modulus up to 300 GPa and strength-to-modulus ratio above 1×10^{-2}.
- *Low-modulus type*: CFs with modulus as low as 100 GPa and low strength. They have an isotropic structure.
- *HT type*: CFs with strength >3 GPa and strength-to-modulus ratio between 1.5 and 2×10^{-2}.

There is overlap between the IM and HT categories, as shown by the above definitions. Commercial continuous CFs are in the form of tows (untwisted bundles) containing typically 1 OW12 OOO fibers (filaments) per tow, or yarns (twisted bundles). They may be sized or unshed. The sizing improves the handleability and may enhance the bonding between the fibers and certain matrices when the fibers are used in composites. HP CFs are widely used in polymer-matrix composites for aircraft, which are lightweight for the purpose of saving fuel. The aircraft Voyager has 90% of its structure made of such composites and achieved a nonstop, unfueled, round-the-world flight in 1986. The use of such composites in passenger aircraft is rapidly increasing. HP CFs are also used in carbon-matrix composites for high-temperature aerospace applications, such as in the Space Shuttle, as the carbon matrix is more temperature resistant than a polymer matrix. These fibers have started to be used in metal matrices, such as aluminum, for aerospace applications, as aluminum is more temperature resistant than polymers. Short GP pitch-based CFs are used to reinforce concrete, because low cost is crucial for the concrete industry. Because this is a large-volume application of CFs, the tonnage of CFs used is expected to increase markedly as this application becomes more widely accepted. GP CFs are also used for thermal insulation, sealing materials, electrically conducting materials, antistatic materials, heating elements, electrodes, filters, friction materials, sorbents, and catalysts.[19]

4.2 PROCESSING OF CARBON FIBERS

4.2.1 Introduction

CFs have high tensile strengths up to 7 GPa with excellent creep resistance, low densities (1.75–1.95 g/cm^3), and high moduli (200–600 GPa).[20] They lack resistance to oxidizing agents, such as hot air and flames, but they are resistant to all other chemical species. The good mechanical properties make CFs attractive for use in composites in the form of woven textiles and continuous or chopped fibers. Such composite parts can be produced through filament winding, tape winding, pultrusion, compression molding, vacuum bagging, liquid molding, and injection molding. The CF industry has been growing continuously with a focus on aerospace, military, construction and medical and sporting goods.[20–23] For the automotive industry, CF-reinforced polymeric composites significantly reduce weight, which is a prerequisite for battery-driven cars.[21]

More than 100 years ago fiber was carbonized by Thomas Edison in his production of incandescent filaments for the world's first commercially viable light bulb. He used cardboard and later bamboo filaments as his starting material to produce carbonized thread. Today CF is manufactured from a precursor fiber, which is generally spun from PAN, pitch or rayon. Approximately 95% of all CF produced is derived from PAN, the remaining 5% being made mostly from pitch with a very small amount made from rayon. Many other polymers have also been investigated for their potential as CF precursors, mainly to reduce the processing cost or improve the mechanical properties of the resultant CFs. However, to date, none of these have become commercial.

Over 98% of CFs are made from PAN as a precursor. Other CF precursor types are described elsewhere.[20,21] The mechanical properties of some commonly used fibers on the market (including PAN-based CF, pitch-based CF and two other typical reinforcing fibers) are listed in Figure 4.3. Besides the continuous growth of CF production in the last few years, the large-volume application of CF in the automotive industry has been impeded by high fiber costs and the lack of high-speed composite fabrication techniques. However, current investigations and future developments might well change the market to establish CFs as a mass product like other synthetic fibers or metals.

Pitch-based CF is categorized as either general performance carbon fiber (GPCF) or high-performance carbon fiber (HPCF). GPCF is made from isotropic pitch[24,25] and has a tensile strength up to 1000 MPa and a tensile modulus of 30–60 GPa. It is used mainly in less mechanically demanding applications, such as reinforced concrete, thermal insulation, electrodes, or activated carbons.[26] The focus of this chapter is on HPCF, which is made from anisotropic or mesophase pitch.

Carbon filaments (to be distinguished from CFs) grow catalytically when a carbonaceous gas is in contact with a small metal particle (the catalyst) at an

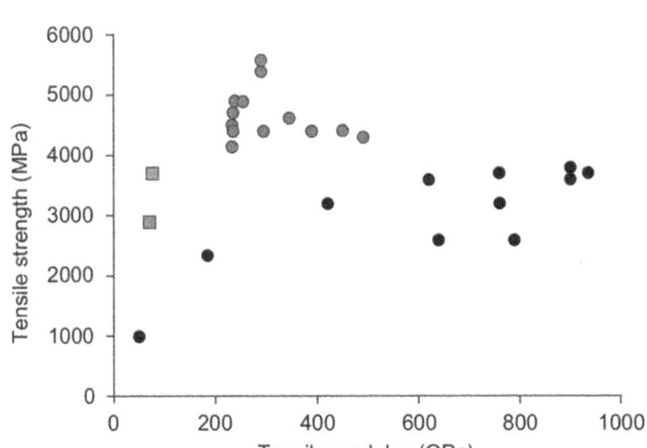

FIGURE 4.3 Tensile strength and modulus of Mitsubishi PAN-based (●) and pitch-based (●) carbon fibers as well as E glass (■) and Kevlar 29 (■), two other typical reinforcing fibers.[27]

elevated temperature.[28] During growth, a carbon filament lengthens such that the filament diameter is equal to the diameter of the catalyst particle that produces it. While the filament lengthens by catalytic growth, noncatalytic chemical vapor deposition of carbon takes place from the carbonaceous gas on the sides of the filament, causing the filament to grow radially (thicken); it thus becomes VGCF.[29]

This chapter describes the methods for fabricating CFs from all three types of precursors. The chemical treatments and coatings needed to enhance the bonding with various matrices in composite materials are addressed in the chapters on composite materials. As CFs are most conveniently used in a woven form, this chapter also covers weaving.

4.2.2 Carbon Fibers Made from Polyacrylonitrile

Global CF consumption and the major CF manufacturers of PAN-based CFs are summarized in Table 4.3, with their estimated capacities sorted by actual suppliers.[12] CFs were classified to the number of filaments, where "1K" means 1000 single filaments. In technical terms, CFs with 1–24K are called "low tow" and 24–320K and more filaments are called "large tow." A second classification describes the mechanical properties. According to a combination of tenacity and modulus of the CFs, a classification as seen in Table 4.4 is used.

4.2.2.1 Polymerization

PAN was first recognized as a suitable precursor material for CFs by Shindo in 1961.[30] It is currently the most important starting materials for the preparation of CFs. The polymerization of acrylonitrile (AN) is of crucial importance particularly since further steps in CF production, such as precursor fiber spinning, stabilization, and carbonization as well as the resulting CF properties, strongly depend

TABLE 4.3
The Name Plate CF Capacity in 2020 in Kilo-Tons

Countries/Regions	Capacity/KT	Percentage
USA	37.3	21.7
Mainland China	36.15	21.1
Japan	29.2	17
Hungary	15.4	9.0
Mexico	10	5.8
Republic of Korea	9.2	5.4
Taiwan, China	8.8	5.1
Germany	5.1	3.0
France	5.1	3.0
British	4.0	2.3
Turkey	3.6	2.1
Russia	2.0	1.2
Spain	0.6	0.3

TABLE 4.4
Classification of CFs[12]

Type of PAN-Based CF	Tensile Strength (GPa)	Tensile Elastic Modulus (GPa)
Low elastic modulus type	>3500	<230
Standard elastic modulus type	>2500	230–265
Intermediate elastic modulus type	>3500	270–315
High elastic modulus type	>2500	315–600
Ultrahigh elastic modulus type	>2500	>600

on the characteristics of the polymer used.[11] AN can be polymerized in bulk, suspension, solution, and emulsion applying free radical, ionic, or atom transfer radical polymerization.[31–35] Solution polymerization and suspension polymerization are currently the most widely used procedures for preparing PAN-based copolymers for CF production. In continuous process control, AN's solution polymerization, and related monomers are probably more satisfactory.[36–41]

Solution polymerization can be carried out with solvents, such as dimethylacetamide (DMAc), dimethylformamide (DMF), dimethylsulfoxide (DMSO), or aqueous sodium thiocyanate solutions, and thus allows immediate manufacturing of spinning dopes.[42,43] Since the produced copolymers have high molecular weight, the polymerization reaction is usually carried out in rather low concentrated solutions. Two important drawbacks limit the application of this polymerization procedure. The conversion of the monomer to polymer reaches 50%–70%, so the resulting solutions may be processed only after complete recovery of the unreacted AN from the spinning dope; however, usually it still contains 0.2–0.3 wt.% of this toxic and carcinogenic monomer.[35]

Another important drawback is that conventional solvent systems used for the polymerization process usually have high transfer constants. Thus, the process of suspension polymerization of AN has the advantage that almost no by-products are produced and that it can be carried out under controlled conditions so that both branching and crosslinking are avoided.[11,33,35–41] Further advantages of the reaction include the simple removal of the polymer by filtration and drying, adjustment of the molecular weights over a very wide range, and control of particle size. Particularly on a large scale, this discontinuous process permits good control over the exothermic nature of the polymerization, allowing up to 90% polymer yield.

4.2.2.2 Preparation of CF Precursors

PAN-based polymers undergo thermally induced cyclization reactions below their melting point and can therefore not be processed in melt with conventional spinning techniques. This makes the melt spinning of pure PAN impossible unless large amounts of solvent additives and plasticizers are added. In such a melt-assisted spinning process, additives are used to reduce the interaction between PAN molecules by decoupling nitrile–nitrile association. This way the melting point of plasticized PAN is reduced to an appropriate range for melt spinning.[44–46] However, until today, no melt-spinning process has reached the CF qualities of wet-spun PAN-based precursor fibers. This is why wet spinning is still the preferred method for CF precursor manufacturing.

The choice of solvents for PAN-based copolymers is of utmost importance in the production of precursor fibers using wet-spinning processes. As already mentioned, the highly polar nitrile groups of AN experience a strong dipole–dipole interaction, making the polymer soluble only in highly polar solvents, such as DMAc, DMF, DMSO, zinc chloride, and sodium thiocyanate,[11,43] but also in ionic liquids.[47] Schildknecht listed a variety of solvent systems in which PAN is soluble.[48]

There are many possible technical designs for wet-spinning processes. The precipitation bath can be situated horizontally or vertically, and different baths for washing and drawing can be applied, followed by further manufacturing operations such as application of lubricants, drying, or heat setting. The most important issue is to obtain fibers having optimum morphological structure, which is very sensitive to the composition and conditions within the coagulation bath. Thus, the coagulation bath is not stationary but circulates with a finite velocity. The flow pattern in the bath is rather complex; the streamlines are usually not parallel to the axis of the spinning line, which creates unequal flow and solidification conditions for the individual filaments. Modern systems used for wet spinning often involve the directed flow of the spinning bath. This modification aims to make the flow pattern in the bath symmetrical and equalize the conditions experienced by the individual filaments to promote homogeneous penetration of the nonsolvent into the bundle of filaments.

4.2.2.3 Heat Treatment

Generally, heat treatment of PAN-based fibers entails three steps: oxidation, carbonization, and graphitization. The precursor fiber is first oxidized within a temperature range between 200°C and 300°C, resulting in an N-containing ladder-type polymer formation. This allows further processing at higher temperatures. After oxidation, the fibers are carbonized at temperatures up to 1200°C–1600°C in an inert atmosphere to obtain a turbostratic carbon structure. Then, to improve the ordering and orientation of the basal planes in the direction of the fiber axis, the fiber can be graphitized by up to 3000°C depending upon the required tensile modulus of the resultant CF. This entire procedure shall be outlined in more detail.

The basic principle used in the conversion of PAN-based precursor fibers into CFs is the treatment of the fibers in an oxidizing atmosphere at temperatures in the range between 200°C and 300°C. Generally, the thermal treatment, that is, oxidation reaction of the homopolymer PAN, is a very difficult reaction to control. Because of a high conversion temperature in the oxidation step, coupled with highly exothermic reaction kinetics, the rapid evolution of heat can cause defects inside the fiber and on the surface of the filaments, particularly in larger fiber tows. However, the exothermic reaction can be influenced by a suitable comonomer composition, whose presence significantly affects the oxidation process.[49–51]

In the next step of the process, the stabilized precursor fiber tows, capable of withstanding high temperatures, are subjected to thermal pyrolysis in an inert atmosphere and converted into CFs. The volatile compounds are removed to give CFs with a carbon yield of about 50 wt.% with respect to the original precursor. At this stage, these CFs contain >98 wt.% carbon, 1–2 wt.% nitrogen, and 0.5 wt.% hydrogen.[12]

The final step in the heat treatment of CFs at temperatures of up to 3000°C is called graphitization. At these temperatures, the ordering and orientation of the small turbostratic crystallites in the direction of the fiber axis take place. This results in higher Young's modulus, but simultaneously leads to the lower tensile strength of the final CF. The HM furnace employs graphite muffle operating in a carefully controlled inert atmosphere. The conditions must be quiescent, since even a low-volume gas flow over the hot graphite element is sufficient to continually remove a molecular layer of graphite and cause severe erosion leading to premature failure.[11]

4.2.3 Carbon Fibers Made from Pitch

Fifty years ago, Brooks and Taylor reported a liquid crystal-like mesophase in pitch, which upon heating to high temperature, formed graphitic carbon.[52] Later, White et al. showed that the mesophase could be oriented by flow and shear and that the structure persisted during the graphitization heat treatment.[53] Singer built on this knowledge, showing that at the correct temperature and viscosity fibers could be drawn; the fibers being highly oriented with the graphite layer planes lying along the fiber axis.[54,55] Heating these fibers resulted in graphitization and the formation of the first high-strength CFs based on pitch. Singer's work, along with that of McHenry and Lewis, formed the basis of Union Carbide's industrial process for making mesophase pitch-based CF.[54,56,57]

4.2.3.1 Pitch Precursors

Pitch is a black tar-like substance comprising hundreds of three to eight ringed aromatic compounds, the exact composition varying depending on the source of the raw material and the method of processing. It is also sometimes referred to as asphalt or bitumen. It occurs naturally in several parts of the world but the most pitch is produced as the by-product of refining petroleum, or the destructive distillation of coal. Petroleum pitch typically has more alkyl-substituted polycyclic aromatic hydrocarbons than coal tar pitch. That is, it has lower aromaticity and a higher degree of substitution of the aromatic ring.[58,59] A synthetic mesophase pitch can also be prepared from ring compounds, such as naphthalene.[60]

Raw pitch is a complex blend of polyaromatic molecules and heterocyclic compounds. As a general rule it is isotropic and is composed of four fractions[61]: (i) saturated aliphatic components of low molecular weight; (ii) naphthalene aromatic compounds, which have low molecular weight with saturated rings;

(iii) polar aromatic compounds of medium molecular weight with some heterocyclic molecules; and (iv) asphaltenes of high molecular weight and a degree of aromaticity. In addition, pitch contains a number of solid impurities. It has a softening point between 60°C and 150°C, which is often too low for melt spinning, and thus pitch needs to be treated to increase the viscosity and softening point. Most of the pitch produced is converted into bitumen for use in road surfacing; however, its use as a precursor for CF is also essential.

4.2.3.2 Preparation of CF Precursors

For the preparation of isotropic pitch suitable for spinning into fibers, the material needs to be processed carefully to avoid the formation of mesophase, which occurs readily during heat treatment at 350°C–450°C.[62] During heating at temperatures <350°C, volatiles are released and dehydrogenation, crosslinking, and condensation reactions occur, leading to increased molecular mass. The combined effect is an increase in the softening temperature. While molten the pitch is filtered to remove the solid particles. The heat treatment can be done using a range of different solvent extraction and thermal methods, which the wiped film evaporator method and the air-blowing method are two that are commonly used.[63,64]

When the isotropic pitch is heated from 350°C to 450°C a mesophase is produced, and the formation rate increases with temperature. The initial step involves condensation and dehydrogenation reactions leading to an increase in the aromatic pitch molecules.[65] These planar polyaromatic molecules then stack together to form a so-called molecular assembly unit. These units subsequently self-assemble into microdomains, which are the anisotropic spheres first discovered by Brooks and Taylor.[52]

4.2.3.3 Heat Treatment

The conversion of mesophase pitch fiber to CF follows a similar process to converting PAN polymer to CF. The first stage is to spin fiber. In the case of mesophase pitch, this involves hot melt spinning, whereas PAN is solution spun from a solvent, such as dimethylacetamide or dimethylsulfoxide. The fiber is then stabilized by heating in an oxidizing atmosphere, before being exposed to higher temperatures in an inert atmosphere to effect carbonization and graphitization. In general, the tension is much lower for mesophase pitch fibers than traditionally used for PAN fibers to accommodate the low strength of the as-spun pitch fibers. The final stage involves surface treatment to protect the surface and prime it to promote adhesion with a matrix polymer.

The stabilized fiber must be further heat-treated to develop the final mechanical properties. This is done in an inert atmosphere, typically in two stages called carbonization and graphitization. The fibers are still relatively weak; however, since the basic alignment is introduced at the spinning step, pitch-based CF can undergo carbonization and graphitization with the fiber in the relaxed state, an advantage over PAN-based fiber.[60,66]

Carbonization is generally carried out at temperatures below 2000°C and may itself involve several stages. Typical carbonizing heat treatment in inert nitrogen may involve 0.5–5 minutes at 700°C followed by 0.5 minutes at 900°C and then final heating in the range from 1500°C to 2000°C.[67] Generally, this diameter reduction is less in pitch-based fibers than in PAN-based fibers because of the higher carbon content of pitch compared to PAN. Partly because of this, but mainly because of the difficulty of spinning fine pitch-based fibers, the pitch-based fibers generally have a greater diameter. If carbonization is carried out too rapidly, then the rapid evolution of this volatiles can result in the disruption of the structure and the creation of defects leading to a loss of strength in the final CF.[68]

Graphitization involves heat treating for 10 seconds to 5 minutes at temperatures above 2000°C, often approaching 3000°C.[57,66,67] While some volatiles are still eluted from the fiber, the main changes are an increase in the axial alignment, size, and perfection of crystalline lamellae.

The exact dwell or soak times and temperatures of heat treatment, along with the choice of pitch and the spinning and stabilization conditions, can be varied to produce a range of grades of CF. Efforts have also been made to reduce the cost of production by reducing the time of heating. Greene et al. reported significant structural evolution and property development for a furnace residence time as brief as 0.7 seconds at 3000°C and used a simple energy consumption analysis to show that this process may have cost benefits.[69]

4.2.4 Carbon Fibers Made from Carbonaceous Gases

Iron is the most commonly used catalyst, though nickel, copper, palladium, and other metals and alloys can be used instead.[70–72] The reactivity between the catalyst and its support (e.g., MgO, SO_2, etc.) should be small, if any. For example, Pd supported on SO_2 suffers from a reaction that forms Pd_2Si and this leads to suppression of the catalytic activity.[72] Iron particles can be obtained from solutions of iron salts or iron organometallics; the iron particle size is of the order of 10 nm. The addition of acetylacetonates of Fe, Co, and Mn to Fe (C_5H_5), (ferrocene) reduces the size of the particles and consequently increases the filament growth rate and the filament yield.[73] In particular, the growth rate and yield are 40 μm/s and 70 wt.%, respectively, when 20 wt.% cobalt acetylacetonate is used along with 80 wt.% ferrocene.[74] Sulfurizing the iron, using thiophene or hydrogen sulfide as the sulfur source, helps penetration of catalyst into the fine pores of the activated carbon particles and allows growth of carbon filaments on to them.[75] This effect of sulfur is due to the molten state of the sulfurized iron.[75] A typical iron concentration is 5×10^{-6} g/cm^2, as only a small percentage of catalyst particles gives filaments.[76] When a catalyst particle becomes completely encased by the vapor deposited carbon, the catalyst is poisoned and the catalytic growth stops.

The mechanism of carbon filament formation is illustrated in Figure 4.4. The carbon-containing gas adsorbs and decomposes on the surface of the catalyst particle. The surface carbon then dissolves in the metal particle and diffuses through the particle from the hotter leading surface, on which hydrocarbon decomposition (an exothermic reaction) occurs, to the cooler rear face, at which carbon

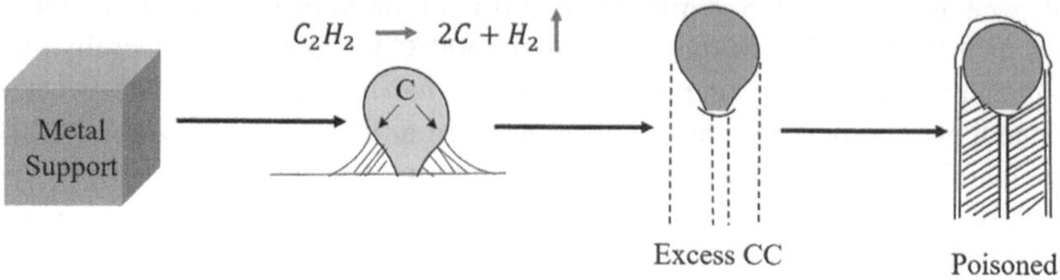

FIGURE 4.4 The most common mechanism of carbon filament formation.[78]

is precipitated from solution (an endothermic reaction). The carbon precipitate constitutes the growing filament. Thus, the catalyst particle is lifted up from its support as the filament grows and it remains at the top of the growing filament, leaving behind a hollow tube (20–500 Å, in diameter) behind it in the center of the filament along the filament axis. Excess carbon, which accumulates at the exposed particle, is transported by surface diffusion around the peripheral surfaces of the particle to form the graphitic skin of the filament. Catalytic filament growth ceases when the leading face is encapsulated by a layer of carbon, which prevents further hydrocarbon decomposition.[77]

The diffusion of carbon through a catalytic particle is the rate-limiting step of carbon filament growth. The observation indicates that the filament growth rate has an inverse square root dependence on the particle size.[79] Thus, too large a particle size will choke off the supply of carbon for the growing precipitate.[80]

When the enveloping carbon has poisoned the catalyst, the catalytic activity can be restored by replacing the hydrocarbon with hydrogen or oxygen and heating for a short time at 700°C. During hydrogasification (Figure 4.5), carbon diffuses in the direction opposite to that during filament growth and is subsequently converted to methane by reacting with adsorbed hydrogen at the front face of the particle. Thus, filament growth and hydrogasification are reversible processes.[81]

When the catalyst particle interacts strongly with its support, it remains on the support as the carbon filament grows above it, as illustrated in Figure 4.8 for the case of Pt/Fe particles. In Figure 4.8, a Pt/Fe particle is modeled as an Fe particle surrounded laterally by Pt, and acetylene (GHZ) is hypothesized to decompose on the Pt apron so that carbon diffuses to the Pt/Fe interface. After this point,

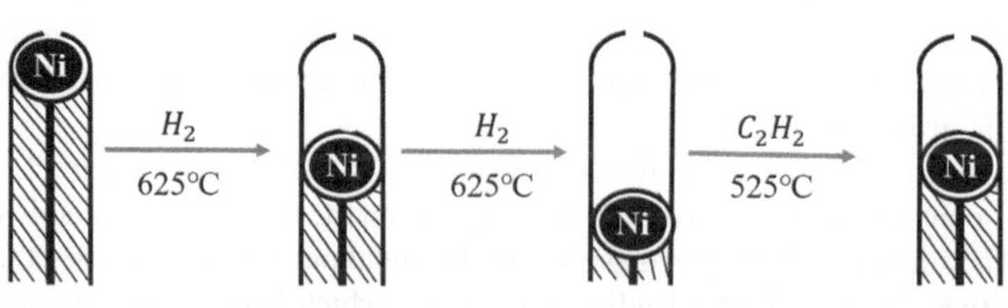

FIGURE 4.5 Reversible gasification/growth of carbon filaments from nickel particles.[78]

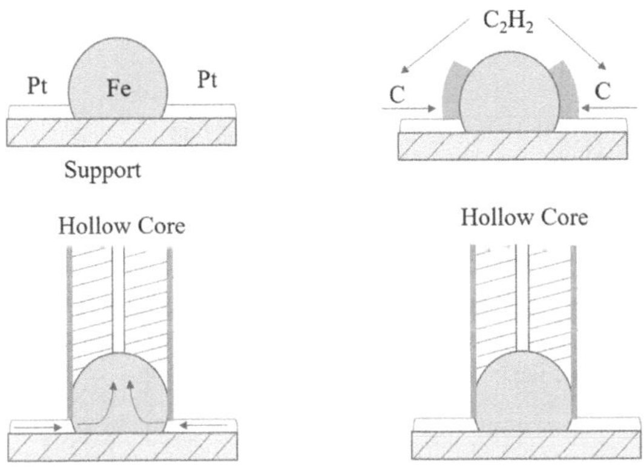

FIGURE 4.6 Mechanism of carbon filament formation by extruded filament growth.[78]

the mechanism is the same as that in Figure 4.8. As the filament appears to be extruded from the catalyst particle, still attached to the support, this mode of growth is known as extruded filament growth; it is not as common as the growth depicted in Figure 4.9.[82]

The growth modes of Figures 4.5 and 4.6 both result in a carbon filament with a duplex structure; this consists of an internal and more reactive (less graphitic) core surrounded by a relatively oxidation-resistant (more graphitic) skin. As a result of this duplex structure, fibers in the form of hollow tubes can be obtained by gasifying the core.[83]

Another way to obtain filaments in the form of hollow tubes involves using an additive (usually an oxide) on the catalyst particle to suppress the filament growth. Not only are the resulting filaments shorter, they tend to be hollow because less material diffuses through the catalyst particle to form the inner core of the filaments.[80]

The growth modes result in whisker like filaments that grow in a single direction. But growth can occur in more than one direction to produce branched, bidirectional, and multidirectional filaments, as illustrated in Figure 4.7, where catalyst particles are indicated by solid figures. Branching does not necessarily require break up of a catalyst particle.[80]

Carbon filaments are generally tubular in shape, but braided filaments and flat ribbons have been reported (Figure 4.8).[81] In particular, coiled fibers with a thickness of 0.1–0.3 μm, a coil diameter of 2–8 μm, and a coil length of 0.1–5 mm have been obtained. They are formed by the intertwining of a pair of coils growing in the same direction simultaneously from a diamond-shaped Ni seed. They can be extended elastically up to about three times the original coil length.[84] On the other hand, for the growth of straight filaments, it helps to preheat the carrier gas before introducing it into the reactor, so as to suppress thermal convection.[85]

The VGCF can be heat-treated in an inert atmosphere up to 3000°C for graphitization. (Carbonization is not necessary.) As the catalyst particles are completely encased in carbon at the conclusion of vapor deposition (thickening), they are not

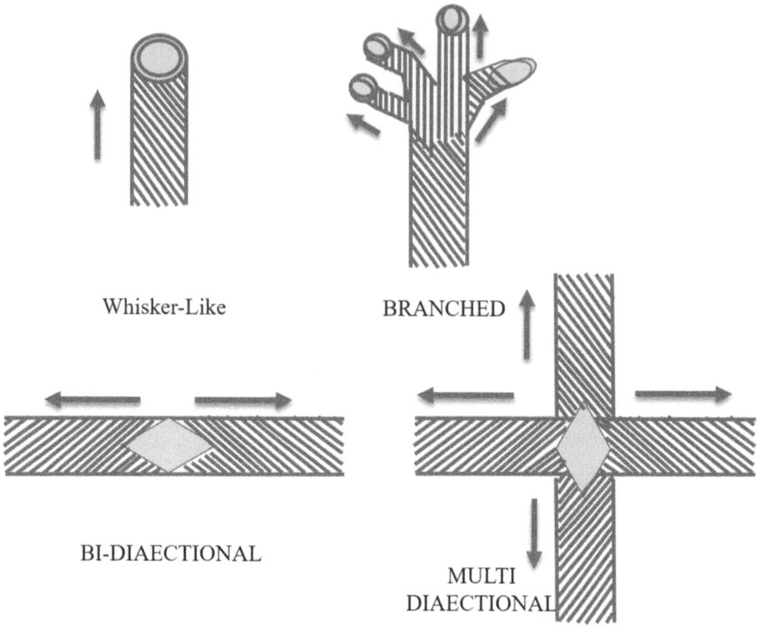

FIGURE 4.7 Schematic representation of different types of growth observed in carbon filaments.[78]

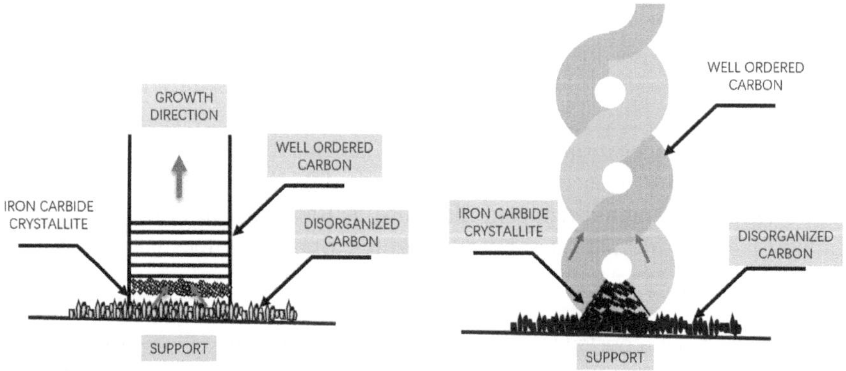

FIGURE 4.8 Schematics of ribbon and braided carbon filament morphologies.[78]

capable of catalytic effects during high-temperature heat treatments. Since the vapor deposited carbon has no metallic impurities and has a uniform cylindrical atomic layer microstructure, VGCF is highly graphitizable.[86] During graphitization, the growth of crystallites in the direction of the fiber axis lags behind the growth of crystallites perpendicular to the fiber axis, as displayed in Figure 4.10.[87]

Instead of a carbonaceous gas, the carbon source may be a plasma generated from an electrical arc produced from a carbon anode. As long as there is a source of carbon, the catalytic growth of carbon filaments is possible.[88]

CFs can be formed from carbonaceous gases without using a catalyst. For example, CFs in the form of tubules (their ends covered by carbon layers instead of the catalyst particle at the tip of a catalytically grown filament) can be grown from carbonaceous gases along with the formation of buckminsterfullerenes. The presence of carbon pentagons, in addition to carbon hexagons, at the end of a tubule allows it to be covered by carbon layers. Carbon tubules are typically smaller in diameter than catalytically grown carbon filaments. The number of carbon layers is around eight in a carbon tubule.[89] As another example, CFs are grown from a carbonaceous gas that causes the deposition of carbon at the tips of growing fibers electrostatically separated from one another.[90] CFs can also grow at the ends of corona wires during negative point-to-plane corona discharges in hydrocarbon atmospheres.[91]

4.2.5 Surface Treatment of Carbon Fibers

Surface treatment of CFs is performed for improving the adhesion between the fibers and the matrix when the fibers are used in a composite. The preferred type of treatment depends on the matrix material, which can be a polymer (a thermoset or a thermoplast), a metal, a carbon, or a ceramic. The details for each matrix are described in the chapter of that title. In general, there are three main methods of surface treatment, namely wet oxidation (e.g., HNO_3, 110°C, 10 minutes to 150 hours), dry oxidation (e.g., air, 02, 500°C–800°C, 30 seconds to 2 hours) and anodic oxidation (e.g., H_2SO_4, K_2SO_4, NaOH, 1–10 minutes).[92] In anodic oxidation, the anodic potential must be below that for the active oxidation reaction, which causes serious damage to the fibers (the anode). Figure 4.11 is a schematic of a typical setup for continuous anodic oxidation of CFs (Figure 4.9).[92–96]

4.2.6 Carbon Fiber Fabrics, Preforms, and Staple Yarns

The brittleness of CFs makes it necessary to coat them with a sizing before weaving. The choice of the sizing material depends on the matrix in which the fibers will be used, as the sizing also serves to enhance the bonding between the fibers and the matrix. The application of a sizing is carried out as the last step in the CF fabrication, i.e., before the fibers are wound on a spool.

Woven fabrics provide the most convenient way to handle continuous fibers. The impregnation of a fabric with a resin is simply performed by immersing the fabric in a resin bath. In contrast, unidirectional fiber prepregs are prepared by immersing a continuous tow of fibers in a resin bath and subsequently winding the tow on to a rotating mandrel. The equipment for preparing unidirectional fiber prepregs is much more complicated than that for preparing fabric prepregs. By using a fabric having a much larger number of yarns (twisted tows) in one direction than in the perpendicular direction, a fabric can be almost unidirectional.

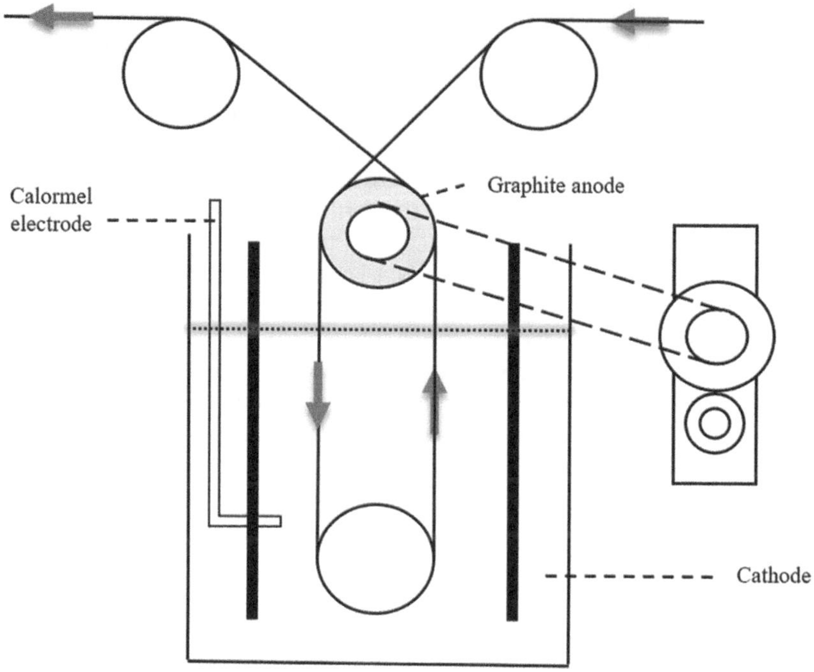

FIGURE 4.9 Schematic of a typical arrangement for continuous anodic oxidation of carbon fiber.[78]

Woven fabrics are defined as interlacing structures with at least two sets of orthogonal yam systems, i.e., warp and fill. Fabrics are usually two-dimensional. If yarns traversing from one fabric plane into at least one other fabric plane exist (by means of structural stitching), the fabric is 3D and is known as a fabric preform.[97,98]

Braided fabrics are defined as interlacing structures that consist of at least two sets of yams, not necessarily orthogonal. For simple two-dimensional biaxial braids, the yarn orientations are generally $\pm\phi$, where ϕ can vary from approximately 15 to 89°. In addition to the angular oriented yarns, yarns oriented along the axial or machine direction of the braid (i.e., $\phi=0$) can be added. These braids are referred to as two-dimensional triaxial braids. If more than one layer of fabric is braided at a time and the layers are interconnected so that yarns from one layer traverse to no less than at least one contiguous layer, the braid is referred to as a 3D braid.[97] Braiding is particularly suitable for fabricating composite tubings from braided seamless sleeves.

The attraction of 3D weaving is the increase in the interlaminar shear strength of the resulting composite, although this increase is accompanied by a decrease in the in-plane tensile modulus and strength.

Fabric preforms provide a way to fabricate composite materials of near-net shapes. Furthermore, they can be used to build structures with a reduced amount of external fasteners. Fastenless preform shapes include Is, Ts, and Js. A reduced amount of fasteners is attractive because the interface between a fastener and a composite is the "weak link" in a structure.[97]

The fabrics and preforms mentioned above are anisotropic. Three-dimensionally isotropic preforms can be prepared by the natural intertwining (self-weaving) of carbon filaments as they grow by catalytic decomposition of a carbonaceous gas. By using a shaped mold for the carbon filament growth, preforms of near-net shapes can be fabricated.[98]

Three-dimensionally or two-dimensionally isotropic preforms can be prepared from short CFs. A binder is used to make the short CF preform. The binder can be carbon (from a pitch or a resin), silica, phosphate, epoxy, etc. The fabrication of a preform involves (i) filtering or pressing a slurry consisting of short CFs, a binder, and a vehicle, (ii) drying, and (iii) heat treatment (in some cases). In the case of a pitch binder, the pitch can be dissolved in toluene or, more safely, methylene chloride, and combined with CFs to form a slurry. Water is the usual vehicle for silica or phosphate binders. Preforms prepared by dry pressing of short CFs suffer from nonuniformity in the fiber distribution. The slurry method is particularly important for preforms with CFs in amounts <25 vol.%.

A staple yarn consists of a series of short fibers that are placed to form a continuous overlapping assembly; the assembly is bound together by twist, which prevents fiber slippage by increasing the interfiber friction. The mean staple length of the fibers is about 100mm (about 4in.) in the Heltra process, which produces staple yarns from continuous CF tows. This length is well above the critical length associated with composite materials. In spite of the discontinuity and twist in the fibers, there is a high translation of tensile properties from the continuous fiber tow to the staple yarn; up to 97% of the tensile modulus and 84% of the tensile strength are retained. Due to the hairiness of a yarn, the use of a yarn in a composite significantly increases the interlaminar shear strength over the continuous fiber counterpart.[99]

4.3 STRUCTURES AND PROPERTIES OF CARBON FIBERS

4.3.1 Structure of Carbon Fibers

The mechanical properties of CF are closely related to its microstructure. From the perspective of composite micromechanics, the tensile strength of CF mostly depends on the meso defects in the material. When the structure is subjected to the external load, these meso defects are easy to cause stress concentration, resulting in the generation and propagation of meso cracks. When the crack propagation exceeds the critical value, the fiber will fracture. Therefore, the tensile modulus of fiber mostly depends on the crystal structure and crystal distribution of the material itself.

After decades of continuous research by researchers, it is now recognized that the microstructure of CF is a polycrystalline disordered graphite structure.[100,101] Inside the fiber, there are not only crystalline graphite crystals, but also amorphous carbon and micropores produced in the production process. Crawford and Johnson proposed a 3D carbon structure model with microcrystalline correlation in 1971 to describe the graphite structure inside the CF, as shown in Figure 4.12.[57]

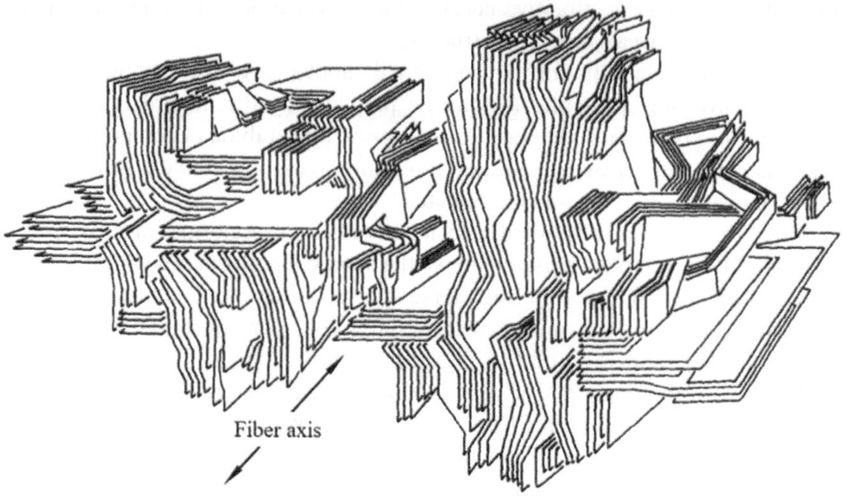

FIGURE 4.10 3D graphite structure in carbon fiber.[57]

Due to the disordered molecular configuration of the precursor PAN, and some amorphous carbon and pores remain in the fiber after fiber molding, the graphite crystal in PAN-based CF is disordered and the overall graphitization degree is low, as shown in Figure 4.11.[57]

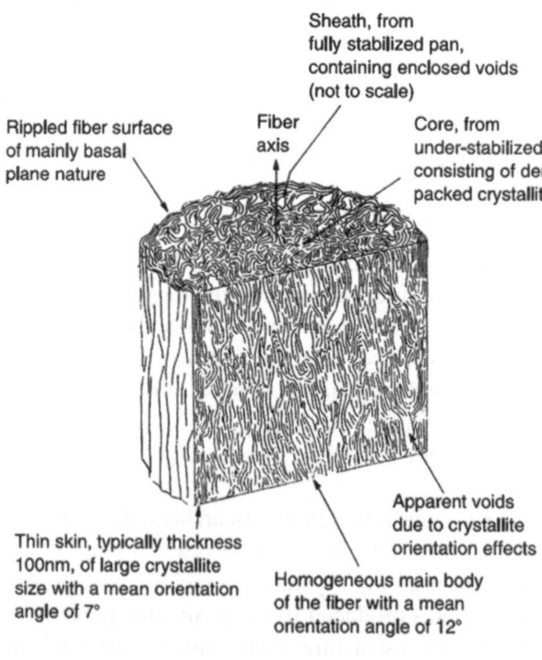

FIGURE 4.11 Microstructure of PAN-based carbon fiber.[57]

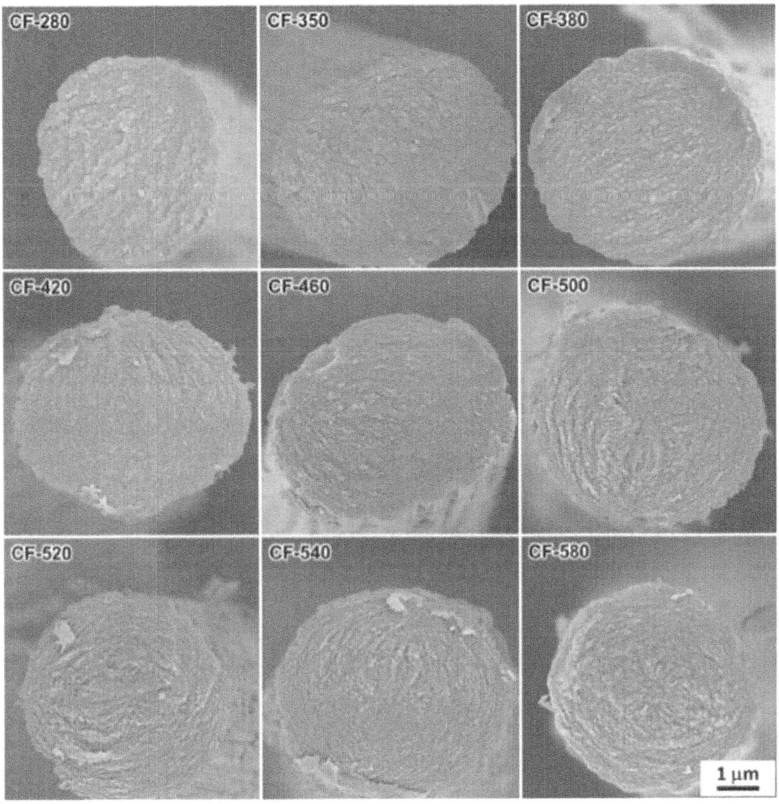

FIGURE 4.12 Cross section of PAN-based carbon fiber.[57]

This means that although the graphite microcrystalline itself has crystal anisotropy, it often presents an isotropic structure macroscopically on the cross section of PAN-based CF, as shown in Figure 4.12.[57]

Compared with PAN-based CFs, the graphite crystals of mesophase pitch-based CFs are more regular and orderly, and the overall graphitization degree is higher. This is because in the process of preparing CF, the anisotropic asphalt mesophase is first formed in the thermal condensation polymerization of asphalt. During the subsequent carbonization and graphitization of the fiber, these asphalt mesophase are easy to form a regular graphite structure, as shown in Figure 4.13.[102] The higher the mesophase pitch content of the precursor, the higher the final graphitization degree of the fiber (the graphitization degree of high-quality mesophase pitch-based CF can reach 98%). The highly graphitized mesophase pitch-based CF shows anisotropy in the section, resulting in different morphology in the section, as shown in Figure 4.14.[102] Large graphite crystals also make it easier for gases to escape from the fibers during carbonization and graphitization. These different morphologies may depend on the spinning temperature,

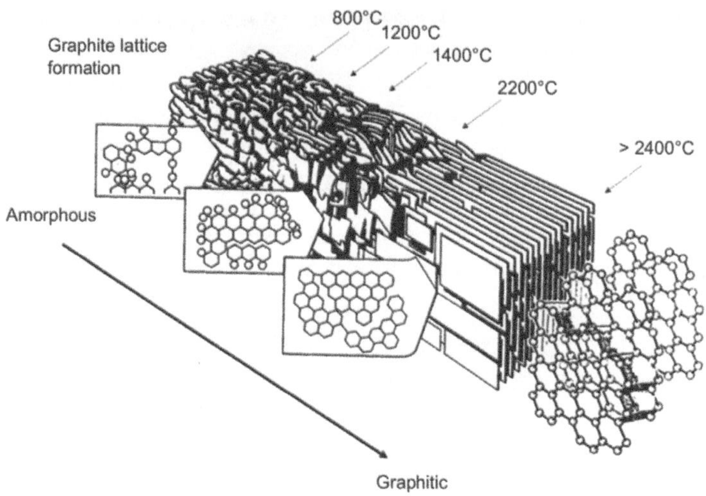

FIGURE 4.13 Graphitization of mesophase pitch.[102]

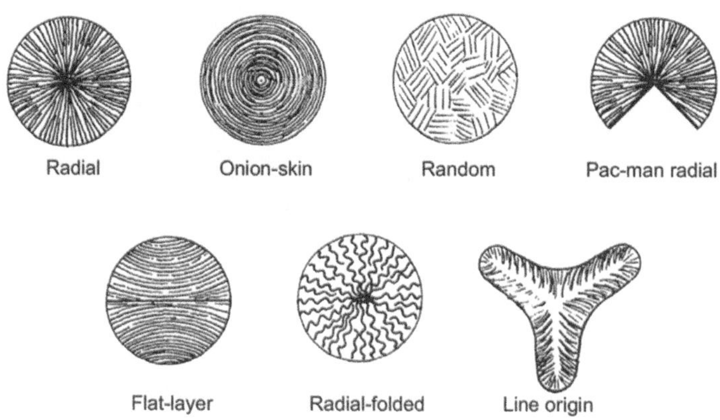

FIGURE 4.14 Cross section of mesophase pitch-based carbon fiber.[103]

spinneret shape, and spinning rate, but these morphologies have little effect on CF's mechanical and thermal properties. The dominant factor is the microcrystalline structure in CF.

Higher graphitization degree and more regular graphite crystal make the tensile modulus of mesophase pitch-based CF easily exceed that of PAN-based CF, but it also means that mesophase pitch-based CF is more brittle and microcosmic cracks are easier to expand, resulting in fiber fracture. Therefore, PAN-based CF's tensile strength is still much higher than that of mesophase pitch-based CF. However, it is worth noting that with the increase of graphitization degree of mesophase pitch-based CF (i.e., the greater brittleness), the strength of the fiber does not decrease but increases very slowly.

The different structure of CF leads to the difference in CF performance. In general, the advantages of CFs include the following: low density; high tensile modulus and strength; low thermal expansion coefficient; thermal stability in the absence of oxygen to over 3000°C; excellent creep resistance; chemical stability, particularly in strong acids; biocompatibility; high thermal conductivity; low electrical resistivity; availability in a continuous form; and decreasing cost (versus time). Conversely, disadvantages of CFs include the following: anisotropy (in the axial versus transverse directions); low strain to failure; compressive strength is low compared to tensile strength; tendency to be oxidized and become a gas (e.g., CO) upon heating in air above about 400°C; oxidation of CFs is catalyzed by an alkaline environment.

As the structure determines each property, the different properties are interrelated. For example, the following trends usually go together: increase in the tensile modulus; decrease in the strain to failure; decrease in the compressive strength; increase in the shear modulus; increase in the degree of anisotropy; decrease in the electrical resistivity; increase in the thermal conductivity; decrease in the coefficient of thermal expansion; increase in density; increase in thermal stability (oxidation resistance); increase in chemical stability; and increase in cost.

4.3.2 Mechanical Properties

Mechanical Properties. Table 4.5 shows the tensile properties of CFs along the fiber axis compared to those of a graphite single crystal along the a-axis, i.e., parallel to the carbon layers.[104] Although the carbon layers in a CF exhibit a strongly preferred orientation parallel to the fiber axis, the layers' alignment is far from perfect, and the crystallite size is finite. Therefore, CFs' tensile modulus and strength are considerably below those of a single graphite crystal. The modulus of HM-type fibers approaches that of a single graphite crystal, but that of HT-type fibers is much below that of a single graphite crystal. The tensile strengths of both HM and HT fibers are very much below that of a single graphite crystal, although the strength of HT is higher than that of HM. There is thus much room for improvement of the tensile strength of CFs. In contrast, there is not much room to improve the tensile modulus.

The tensile properties of some commercial CFs of the HP grade are shown in Table 4.6.[106] For the same precursor material (PAN or mesophase pitch), the tensile strength, modulus, and strain to failure vary over large ranges.

The tensile modulus is governed by the preferred orientation of the carbon layers along the fiber axis, so it increases with decreasing interlayer spacing (d_{002}) and with increasing L_c, and L_a, as shown in Table 4.7.[108] for a series of mesophase pitch-based CFs produced by du Pont.

Comparison of the du Pont fibers (Table 4.7) with the Amoco fibers, both of which are based on mesophase pitch, indicates the superior tensile strength of the du Pont fibers. Unfortunately, the du Pont fibers are not commercially available, whereas the Amoco fibers are.

TABLE 4.5
Considerations Concerning Young's Modulus (E) and the Tensile Strength (σ) of Carbon Fibers

	Theoretical Values for Graphite Single Crystal	CFs		Future Trends
		HT Type	HM Type	
Young's modulus €	$E = 1000\,\text{GPa}$	$E = 250\,\text{GPa}$	$E = 700\,\text{GPa}$	Further increase not necessary
Tensile strength (σ)	$\sigma_{\text{theor.}} = E/10 = 100\,\text{GPa}$	$\sigma_{theor.} = 25GPa$ $\sigma_{\text{exp.}} = 5\,\text{GPa}$ $= 20\%$ of $\sigma_{\text{theor.}}$	$\sigma_{\text{theor.}} = 70\text{GPa}$ $\sigma_{\text{exp.}} = 3\,\text{GPa} = 4\%$ of $\sigma_{\text{theor.}}$	Further improvement expected

Source: From Ref.[105]

TABLE 4.6
Tensile Modulus, Strength, and Strain to Failure of CFs

Manufacturer	Fiber	Modulus (GPa)	Strength (GPa)	Strain to Failure (%)
		PAN-Based, High Modulus (Low Strain to Failure)		
Celanese	Celion GY-70	517	1.86	0.4
Hercules	HM-S Magnamite	345	2.21	0.6
Hysol Grafil	Grafil HM	370	2.75	0.7
Toray	M50	500	2.50	0.5
		PAN-Based, Intermediate Modulus (Intermediate Strain to Failure)		
Celanese	Celion 1000	234	3.24	1.4
Hercules	IM-6	276	4.40	1.4
Hysol Grafil	Ap0110 IM 43-600	300	4.0	1.3
Toho Beslon	Sta-grade Besfight	240	3.73	1.6
Amoco	Thornel 300	230	3.10	1.3
		PAN-Based, High Strain to Failure		
Celanese	Celion ST	235	4.34	1.8
Hercules	AS-6	241	4.14	1.7
Hysol Grafil	Apollo HS 38-750	260	5.00	1.9
Toray	T 800	300	5.70	1.9
		Mesophase Pitch-Based		
Amoco	Thornel P-25	140	1.40	1.4
	P-55	380	2.10	0.5
	P-75	500	2.0	0.4
	P-100	690	2.20	0.3
	P-120	820	2.20	0.2

Source: From Ref.[107]

TABLE 4.7
Mechanical Properties of Pitch-Based CFs and Their Structural Parameters as Determined by X-Ray Diffraction: d_{002}, the Interlayer Spacing; L_c, the Out-of-Plane Crystallite Size; and L_a, the In-Plane Crystallite Size Parallel to the Fiber Axis

Fibers	Tensile Modulus (GPa)	Tensile Strength (GPa)	d_{002}(nm)	L_c(nm)	L_a(nm)
E-35	241	2.8	0.3464	3.2	7.2
E-55	378	3.2	0.3430	8.2	16.2
E-75	516	3.1	0.3421	10.7	22.4
E-105	724	3.3	0.3411	17.3	46.1
E-120	827	3.4	0.3409	18.9	51.4
E-130	894	3.9	0.3380	24.0	180.4

Source: From Ref.[109]

For a high-modulus CF (e.g., HM70), the stress–strain curve is a straight line up to failure; as the modulus decreases there is an increasing tendency for the slope to increase with increasing strain. This effect occurs because the fiber is increasingly stretched as the strain increases; the carbon layers become more aligned and the modulus therefore increases. It forms the basis of a process called stress graphitization.

Flaws strongly influence the tensile strength, so it increases with decreasing test (gage) length and with decreasing fiber diameter. There are two types of flaws, namely surface flaws and internal flaws. The surface flaws control the strength of CFs that have not been heat-treated above 1000°C–1200°C; the internal flaws control the strength of CFs that have been heat-treated above 1000°C–1200°C.[110] Upon etching a fiber, the amount of surface flaws is decreased, causing the fiber strength to increase. The minimum practical gage length is 0.5 mm,[111] even though the ultimate fragment length of a stressed single fiber composite is 0.3 mm[110] and it is the ultimate fragment length (also called the critical length) that determines the composite strength. Table 4.9[111] shows the tensile strengths of CFs (Hercules AS-4, PAN based) at different gage lengths, as determined by traditional tensile testing and by in situ fiber strength testing. The latter testing method involves embedding a single fiber in a matrix (e.g., epoxy) and pulling the unembedded ends of the fiber to increasing strain levels up to approximately three times greater than the failure strain of the fiber. While the strain is gradually increased, the number of breaks in the fiber is counted in situ.[111] The fiber eventually breaks into fragments of a length equal to the critical length, which is related to the tensile properties of the fiber and the interfacial shear strength between the fiber and the matrix. Table 4.8 shows that the fiber strength determined by either method increases with decreasing gage length. The latter method is applicable to very

TABLE 4.8
Tensile Strengths of AS-4 Fibers at Different Gage Lengths as Determined by Traditional Tension Testing and by In Situ Fiber Strength Testing in Epoxy and Solvent-Deposited Polycarbonate Matrices

		In Situ Strength (MPa)	
Gage Length (mm)	Conventional Tension Test (MPa)	Epoxy	Polycarbonate
25.4	3215±966	3188±704	2698±518
8.0		3850±738	3347±725
4.0		4264±787	3733±752
2.0		4720±856	4175±773
1.0	5285±1731	5223±911	4582±814
0.55	5644±994	5693±945	4996±869
0.3		6189±973	5437±883

Source: From Ref.[78]
Tensile strengths appear in MPa followed by standard deviations.

short gage lengths, but it has the disadvantage of being sensitive to the fiber prestrain resulting from the specimen preparation technique. The difference between the in situ fiber strengths for epoxy and polycarbonate matrices (Table 4.8) is due to a difference in the fiber prestrain.

The compressive strength is much lower than the tensile strength, as shown in Table 4.9.[112] The compressive strength ratio to the tensile strength is smaller for graphitized fibers than for carbonized fibers. Pitch-based CFs have even lower compressive strength than PAN-based fibers. Moreover, the compressive failure strain is much lower for pitch-based CFs than PAN-based CFs, as shown in Table 4.10.[113] These differences between pitch-based and PAN-based fibers are

TABLE 4.9
Tensile and Compressive Strength of CFs

	Pitch-Based		PAN-Based	
Carbon Fiber	Carbonized Fiber HTX	Graphitized Fiber HTX	Carbonized Fiber T-300	Graphitized Fiber M40
Tensile strength $(\sigma_f)_{ten.}$ (GPa)	3.34	4.33	3.50	2.88
Estimated compressive strength $(\sigma_f)_{comp.}$ (GPa)	1.25	0.54	2.06	0.78
$(\sigma_f)_{comp.}/(\sigma_f)_{ten.}$ (%)	37.4	12.5	58.9	27.1

Source: From Ref.[19]

TABLE 4.10
Compressive Failure Strains for Pitch-Based Fibers and PAN-Based CFs

Precursor	Fiber	Modulus (GPa)	Mean Failure Strain (%)	Standard Deviation (%)	No. of Specimens
Pitch	HM	519	0.346	0.074	39
	HT	244	0.981	0.064	9
	UHM	662	0.163	0.063	9
	P-75	517	0.248	0.04	8
PAN	AS-1	228	2.59	0.181	8
	AS-4	241	2.27	0.256	8
	IM-6	276	2.35	0.132	8
	IM-7	303	2.11	0.317	8
	T-700	234	2.66	0.111	8
	T-300	230	2.36	0.125	8
	GY-30	241	2.65	0.207	8

Source: From Ref.[115]

TABLE 4.11
Properties of Some PAN-Based CFs

Fiber	Tensile Modulus (GPa)	Tensile Strength (GPa)	Shear Modulus (GPa)
T-300	230	3.5	17.0
M30	290	3.9	17.0
T-50	390	2.4	
M-40	400	2.7	15.8
M-46	450	2.35	14.8
GY-70	520	1.8	
T-800H	190	5.6	18.1
M-40J	390	4.3	17.0
M-46J	450	4.2	17.0
M-60J	585	3.8	

Source: From Ref.[108]

consistent with the difference in the compressive failure mechanism. Pitch-based fibers of HM typically deform by a shear mechanism, with kink bands formed on a fracture surface at 45″ to the fiber axis. In contrast, PAN-based fibers typically buckle on compression and form kink bands at the innermost part of the fracture surface, which is normal to the fiber axis.[114] The difference in compressive behavior between pitch-based and PAN-based CFs is attributed to the strong preferred orientation of the carbon layers in pitch-based fibers and the more random microstructure in PAN-based fibers (Table 4.11). The oriented layer microstructure causes the fiber to be susceptible to shearing.[113] Thus, the compressive strength decreases with increasing tensile modulus.

The shear modulus of CFs decreases with increasing L_c and with increasing L_a.[116] This is expected, since increases in L_c and L_a imply a greater degree of carbon layer preferred orientation. A decrease in the shear modulus is accompanied by a decrease in the compressive strength.

The values of the torsional modulus of various commercial CFs are listed in Table 4.12.[110] The torsional modulus is governed mostly by the cross-sectional microstructure. Mesophase pitch-based CFs have low torsional modulus because they have an appreciable radial cross-sectional microstructure, which facilitates interlayer shear. Hence, the torsional modulus of mesophase pitch-based CFs is even lower than that of isotropic pitch-based CFs. On the other hand, PAN-based CFs have high torsional modulus because they have an appreciable degree of circumferential microstructure.[110]

Electrical Properties. The electrical resistivity of the mesophase pitch-based CFs of Table 4.14 is shown as a function of temperature from 2 to 300 K.[108] The resistivity decreases with increasing temperature for each type of fiber. This is because the carrier density increases with temperature, just as for carbons and graphites in general. At a given test temperature, the resistivity decreases with

TABLE 4.12
Torsional Modulus and Young's Modulus of Various Carbon Fiber Types

Carbon Fiber Type	Young's Modulus (GPa)	Torsional Modulus (GPa)
	Pitch Mesophase	
PM-A	184	13.5
PM-B	262	9.0
		9.2
PM-C	364	10.8
		9.7
PM-C (1700)	400	10.4
	Isotropic Pitch	
KCF-200	80	16.4
KCF-2700	~80	18.0
	PAN	
T-400	226	21.4
AS	215	21.0
HM-S(H)	~370	20.2
Modmor I	~400	28.2
HM-S(C)	~380	35.3
	Rayon	
T-11 (carbon)	79	15.3
T-12 (graphite)	83	22.8
T-25	176	12.3
T-70	390	12.3
T-75	540	13.8
T-100	680	16.9

Source: From Ref.[17]

increasing tensile modulus. This is because an increase in the tensile modulus is accompanied by a decrease in the concentration of defects, and defects cause carrier scattering.

An effective way to decrease the resistivity of CFs by a factor of up to 10 is intercalation. Intercalation is the formation of layered compounds in which foreign atoms (called the intercalate) are inserted between the carbon layers. The intercalate acts as an electron acceptor or an electron donor, thus doping the CFs. This doping causes the carrier concentration to increase, thereby decreasing the electrical resistivity. Intercalation is only possible in relatively graphitic CFs. For example, bromination (i.e., intercalation with bromine, an acceptor) causes a weight uptake of 18%–20% for Amoco's Thornel P-100 and P-75 fibers, but 0% for P-55; it causes a resistivity decrease of 73%–79% for P-120, P-100, and

P-75 fibers, but just 4% for P-55.[117] For the case of brominated P-100-4 (P-100-4 is even more graphitic than P-loo), a resistivity of 11.0pfI.cm has been reported.[6] On the other hand, the severity of the intercalation reaction in highly graphitic fibers can cause physical damage to the fibers, so that the mechanical properties and oxidation resistance are degraded.[118] Therefore, there is an optimum degree of graphitization, which corresponds to that of P-100 fibers for the case of bromine as the intercalate.[117] The low electrical resistivity of intercalated CFs makes these fibers useful in composites for electromagnetic interference shielding.

A way to decrease the electrical resistivity and increase the thermal conductivity of CFs is to coat the fibers with a metal that is more conductive than the fibers. All types of CFs are higher in electrical resistivity than metals, therefore all metal-coated CFs are more electrically conductive than the corresponding bare CFs. However, the thermal conductivities of the highly graphitic CFs, such as pitch-based Thornel P-100, P-120, and K1100X fibers of Amoco, are even higher than copper. Thus, metal-coated CFs are superior to the corresponding bare fibers in thermal conductivity only for the less graphitic fibers, which constitute the vast majority of CFs used in practice anyway.

Table 4.13[119] shows the effect of a 0.35 μm thick electrodeposited nickel coating on the properties of an originally 7.0 μm thick PAN-based AS-4 CF (Hercules), which is not graphitic. The nickel coating causes the density to increase, the electrical resistivity to decrease, the thermal conductivity to increase, the thermal expansion coefficient to increase, and the tensile modulus, strength, and elongation to decrease. The largest effect is the decrease in the electrical resistivity.

Thermal Conductivity. The longitudinal thermal conductivity, thermal expansion coefficient, density, and specific thermal conductivity (conductivity/density) of Amoco's mesophase pitch-based CFs are shown in Table 4.14.[121] The thermal conductivities of P-100, P-120, and KllOOX fibers are all higher than that of copper, while the thermal expansion coefficients and densities are much lower than

TABLE 4.13
Effect of Nickel Coating on the Properties of PAN-Based AS-4 (Hercules) Carbon Fiber

Property	Bare Fiber	Ni-Coated Fiber
Diameter (μm)	7.0	7.8
Density (g/cm^3)	1.80	2.97
Electrical resistivity (10^{-6} Ω cm)	1530	7
Thermal conductivity (W/m/K)	7.2	10.7
Thermal expansion coefficient (10^{-6}/°C)	−1.7	−0.8
Tensile modulus (GPa)	234	210
Tensile strength (MPa)	3582	2582
Tensile elongation (%)	1.53	1.33

Source: From Ref.[120]

TABLE 4.14
Thermal Properties of the Most Advanced Pitch-Based CFs

Material	Longitudinal Thermal Conductivity (W/m/K)	CTE (10^{-6}/K)	Density (g/cm³)	Specific Conductivity (W cm³/m/K/g)
Al 6063	218	23	2.7	81
Copper	400	17	8.9	45
P-100	520	−1.6	2.2	236
P-120	640	−1.6	2.1	305
3.5K1100X	1100	−1.6	2.2	500
K1100X/Al (55 vol.%)	634	0.5	2.5	236
K1100X/epoxy (60 vol.%)	627	−1.4	1.8	344
K1100X/Cu (46 vol.%)	709	1.1	5.9	117
K1100X/C (55 vol.%)	696	−1.0	1.8	387

Source: From Ref.[112]

those of copper. Thus, the specific thermal conductivity is exceptionally high for these CFs. In general, the thermal expansion coefficient of CFs decreases with increasing tensile modulus. Table 4.15 also shows that use of KllOOX fibers in Al, epoxy, Cu and C matrices results in composites of high thermal conductivity.

Oxidation Resistance. The oxidation resistance of CFs increases with the degree of graphitization. In the case of Kureha isotropic pitch-based CFs, weight loss starts at about 400°C and is gradual up to 500°C; from 500°C to 600°C, the weight loss is sharp; complete weight loss takes place at 620°C. In the case of Kureha isotropic pitch-based CFs graphitized at 2700°C, weight loss starts at about 400°C but is gradual up to 700°C; beyond 700°C, the weight loss is sharp; complete weight loss takes place by 850°C. In the case of Amoco mesophase pitch-based CFs, weight loss starts at about 500°C and is only about 2% up to 600°C; complete weight loss takes place at about 880°C. The apparent activation energies range from 112 to 205 kJ/mol for various pitch-based CFs.[122] The oxidation has been modeled to yield information on the oxidation kinetics, the activation energy, and the rate-determining step.[123]

Severe oxidation causes CFs to lose weight due to the evolution of CO or CO_2 gases. However, slight oxidation may cause CFs to gain weight slightly due to the formation of chemical bonds to various oxygen-containing functional groups on the surface of the fibers. The oxygen-containing groups (or adsorbed oxygen) increase the polar component of the surface free energy and hence result in enhanced electrochemical response (relevant for fiber electrodes) and improved fiber-matrix bonding (when the fibers are used in a composite).

Although oxidation due to molecular oxygen is most common on earth, oxidation due to atomic oxygen is important in aerospace applications. Atomic oxygen is more reactive than molecular oxygen and is the dominant atmospheric element in low earth orbit.[124]

4.4 CARBON FIBER COMPOSITES

Fiber-reinforced composite materials represent a radical approach to designing structural materials when compared with traditional materials, such as metals and ceramics. Although the living world, both the structures of plants and animals, is based on the reinforcement of matrix materials by flexible filaments, it was not until the second half of the 20th century that fiber-reinforced composites began to be really developed. This was possible as organic resins had begun to be produced in the 1930s and in the same decade the first synthetic fibers were made. Composite materials, today, find uses in all areas and in some have become the dominant form of structural material used against which other materials are judged.[125] This is undoubtedly the case for the aerospace industry and will increasingly become the case for other industrial sectors as the 21st century progresses. Matter in the form of a filament can possess extraordinary mechanical properties of strength and stiffness. Composite materials, or at least most of them, exploit this fact. The reasons why fibers are so much stronger and often stiffer than the same material in bulk form is discussed in Chapter 2 but for the present let us just remember that a glass fiber can be hundreds of times stronger than the same glass in bulk form. This is because the fine fibers contain far fewer defects than does the bulk material so an assembly of fibers making up a given volume will be much stronger than the same volume of the same glass in bulk form. What is true for glass fibers is also true for other fiber-form materials, so fibers are really an extraordinary form of matter. If we reconsider the assembly of fibers mentioned above, it should be obvious that the strength we are considering is parallel to the fibers' axes. The strength will be negligible if we pull the bundle at right angles to the fibers. The fiber assembly is highly anisotropic. This would not be the case for the bulk material, which has the same characteristics in all directions and is therefore isotropic. This immediately illustrates the difference between composites and most traditional materials. Composites, at the level of layers of fibers, are inherently anisotropic, whereas most traditional structural materials are isotropic.

4.4.1 Polymer-Matrix Composites

Polymer-matrix composites are much easier to fabricate than metal-matrix, carbon-matrix, and ceramic-matrix composites, whether the polymer is a thermoset or a thermoplast.[126] This is because of the relatively low processing temperatures required for fabricating polymer-matrix composites. For thermosets, such as epoxy, phenolic, and furfuryl resin, the processing temperature typically ranges from room temperature to about 200°C; for thermoplasts, such as polyimide (PI), polyethersulfone (PES), polyetheretherketone (PEEK), polyetherimide (PEI), and polyphenyl sulfide (PPS), the processing temperature typically ranges from 300°C to 400°C. Thermosets (especially epoxy) have long been used as polymer matrices for CF composites. During curing, usually performed in the presence of heat and pressure, a thermoset resin hardens gradually due to the completion of polymerization and the crosslinking of the polymer molecules. Thermoplasts have recently become important because of their greater ductility and processing speed compared to thermosets, and the recent availability of thermoplasts that can withstand high temperatures. The higher processing speed of thermoplasts is due to the fact that thermoplasts soften immediately upon heating above the glass transition temperature (T) and the softened material can be shaped easily. Subsequent cooling completes the processing. In contrast, curing a thermoset resin is a reaction that occurs gradually.

Surface treatments of CFs are essential for improving the bonding between the fibers and the polymer matrix. They involve oxidation treatments and the use of coupling agents, wetting agents, and/or sizings (coatings). CFs need treatment both for thermosets and thermoplasts. As the processing temperature is usually higher for thermoplasts than thermosets, the treatment must be stable to a higher temperature (300°C–400°C) when a thermoplast is used.[127]

CF polymer-matrix composites have the following attractive properties:

1. low density (40% lower than aluminum),
2. HS (as strong as high-strength steels),
3. high stiffness (stiffer than titanium, yet much lower in density),
4. good fatigue resistance (a virtually unlimited life under fatigue loading),
5. good creep resistance,
6. low friction coefficient and good wear resistance (a 40 wt.% short CF nylon-matrix composite has a friction coefficient nearly as low as Teflon and unlubricated wear properties approaching those of lubricated steel),
7. toughness and damage tolerance (can be designed by using laminate orientation to be tougher and much more damage tolerant than metals),
8. chemical resistance (chemical resistance controlled by the polymer matrix),
9. corrosion resistance (impervious to corrosion),
10. dimensional stability (can be designed for zero coefficient of thermal expansion),
11. vibration damping ability (excellent structural damping when compared with metals),
12. low electrical resistivity,
13. high electromagnetic interference (EMI) shielding effectiveness, and
14. high thermal conductivity.

CF polymer-matrix composites can be classified according to whether the matrix is a thermoset or a thermoplast. Thermoset-matrix composites are by tradition far more common, but thermoplast-matrix composites are under rapid development. The advantages of thermoplast-matrix composites compared to thermoset-matrix composites include the following:

1. lower manufacturing cost,
2. no cure,
3. unlimited shelf-life,

4. reprocessing possible (for repair and recycling),
5. less health risks due to chemicals during processing,
6. low moisture content,
7. thermal shaping possible,
8. weldability (fusion bonding possible).
9. better performance:
10. high toughness (damage tolerance),
11. good hot/wet properties, and
12. high environmental tolerance.

The disadvantages of thermoplast-matrix composites include the following:

1. limitations in processing methods,
2. high processing temperatures,
3. high viscosities,
4. stiff and dry prepregs when a solvent is not used (Le., not drapeable or tacky), and
5. fiber surface treatments less developed.

CF polymer-matrix composites can be classified according to whether the fibers are short or continuous. Continuous fibers have much more effect than short fibers on the composite's mechanical properties, electrical resistivity, thermal conductivity, and other properties, too. However, they give rise to composites that are more anisotropic. Continuous fibers can be in unidirectionally aligned tape or woven fabric form.

CF polymer-matrix composites are predominantly used for the aerospace industry, but the decreasing price of CFs is widening the applications of these composites to include the automobile, marine, sports, biomedical, construction, and other industries.

One area of aerospace applications is space vehicles. The United States Space Shuttle uses CF epoxy-matrix composites for its payload bay door and remote manipulator arm[128]; its solid rocket motor cases also use epoxy-matrix composites; its booster tail and fins use polyimide-matrix composites. Satellite structures and solar panels also use CF polymer-matrix composites.[1] Most space applications utilize standard aerospace grade CFs (tensile strength 3550 MPa, tensile modulus 235 GPa) combined with a 177°C cure multifunctional epoxy resin matrix. Filament wound rocket motor cases employ a 121°C cure, modified bis-A-epoxy as the resin matrix. Stiffness requirements of some satellite applications dictate the use of high-modulus CFs (350 GPa).[129] Thermoplast matrices, such as PEEK and PES, are gaining attention for space applications.

A second area is military aircraft. Examples include Gripen, EFA, French Rafale, and U.S. B-2, which use the 177°C cure toughened thermoset-matrix resins along with intermediate-modulus (295 GPa), or high-strength (5590 MPa), intermediate-modulus CFs.[12] The U.S. Advanced Tactical Fighter is planned to use thermoplastics or toughened bismaleimide as a primary structural material. Older military aircraft are being modified with epoxy-matrix composite wings.

Helicopters are a third area, both for military and commercial use. For example, the all-composite MBB BK117 helicopter is two-thirds CF epoxy-matrix composite, one-third aramid fiber and glass fiber epoxy-matrix composite. Both 121°C and 177°C cure epoxy resins are used.

A fourth area is concerned with primary and secondary structures in commercial aircraft. Examples of primary structural applications include the Airbus A310/A320 vertical tail fin boxes (121°C cure toughened epoxy), the A320 horizontal tail fin (177°C cure epoxy) and the ATR 72 external wing box (177°C toughened epoxy).

A fifth area of aerospace applications is commercial aircraft engines. Outer and front sections of the engine are subjected to lower temperatures and can utilize an epoxy matrix. For example, the front fan ducts on Rolls-Royce engines and the blocker doors and transcowls on General Electric's CF6-8OC2 engines use 177°C cure epoxy. Engine rear section components operate at higher temperatures; this necessitates polyimide matrices such as PMR-15, which is used for thrust reversers and bypass ducts. Thermoplasts such as PEEK are being considered for engine applications.

4.4.2 Metal-Matrix Composites

CF metal-matrix composites are gaining importance because the CFs reduce the coefficient of thermal expansion, increase the strength and modulus, and decrease the density.[130] If a relatively graphitic kind of CF is used, the thermal conductivity can be also enhanced. Compared to the metal itself, a CF metal-matrix composite is characterized by a higher strength-to-density ratio (i.e., specific strength), a higher modulus-to-density ratio (i.e., specific modulus), better fatigue resistance, better high-temperature mechanical properties (a higher strength and a lower creep rate), a lower a, and better wear resistance. Compared to CF polymer-matrix composites, a CF.

Metal-matrix composite is characterized by higher temperature resistance, higher fire resistance, higher transverse strength and modulus, the lack of moisture absorption, a higher thermal conductivity, a lower electrical resistivity, better radiation resistance, and absence of outgassing. On the other hand, a metal-matrix composite has the following disadvantages compared to the metal itself and the corresponding polymer-matrix composite: higher fabrication cost and limited-service experience.

The fabrication of metal-matrix composites often involves the use of an intermediate, called a preform, in the form of sheets, wires, cylinders, or near-net shapes. The preform contains the reinforcing fibers usually held together by a binder, which can be a polymer (e.g., acrylic, styrene), a ceramic (e.g., silica, aluminum metaphosphate), or the matrix metal itself. For example, continuous fibers are wound around a drum and bound with a resin, and subsequently the wound fiber cylinder is cut the drum and stretched out to form a sheet. During subsequent composite fabrication, the organic binder evaporates. As another example, short fibers are combined with a ceramic or polymeric binder and a liquid carrier (usually water) to form a slurry; this is then filtered under pressure or wet pressed to form a wet "cake," which is subsequently dried to form a preform. In the case of using the matrix metal as the binder, a continuous fiber bundle is immersed in the molten matrix metal so as to be infiltrated with it, thus forming a wire preform; alternately, fibers placed on a matrix metal foil are covered and fixed in place with a sprayed matrix metal, thus forming a sprayed preform.

The most popular method for fabricating CF metal-matrix composites is the infiltration of a preform by a liquid metal under pressure.[131] The low viscosity of liquid metals compared to resins or glasses makes infiltration very appropriate for metal-matrix composites. Nevertheless, pressure is required because of the difficulty for the liquid metal to wet the CFs. The pressure can be provided by a gas (e.g., argon).

A second method for fabricating CF metal-matrix composites is diffusion bonding. In this method, a stack of alternating layers of CFs and metal foils is hot pressed (at, say, 24 MPa for 20 minutes) to cause bonding in the solid state. This method is not very suitable for fiber cloths or continuous fiber bundles because of the difficulty for the metal to flow to the space between the fibers during diffusion bonding. In contrast, the infiltration method involves melting the metal, so metal flow is relatively easy, making infiltration a more suitable method for fiber cloths or continuous fiber bundles. A variation of diffusion bonding involves the hot pressing of metal-coated CFs without the use of metal foils. In this case, the metal coating provides the metal for the metal-matrix composite. In general, the diffusion bonding method is complicated by the fact that the surface of the metal foil or metal coating tends to be oxidized and the oxide makes the bonding more difficult. Hence, a vacuum is usually required for diffusion bonding.

A third method of fabricating CF metal-matrix composites involves hot pressing above the solidus of the matrix metal. This method requires lower pressures than diffusion bonding, but the higher temperature of the pressing tends to cause fiber degradation, resulting from the interfacial reaction between the fibers and the matrix metal. A way to alleviate this problem is to insert a metal sheet of a lower solidus than the matrix metal alternately between wire preform layers then to hot press at a temperature between the two solidus temperatures. A combination of the second and third methods involves first heating CFs laid up with matrix metal sheets between them in vacuum in a sealed metal container above the liquidus of the matrix metal, then immediately hot pressing the container at a temperature below the solidus of the matrix metal.

A fourth method of fabricating CF metal-matrix composites involves the plasma spraying of the metal on to continuous fibers. As this process usually results in a composite of high porosity, subsequent consolidation (e.g., by hot isostatic pressing) is usually necessary. Compared to the other methods, plasma spraying has the advantage of being able to produce continuous composite parts, though the subsequent consolidation step may limit their size.

Slurry casting, a fifth method, is complicated by the tendency for the CFs (low in density) to float on the metal melt. To overcome this problem, which causes nonuniformity in the fiber distribution, compocasting is necessary. Compocasting (rheocasting) involves vigorously agitating a semi solid alloy so that the primary phase is nondendritic, thereby giving a fiber-alloy slurry with thixotropic properties.

The infiltration method can be used to produce near-net shape composites, so that subsequent shaping is not necessary. As infiltration using a gas pressure involves a smaller rate of pressure increase compared to using a piston, infiltration is more suitable than squeeze casting for near-net shape processing. If shaping is necessary, it can be achieved by plastic forming (e.g., extrusion, swaging, forging, and rolling) for the case of short fibers. Plastic forming tends to reduce the porosity and give a preferential orientation to the short fibers. These effects result in improved mechanical properties. For continuous fiber metal-matrix composites, shaping cannot be achieved by plastic forming and cutting is necessary.

4.4.3 Carbon-Matrix Composites

CF carbon-matrix composites, also called carbon-carbon composites, are the most advanced form of carbon. In the absence of oxygen which oxidizes carbon from around 400°C, carbon-carbon composites can resist temperatures in excess of 3000°C.[132] Carbon-carbon composites can be made using carbon particles but the composites which interest us here are made from CFs. The CFs used for carbon-carbon composites are usually continuous and woven. Both two-dimensional and higher-dimensional weaves are used, though the latter has the advantage of an enhanced interlaminar shear strength. The carbon matrix is derived from a pitch, a resin, or a carbonaceous gas. Such as pitch, which contains a very high (~90%) carbon content, can be forced into the preform under high pressure (1000 atm or 100 MPa) and high temperature of the order of 1000°C.[133] The high-pressure infiltration of pitch produces a composite with little porosity. Several infiltration cycles are necessary, usually up to six, but the density which can be obtained is close to that which is of a fully dense carbon of around 2 g/cm^3.

The fabrication of carbon-carbon composites is carried out by using four main methods, namely (i) liquid phase impregnation (LPI), (ii) hot isostatic pressure impregnation carbonization (HIPIC), (iii) hot pressing, and (iv) chemical vapor infiltration (CVI).[134]

All of the methods (except, in some cases, CVI) involve firstly the preparation of a prepreg by either wet winding continuous CFs with pitch or resin (e.g., phenolic), or wetting woven CF fabrics with pitch or resin. For highly directional carbon-carbon composites, fabrics which have a greater number of fibers in the warp direction than the fill direction may be used. The pitch or resin needs to be pyrolyzed or carbonized by heating at 350°C. Due to the shrinkage of the pitch or resin during carbonization (which is accompanied by the evolution of volatiles), additional pitch or resin is impregnated in the case of LPI and HIPIC. In LPI carbonization and impregnation are carried out as distinct steps, whereas in HIPIC carbonization and impregnation are performed together as a single step.

During hot pressing, graphitization may occur even for thermosetting resins, which is known as stress graphitization. Subsequently, further graphitization may be performed by heating without applied pressure at 2200°C–3000°C. Composites made by hot pressing have flattened pores in the carbon matrix and the part thickness is reduced by about 50%. Excessive pressure (say, 5 MPa) causes the formation of vertical cracks.

In CVI (also called CVD, chemical vapor deposition), gas-phase impregnation of a hydrocarbon gas (e.g., methane, propylene) into a CF preform takes place at 700°C–2000°C, so that pyrolytic carbon produced by the cracking of the gas is deposited in the open pores and surface of the preform. There are three CVI methods, namely the isothermal method, the temperature gradient method, and the pressure gradient method.

Carbon-carbon composites with high thermal conductivity are important for first wall components for nuclear fusion reactors, hypersonic aircraft, missiles and spacecraft, thermal radiator panels, and electronic heat sinks. Of the world market in carbon-carbon composites, 79% resides in the United States, 20% in Europe and the former USSR, and 1% in Japan. And future applications include air-breathing engine components, hypersonic vehicle airframe structures, space structures, and prosthetic devices.

In the isothermal method, the gas and sample are kept at a uniform temperature. As carbon growth in the pores will cease when they become blocked, there is a tendency for preferential deposition on the exterior surfaces of the sample. This causes the need for multiple infiltration cycles.

In the temperature gradient method, the sample is supported by an inductively heated mandrel (a susceptor) so that the inside surface of the sample will be at a higher temperature than the outside surface. Due to the temperature gradient, the deposition occurs first at the inside surface of the sample and progresses toward the outside surface, thereby avoiding the crusting problem.

In the pressure gradient method, the hydrocarbon gas impinges on to the inside surface of the sample, so the gas pressure is higher at the inside surface than the outside surface. The pressure gradient method is not as widely used as the isothermal method or the temperature gradient method.

Both the temperature gradient method and the pressure gradient method are limited to single samples, whereas the isothermal method can handle several samples at once. However, the isothermal method is limited to thin samples due to the crusting problem. A drawback of CVI is the low rate of deposition resulting from the use of a low gas pressure (1–150 torr), which favors a long mean free path for the reactant and decomposed gases; a long mean free path enhances deposition into the center of the sample. A diluent gas (e.g. He, Ar) is usually used to help the infiltration. For example, a gas mixture containing 3–10 vol.% propylene (C3H6) in Ar was used for CVI at 760°C–800°C. Hydrogen is often used as a carbon surface detergent. An attraction of CVI is that CVI carbon is harder than char carbon from pitch or resin, so that CVI carbon is particularly desirable for carbon-carbon composites used for brakes and friction products. The methods of CVI and LPI may be combined by first performing CVI on dry wound CFs then performing LPI. The CVI step serves to make the CFs rigid prior to impregnation.

The main disadvantages of carbon-carbon composites lie in the high fabrication cost, the poor oxidation resistance, the poor interlaminar properties (especially for two-dimensionally woven fibers), the difficulty of making joints, and the insufficient engineering data base.

Applications for carbon-carbon composites include aircraft brakes, heat pipes, reentry vehicles, rocket motor nozzles, hip replacements, biomedical implants, tools and dies, engine pistons, tiles for plasma facing armor, and electronic heat sinks.

4.4.4 Ceramic-Matrix Composites

Ceramic-matrix CF composites are gaining increasing attention because (i) the good oxidation resistance of the ceramic matrix (compared to a carbon matrix) makes the composites attractive for high-temperature applications (e.g., aerospace and engine components) and (ii) the continuously decreasing price of CFs makes the use of CF-reinforced concrete economically feasible.[134]

Ceramic-matrix CF composites that have been reported fall into the following categories: (i) CF-reinforced concretes and gypsums, (ii) CF-reinforced glasses, and (iii) other CF-reinforced ceramics (MgO, $A1_2O_3$, SiC, mullite).[12] The choice of the ceramic matrix depends on the application and on the chemical compatibility of the matrix with the CFs. As an alkaline environment catalyzes the oxidation of carbon, an alkaline matrix is not preferred. Nevertheless, CFs are used in concrete, which is alkaline, because competitive fibers, such as glass fibers and polymer fibers, are even less stable in concrete. There has been no report of CF degradation in concrete. Matrices that react with carbon to form carbides are to be avoided, especially in high-temperature applications, as such reactions will degrade the CFs. The fibers which are used in ceramic-matrix composites are based on SiC or alumina. They are fine with diameters of <15 μm and continuous. In the former case, the matrix material is usually SiC, deposited by techniques analogous to those used to make carbon-carbon composites, or mullaite produced by powder slurry infiltration or a sol–gel process in the latter case.

Chemical vapor infiltration (CVI) is used to introduce the matrix material into a fibrous preform at a temperature in the region of 1000°C. Silicon carbide reinforced with silicon carbide-based fibers (SiC–SiC) is prepared by a reaction in which methyltrichalorosilane decomposes in the presence of hydrogen. As with the carbon-carbon composites, the closing of pores in the preform is a limitation to achieving maximum density, so close control of temperature and gas flow is important. However, infiltration times can be as long as several days or weeks. The final product has then to be surface ground to achieve the finish required. Typically, densities of the order of 80%–90% of the theoretical maximum can be achieved.

An alternative approach is to infiltrate the fibrous preform with a polymer, such as polycarbosialane, as is used in the production of the fibers. Deposition of the SiC onto the fibers allows the SiC–SiC CMC to be produced, although repeated infiltration cycles are necessary. The use of polyasilazanes allows a matrix of silicon nitride (Si_3N_4) to be produced. Long processing times are again a difficulty and with the high cost of the precursor materials contribute to the high cost of manufacture, even though the lower temperatures involved, which are <1000°C, are an advantage.

Slurry infiltration can be used to make composites with glass matrices. The glasses which are used for this process are usually lithium aluminum silicates and magnesium aluminum silicates. The glass in powder form is made into a slurry which is used to impregnate the ceramic fibers. As the glasses have relatively low melting points of around 1200°C, densification can occur when pressure is applied during sintering. These vitreous ceramic composites differ from the SiC and Al_2O_3 matrix composites as the elastic moduli of the matrix material is lower than that of the reinforcing fiber.

MgO, Al_2O_3, SiC, Si_3N_4, mullite, p′-sialon, and ZrO_2 are other ceramics that have been used as matrices for CF composites. These composites are made by a number of methods, including (i) sintering the ceramic-matrix powder and the CFs, (ii) sintering the ceramic precursor and the CFs, (iii) polymer infiltration and pyrolysis, (iv) infiltrating the CF preform with a ceramic sol then subsequently gelling and sintering, (v) infiltrating the CF preform with a molten solution (e.g., Si) which will react with the CFs to form the desired ceramic (e.g., SiC), and (vi) CVI of the CF preform.

4.4.5 Hybrid Composites

Hybrid composites usually refer to composites containing more than one type of filler and/or more than one type of matrix. They are commonly used for improving the properties and/or lowering the cost of conventional composites. The possibility of using more than one type of filler or more than one type of matrix should always be considered in composite design.[135]

When more than one type of filler is used in a CF composite, the second type of filler may be a different type of fiber; it may even be whiskers or particles. When both fillers are fibers, one or both of the fillers can be short or continuous. The addition of a second type of fiber, such as glass or aramid fibers, can serve to increase the toughness of the CF composite. The two types of fibers may be in different laminae or the same lamina in the composite laminate. If they are in the same lamina, they may be cowoven, twisted, stapled, randomly mixed, or bound by a binder. When the second filler is particles, the particles are usually intermixed with the fibers within a lamina by incorporating the particles or the particle precursor in the matrix prior to impregnation of the matrix precursor into the fiber preform. However, they are sometimes located instead between adjacent fiber laminae, such that the matrix permeates the space between adjacent fibers as well as the space between adjacent particles. The particulate filler can serve to enhance the mechanical or electrical properties of the composite.

When two types of matrix are used in a composite, one of two cases applies. In the first case, one type of matrix is in contact with the filler while the second type of matrix is not; in other words, the second type of matrix is just in contact with the first type of matrix. This structure is obtained by infiltrating the CF preform with the precursor of the first type of matrix, and subsequently with the precursor of the second type of matrix. For example, the first type of matrix can be a thermoset (a lower-viscosity resin) while the second type of matrix can be a thermoplast (a higher-viscosity resin); the first type of matrix can be carbon (which is not oxidation resistant) while the second type of matrix can be SiC (which is more oxidation resistant).

The second type of matrix is the one that is at the exterior surface of the composite. In the second case, the two types of matrix are intimately mixed in a random fashion so that the mixture serves as the matrix. An example is the use of carbon and SiC together as the matrix by pyrolyzing a mixture of a carbon precursor (e.g., pitch) and a SiC precursor (e.g., a polycarbosilane). The interest behind the use of SiC and C together as the matrix lies in oxidation protection of the carbon composite. The sandwich composites are a special class of hybrid composites (not described above). A sandwich composite has an interlayer (interleaf) between adjacent CF laminae in the laminate, though an interlayer does not necessarily reside between every pair of fiber laminae. The interlayer itself may be a composite or not a composite. An example of an interlayer that is itself a composite is a honeycomb, such as an aluminum honeycomb formed by brazing aluminum sheets together in such a way as to form a structure that looks like a honeycomb (akin to corrugated paper). Honeycombs are attractive for their very low density, which is due to their very high porosity. Examples of interlayers that by themselves are not composites include aluminum and polymers, which serve to improve the toughness and/or control the thermal expansion coefficient.

Carbon-carbon composites contain CFs in a carbon matrix that is often not fully dense. The filling of the pores in a carbon-carbon composite with SiC, TiC, or other ceramic materials are more oxidation resistant than carbon yields a hybrid composite with more than one type of matrix. CVI has been used for the deposition of SiC and TiC. In particular, the thermal gradient CVI method has been used for the codeposition of C and SiC at 1100°C–1400°C, with propane (C,H) and methyltrichlorosilane (CH_3SiC_{13}) as the sources of carbon and silicon (carbide-forming element) respectively. Hybrid composites containing 12–29 wt.% SiC have been prepared, such that a CH_3SiC_{13}/N_2 ratio (N_2 being the carrier gas) increase gives rise to a SiC weight fraction increase. At low deposition temperatures, SiC codeposition does not significantly affect the microstructure of the carbon matrix. Since SiC is stiffer than carbon, the composites' mechanical properties (e.g., Young's modulus) are improved by SiC codeposition. However, at high deposition temperatures (1300°C–1400°C), due to the increase of the degree of crystalline order of the carbon matrix by the SiC codeposition, a matrix of columnar structure and large crystallite size is formed, resulting in no improvement of the mechanical properties by the SiC codeposition. Impregnation of a SiC precursor (polycarbosilane) into a carbon-carbon composite has also been achieved by the use of supercritical fluids, which typically exhibit densities approaching that of their liquid phase but have no surface tension and very low viscosity. The supercritical fluids dissolve, transport, and precipitate ceramic precursors within the pores of the carbon-carbon composite. Titanium carbide is attractive because of its high thermal stability compared to SiC. Furthermore, TiC has a CVI infiltration capability much higher than those of HfC and TaC. Due to the lower viscosity of epoxy resins than thermoplastics and the better fiber-matrix adhesion for epoxy than thermoplastic matrices, CF preforms may be partially impregnated with epoxy and subsequently fully impregnated with a thermoplastic. This results in a hybrid composite with more than one matrix.

The HS, low electrical resistivity, and low thermal expansion coefficient of CFs are exploited in a sandwich composite in which a high-T, ceramic superconductor is sandwiched by a tin-matrix unidirectional CF composite fabricated by squeeze casting. Tin (preferably 96Sn-4Ag) is used as the matrix because of the possibility of diffusion bonding it to the superconductor at a temperature low enough to avoid degradation of the superconducting properties of the superconductor. The sandwiching enables the superconductor to be packaged for mechanical durability, as the superconductor by itself is mechanically weak, especially under tension.

4.5 TRENDS OF CARBON FIBER COMPOSITES APPLICATIONS AND PROSPECTS

The use of CFs in composites is predicted to grow in future due especially to their decreasing price, lightweight, specific strength and specific stiffness.

In 2020, the global CFs composite market reached US$ 18.68 B (Figure 4.15).

The CFs are widening the applications of these composites to include the automobile, marine, sports, biomedical, construction, and other industries.

4.5.1 Aerospace

The development trend of the market is as follows: The demand for CF in 2020 is 16,450 tons (Figure 4.16).

The segment market share (tons) of the aerospace market is as follows:

According to data from the Teal Group, an aviation industry analysis agency, global air passenger traffic in 2020 has dropped by ~63% compared to 2019; Boeing and Airbus have significantly reduced the production capacity of aircraft, as displayed in Figure 4.17.

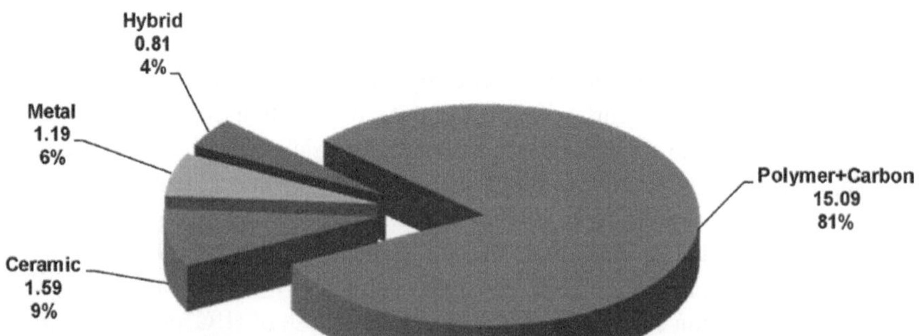

FIGURE 4.15 2020 Global demand for composites by matrix (Billion US$).

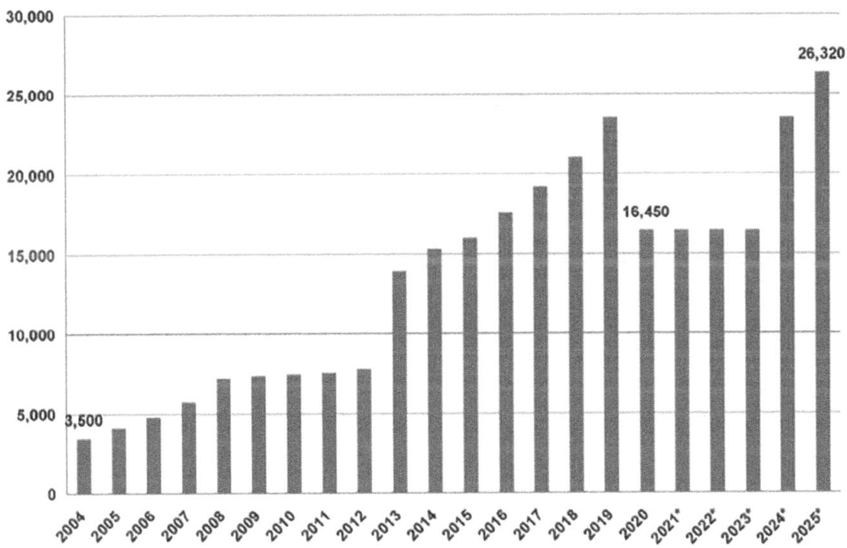

FIGURE 4.16 Carbon fiber demand in Aerospace-Trend (MT).

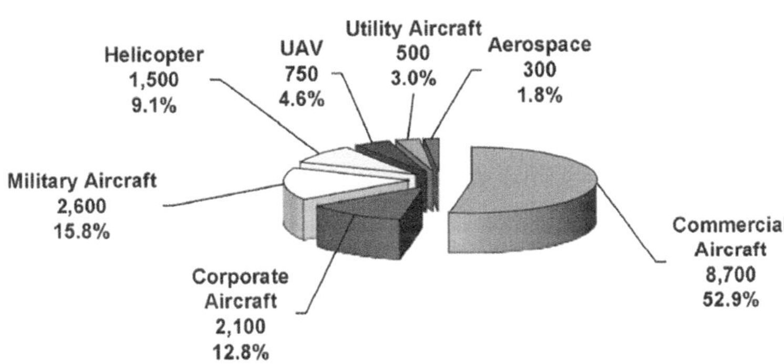

FIGURE 4.17 2020 Carbon fiber demand in aerospace segment (MT).

In June 2020, Toray Composite Materials America announced that it would immediately suspend operations at its plant in Spartanburg, South Carolina, and greatly reduce the production capacity of its Tacoma prepreg plant. All these actions will result in a reduction of about 25% of the workforce in U.S. factories; Hexcel's first-quarter net profit dropped by more than 40% year-on-year and announced a new round of layoffs and pay cuts plans; According to our comprehensive judgment, CF for aerospace use has decreased by 30% compared with the same period last year.[136]

With regard to the recovery of the aerospace market, based on the impact of the epidemic and the complex and slow characteristics of the aerospace system, we predict that the demand will return to the level of 2019 in 2024.

4.5.2 Wind Blades

In 2020, the development of wind power market developed rapidly, and the whole industry chain is full of profits (Figure 4.18).

In 2020, from the perspective of wind power demand for CF, Vestas still be the main force; other giants, such as Siemens Gamesa, GE-LM and Nordex, have adopted CF pultrusion manufacturing and testing prototypes in their new models.

A few years ago, "the ball" was at the foot of the domestic wind power giants, and the CF industry kept urging it: "Kick it out!" But starting from 2021, "the ball" is sure to be at the foot of the CF industry. Wind power giants will push hard: "kick it out! How dare we plan the blade demand of tens of thousands of tons if you can't produce the capacity of carbon fiber?" The whole CF industry, get ready to meet this "trouble of happiness"!

4.5.3 Sports Leisure

The development of demand for sports leisure market over the past decade is shown as below; the demand in 2020 is 15,400 tons (Figure 4.19).

The segment shares of the sports market are as follows (Figure 4.20).

Under the epidemic situation, the CF equipment for group sports, such as (ice) Hockey sticks and ski poles, have declined substantially; On the contrary, the equipment for personal sports and leisure increased, mainly including golf shafts, bicycles and fishing rods. In addition, healthy and green travel has been popular in Europe and the United States in recent years and the demand for e-bikes has also increased greatly.

4.5.4 Automobile

In previous reports, we repeatedly reminded the concept of lightweight value of whole life cycle; From F1 racing car to super car and super luxury car, the CF composite is expanding its regular and trend of application. In the calculation of lightweight value, in addition to the conventional value of energy saving and cost

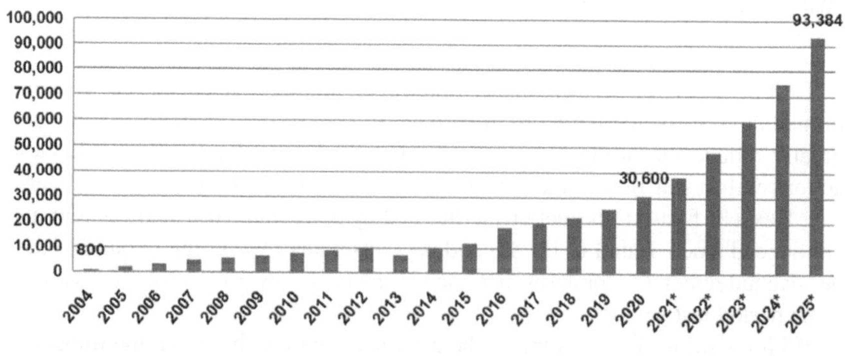

FIGURE 4.18 Carbon fiber demand in wind blades trend (MT).

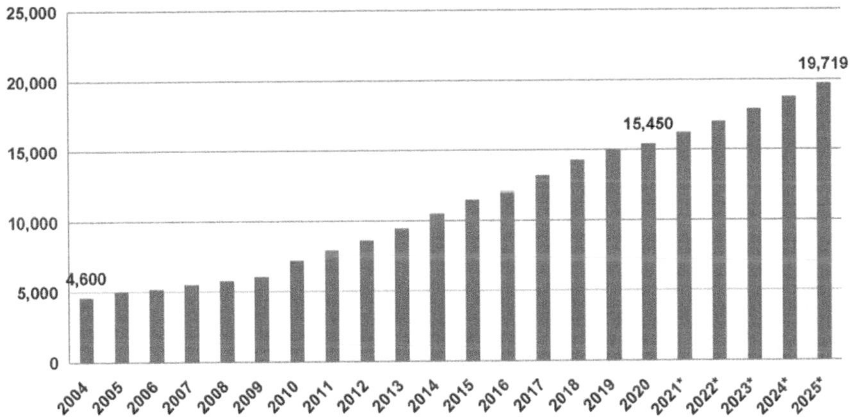

FIGURE 4.19 Carbon fiber demand in sports/leisure trend (MT).

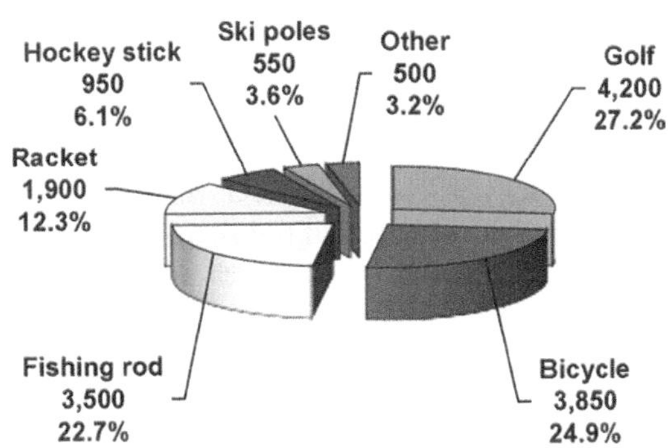

FIGURE 4.20 2020 Carbon fiber demand in sports/leisure trend (MT).

reduction, the cost of green environmental protection is a large variable cost due to the trend (Figure 4.21).

4.5.5 Others

4.5.5.1 Rail Transit

In September, 2020, Kawasaki delivered four railway vehicle bogies with CF-reinforced resin matrix composite springs to Japan's Shikoku Railway Corporation (hereinafter referred to as JR Shikoku). In January, 2021, the first high-temperature superconducting high-speed maglev engineering prototype vehicle and test line with fully independent intellectual property rights were officially put into use in Jiuli Campus of Southwest Jiaotong University. The prototype vehicle adopted full CF lightweight body and low resistance head shape, with an overall weight about 12 tons and a suspension height about 10 mm.

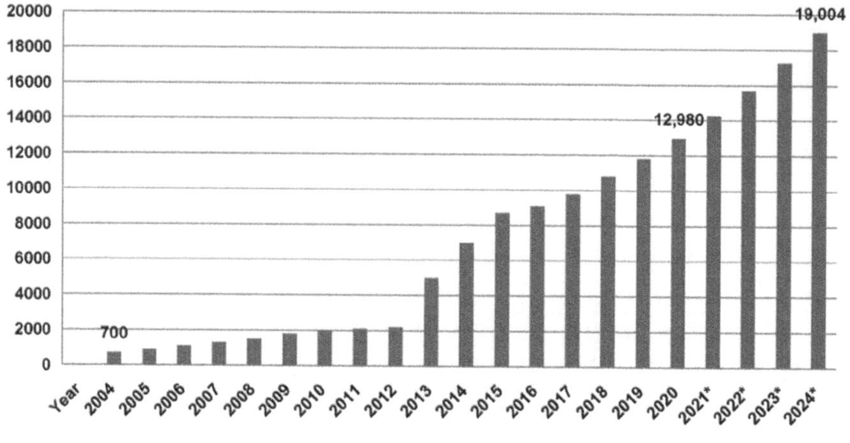

FIGURE 4.21 Carbon fiber demand in auto trend (MT).

In the aspect of developing composite of rail transit, China had an incomparable and comprehensive advantage over other countries in the world. The main shortcoming was still the supply of low-cost CF in the industrial chain and the system capability of research and development. Among them, there was not only the problem of application technology of rail transit system but also the problem of innovation technology of composite field. Encouragingly, the associated research and development between the rail transit system and composite system was more and closer in 2020. And this research and development association would be more and more powerful. The rail transit system was also preparing its own composite scientific research system.

In point of the development of CF industry, the next 3–5 years is the gold year for the development of low cost and large tow CF. Tens of thousands of tons of production capacity of CF will supply the market. Meanwhile, if rail transit and related enterprises of CF and composite strengthen the efforts to the research and development and engineering practice on this basis. I think this will build a wonderful future for the above two industries.

4.5.5.2 Functional Carbon Fiber Materials

The main functional applications of CF besides the traditional flame-retardant materials are pre-oxygen wire and pre-oxygen felt and so on. The traditional thermal field materials have applications to carbon felt, graphite hard and soft felt, and electric blanket, but also the traditional antistatic and electromagnetic shielding materials, such as short-cut CF and grinding reinforced plastic.

The new functional materials of CF mainly include:

A. Gaseous diffusion layer (GDL) of fuel cell;
B. Carbon felt (graphite felt) of the electrode material of flow cell;
C. Carbon felt plate to replace lead plate electrode of lead-acid battery;
D. Microporous CF material to replace the energy density of lithium-ion battery;

E. Flame retarded, silent, hydrophobic and heat preserved multifunctional material prepared by pre-oxidized felt and silica aerogel. It can be used for the isolation between batteries for electric vehicle and all kinds of skillful or functional buildings, as well as the silence of submarines and tanks, etc.

4.6 OBSERVATION AND PONDERATION

4.6.1 World Map of Carbon Fiber

Seven traditional giants of CF (Thereinto Zoltek was purchased by Toray and now six giants) have basically completed the layout of international industrial bases. Especially Toray, it had completed the layout of small and large tow CF in Japan, Korea, the United States, France, Hungary, and Mexico in the past decades. These giants still dominated the global CF in terms of production technology and cost as well as application technology and ecology (Figure 4.22).

CF is different from the conventional chemical fiber products in terms of the complexity of preparation technology, the technical complexity of industrial chain, or the difficulty of application technology development and ecological construction. It is unpractical at least for the next 20 years that attempting to develop CF by undertaking industrial transfer of developed countries. How does CF of emerging countries catch up with the developed countries? That is what we are going to focus on this time.

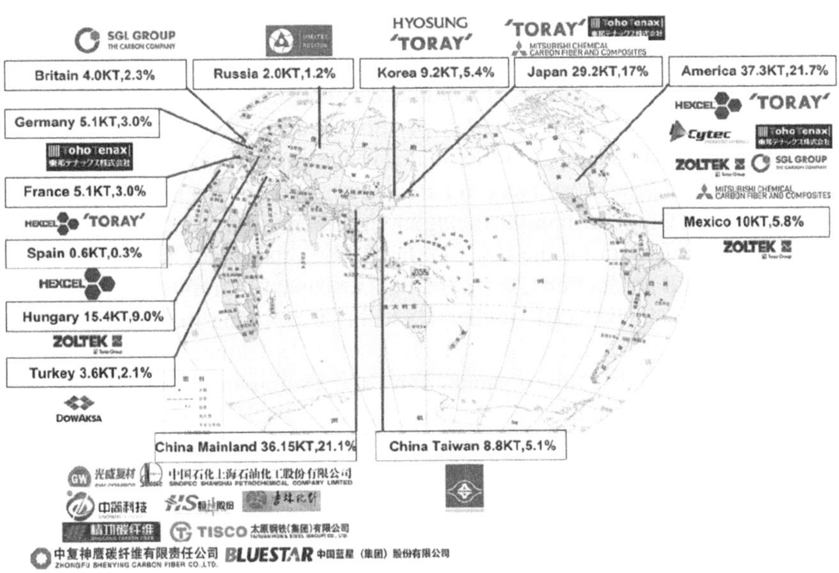

FIGURE 4.22 World map of carbon fiber.

4.6.2 Development Direction of Composite Process and Application Technology

Except the need for low-cost CFs, the previously mentioned explosive growth of the application market has also brought new requirements for the development of composite process and application technology. Its core is to develop a low-cost industrial chain from CF–resin–composite process–application technology. Only the healthy development of this industrial chain can turn the foregoing application prospects into reality.

4.7 SUMMARY AND PERSPECTIVE

Carbonaceous filler-reinforced polymer-matrix composites have arrested salient attention, the relative maturity of composites take-up in sectors, such as aerospace, and the potential for significant growth in advancing markets is on-going. The expansion of the CF-reinforced polymer market is driving large investments in advanced industrial facilities. Despite the promising features of CF-reinforced polymer-based products, the penetration of processing technologies in some industries is hindered by significant challenges. In volume-intensive industries, such as the automotive, poor economies scale and learning increase costs, keeping most applications in niche or premium markets. Coupled with a lack of awareness about composite applications, proper design standards, and an enduring metals-based design culture, CF-reinforced polymer technologies are limited for widespread utilization across sectors. Scattered knowledge about composites processing technologies affect the performance of supply chains, since the lead time for technology adoption, namely the acquisition of equipment and installations, may be less than the time needed to assimilate technical knowledge.

The broad CF market segmentation includes the product type gamut, application spectrum, end-user industry landscape, significant geographical regions, and the top market contenders. The unbiased industry expert opinions on the current market scenario, past market performance, production and consumption rates, demand and supply ratio, and revenue generation forecasts over the estimated period. The most uncertain situation is that of CF supply, made complicated by the Chinese manufacturing environment. Tony and Dan are confident of CF supply data for the US, EU, Japan and other non-Chinese locations and estimate real production (70% of nameplate capacity) from all non-Chinese locations to be 94,850 metric tonnes (MT) in 2021, rising to 101,500 MT by 2023 with further expansion in 2026 not known. In China, where real, local production is more difficult to gauge, Tony and Dan assumed actual domestic output to be 55% of nameplate capacity, which puts 2021 total supply from within China at 19,250 MT, rising to 23,650 MT by 2023, with further expansions through 2026 to be expected. Taken together, these numbers suggest 2021 total global CF supply at 114,100 MT, and by 2023 global CF supply at 125,150 MT.

As might be expected, an already complicated CF supply situation is made doubly so when demand is factored in. There are three end markets in particular:

Wind, aerospace/defense and infrastructure/energy – that are expected to drive substantially increased use of CF over the next 5 years. In fact, the wind energy industry alone, already the largest consumer by volume of CF in the world, is expected to almost double its CF demand, from 28,000 MT in 2021 to 54,000 MT in 2026. This will be driven by rapidly increasing installations of onshore and offshore wind capacity, particularly in China. Similarly, aerospace/defense CF demand is expected to increase from 16,500 MT in 2021 to 29,100 MT in 2026, driven primarily by Boeing and Airbus. Tony and Dan admit, however, that this increase is somewhat uncertain given that demand for the composites-intensive Boeing 787 and Airbus A350 is not expected to return to pre-COVID-19 levels until 2025. In 2021, global demand for all CF is 113,800 MT. By 2026, that figure is expected to increase to 180,400 MT. Recall that Tony and Dan estimate CF supply from existing facilities in 2026 reaching only 125,150 MT, which creates a *potential* CF shortfall of 55,250 MT by 2026, with signs of this supply deficit appearing as soon as 2023 or 2024.

The question is whether or not this forecasted CF shortfall is an actual problem or just a potential problem. We have seen shortage estimates like this before in the last 15 years, and CF suppliers, seeing the same thing, have typically responded by adding capacity. The challenge, as always, is that it takes a lot of time – 2 years at least – and money to construct and commission a new CF line. In addition, two of the suppliers of the type of CF used in pultruded spar caps for wind blades, Zoltek and DowAksa, have already announced capacity expansions. Others, like Toray, Hexcel, Teijin, and Solvay, primarily serve the HP market and might be reluctant to announce new capacity without a signal from Boeing or Airbus about the launch of a clean-sheet aircraft program. In any case, the next few years promise to be interesting and challenging for producers and consumers of CF.

REFERENCES

1. Das, T.K., P. Ghosh, and N.C. Das, Preparation, development, outcomes, and application versatility of carbon fiber-based polymer composites: A review. *Advanced Composites and Hybrid Materials*, 2019. **2**(2): 214–233.
2. Shen, T., et al., Crystalline and pseudo-crystalline phases of polyacrylonitrile from molecular dynamics: Implications for carbon fiber precursors. *Polymer*, 2018. **155**: 13–26.
3. Marsh, H. and F.R. Reinoso, *Activated Carbon*. Elsevier, Amsterdam, Netherlands, 2006, 23.
4. ASBURY. Crystalline graphite is not simply a bunch of graphene layers piled one on top of the other, but is highly ordered structure. 2019; Available from: https://asbury.com/resources/education/graphite-101/structural-description/.
5. Gray, D., A. McCaughan, and B. Mookerji, Crystal structure of graphite, graphene and silicon. *Physics for Solid State Applications*, 2009. **6**: 730.
6. Wang, Z., X. Zhu, and M. Wang, A prediction of a new porous metallic carbon allotrope with an sp2 hybridized network: cP-C24. *Solid State Sciences*, 2020. **105**: 106247.
7. Painters, P. Crystal structure of graphene. 2022; Available from: https://www.proservpainters.com/jepdmv/crystal-structure-of-graphene.html.

8. Mo, Y., & Schleyer, P. V. R. An energetic measure of aromaticity and antiaromaticity based on the Pauling–Wheland resonance energies. *Chemistry–A European Journal*, 2006. 12(7): 2009–2020.
9. Melo, G.R., Carbon based nano-composite materials for energy storage applications. 2015: The University of Texas at El Paso.
10. Xue, Y., et al., Diamond and carbon nanostructures for biomedical applications. *Functional Diamond*, 2022. **1**(1): 221–242.
11. Morgan, P., *Carbon Fibers and Their Composites*. CRC Press, Boca Raton, FL, 2005.
12. Chung, D., *Carbon Fiber Composites*. Elsevier, Amsterdam, 2012, 31.
13. Askeland, D.R., et al., *The Science and Engineering of Materials*. Cengage Learning, Boston, MA, 2003.
14. Fridman, L. and S. Grebennikov, Theoretical aspects in the preparation and use of fibrous carbon adsorbents. *Fibre Chemistry*, 1990. **22**(6): 363–367.
15. Mochida, I. and S. Kawano, Capture of ammonia by active carbon fibers further activated with sulfuric acid. *Industrial & Engineering Chemistry Research*, 1991. **30**(10): 2322–2327.
16. Park, S.-J. and G.-Y. Heo, Precursors and manufacturing of carbon fibers, in *Carbon Fibers*. Springer, Berlin, Germany, 2015, 31–66.
17. Donnet, J.-B. and R.C. Bansal, *Carbon Fibers*. CRC Press, Boca Raton, FL, 1998, 15.
18. Waterbury, M., et al., In situ observation of the single-fiber fragmentation process in metal-matrix composites by ultrasonic imaging. *Composites Science and Technology*, 1994. **52**(2): 261–266.
19. Ohsawa, T., et al., Axial compressive strength of carbon fiber. *Journal of Applied Polymer Science*, 1990. **39**(8): 1733–1743.
20. Balasubramanian, M., et al., Evolution of structure and properties in the formation of pan-based carbon-fibers. *Abstracts of Papers of the American Chemical Society*, 1986. **192**: 29.
21. Frank, E., et al., Carbon fibers: Precursor systems, processing, structure, and properties. *Angewandte Chemie-International Edition*, 2014. **53**(21): 5262–5298.
22. Gao, X.Q., et al., The fabrication of chopped carbon fiber-carbon composites and their thermal/electrical conductivity and microstructure. *New Carbon Materials*, 2005. **20**(1): 18–22.
23. Safonova, A.M., et al., Structure formation and properties of nickel-carbon fibers. *Journal of Applied Chemistry of the Ussr*, 1991. **64**(11): 2289–2292.
24. Mora, E., et al., A study of pitch-based precursors for general purpose carbon fibres. *Carbon*, 2002. **40**(14): 2719–2725.
25. Wazir, A.H. and L. Kakakhel, Preparation and characterization of pitch-based carbon fibers. *New Carbon Materials*, 2009. **24**(1): 83–88.
26. Hassan, M.F., et al., Recent trends in activated carbon fibers production from various precursors and applications: A comparative review. *Journal of Analytical and Applied Pyrolysis*, 2020. **145**: 1–22.
27. Ohno, H., High performance pitch based carbon fiber and its applications. *Sen-I Gakkaishi*, 2014. **70**(5): 151–155.
28. Tibbetts, G.G., Vapor grown carbon-fibers: Status and Prospects. *Carbon*, 1989. **27**(5): 745–747.
29. Bradley, J.R. and G.G. Tibbetts, Improved yield of carbon-fibers by pyrolysis of natural-gas in stainless-steel tubes. *Carbon*, 1985. **23**(4): 423–430.
30. Shindo, A., *Report of the Government Industrial Research Institute*. Japan: Osaka, 1961, 23: 1–8.
31. Ashina, Y., I. Oshima, and K. Sekine, *Temperature Control in Suspension Polymerisation*. Nitto Chemical Industry Co., Mitsubishi Rayon Co, Ibaraki, Osaka, 1976.

32. Bacon, R., Kinetic studies in the chemistry of rubber and related materials. III. Thermochemistry and mechanisms of olefin oxidation. *Transactions of the Faraday Society,* 1946, 42, 14–55.
33. Bero, M. and T. Rosner, Polymerisationskinetik von Acrylnitril in wassriger losung. *Macromolecular Chemistry,* 1970, **136**(1): 1–10.
34. Fritzsche, P. and J. Ulbricht, Das Redoxsystem Peroxydisulfat-Hydrogensulfit-Metallionen als Initiator der Acrylnitrilpolymerisation. *Faserforsch U Textiltechn.* 1964. **14**: 93–100.
35. Sandler, S.R., *Polymer Synthesis.* Academic Press: New York and London, 1974, 1:8.
36. Milczarska, T. and J. Szafko, Kinetics studies on the free-radical polymerization of styrene in the presence of asphalts. *Angewandte Makromolekulare Chemie,* 1994. **218**: 171–182.
37. Peebles, L.H., Polyacrylonitrile prepared in ethylene carbonate solution.3. molecular parameters. *Journal of Polymer Science Part a-General Papers,* 1965. **3**(1): 361–372.
38. Peebles, L.H., Polyacrylonitrile prepared in ethylene carbonate solution.2. kinetics at high conversion. *Journal of Polymer Science Part a-General Papers,* 1965. **3**(1): 353–360.
39. Peebles, L.H., Polyacrylonitrile prepared in ethylene carbonate solution.i. kinetics at low conversion. *Journal of Polymer Science Part a-General Papers,* 1965. **3**(1): 341–352.
40. Peebles, L.H., Kinetic model of persulfate-bisulfite-initiated acrylonitrile polymerization. *Journal of Applied Polymer Science,* 1973. **17**(1): 113–128.
41. Szafko, J. and E. Turska, Free-radical polymerization of some monomers in dimethylformamide. *Makromolekulare Chemie,* 1972. **156**: 297–306.
42. Szafko, J. and E. Turska, Copolymerization of acrylonitrile and methyl-esters of alpha-substituted acrylic acids. *Makromolekulare Chemie,* 1972. **156**: 311–320.
43. Monobe, K., Fundamental approach for fiber formation of crystalline polymers. Abstracts of Papers of the American Chemical Society, 1979: 124–124.
44. Daumit, G.P., Formation of melt-spun acrylic fibers which are well suited for thermal conversion to high strength carbon fibers. Basf Aktiengesellschaft, 1990.
45. Daumit, G.P., Formation of melt-spun acrylic fibers which are particularly suited for thermal conversion to high strength carbon fibers. Basf Aktiengesellschaft, 1990.
46. Daumit, G.P., Formation of melt-spun acrylic fibers possessing a highly uniform in ternal structure which are particularly suited for thermal conversion to quality carbon fibers. Basf Aktiengesellschaft, 1990.
47. Ingildeev, D., et al., Novel cellulose/polymer blend fibers obtained using ionic liquids. *Macromolecular Materials and Engineering,* 2012. **297**(6): 585–594.
48. Schildknecht, C.E., *Vinyl and Related Polymers: Their Preparations, Properties, and Applications in Rubbers, Plastics, Fibers, and in Medical and Industrial Arts.* Wiley & Sons: New York, 1952, 39.
49. Tsai, J.S. and C.H. Lin, Effect of comonomer composition on the properties of polyacrylonitrile precursor and resulting carbon-fiber. *Journal of Applied Polymer Science,* 1991. **43**(4): 679–685.
50. Tsai, J.S. and C.H. Lin, The effect of the side-chain of acrylate comonomers on the orientation, pore-size distribution, and properties of polyacrylonitrile precursor and resulting carbon-fiber. *Journal of Applied Polymer Science,* 1991. **42**(11): 3039–3044.
51. Tsai, J.S. and C.H. Lin, The effect of molecular-weight on the cross-section and properties of polyacrylonitrile precursor and resulting carbon-fiber. *Journal of Applied Polymer Science,* 1991. **42**(11): 3045–3050.

52. Brooks, J.D. and G.H. Taylor, The formation of graphitizing carbons from the liquid phase. *Carbon*, 1965. **3**: 185–193.
53. White, J.L. and M. Buechler, Microstructure formation in mesophase carbon-fibers and other graphitic materials. *ACS Symposium Series*, 1986. **303**: 62–84.
54. Singer, L.S., The mesophase and high modulus carbon fibers from pitch. *Carbon*, 1978. **16**(6): 409–415.
55. Singer, L.S., Carbon-fibers from mesophase pitch. *Fuel*, 1981. **60**(9): 839–847.
56. Liedtke, V. and K.J. Huttinger, Mesophase pitches as matrix precursor of carbon fiber reinforced carbon.1. Mesophase pitch preparation and characterization. *Carbon*, 1996. **34**(9): 1057–1066.
57. Mochida, I., Y. Sone, and Y. Korai, Preparation and properties of carbonaceous mesophase.2. highly soluble mesophase from ethylene tar modified using aluminum-chloride as a catalyst. *Carbon*, 1985. **23**(2): 175–178.
58. Chun, I.S., et al., Carbon nanofibers from polyacrylonitrile and mesophase pitch. *Journal of Advanced Materials*, 1999. **31**(4): 7–14.
59. Yang, H., et al., Improving graphitization degree of mesophase pitch-derived carbon fiber by solid-phase annealing of spun fiber. *Carbon*, 2003. **41**(3): 397–403.
60. Mochida, I., et al., Microstructure of mesophase pitch-based carbon fiber and its control. *Carbon*, 1996. **34**(8): 941–956.
61. Barraza-Burgos, J.M. and J.M. Ospina-Espinosa, Effect of pitch petroleum oxidation on mesophase production. *Ct&F-Ciencia Tecnologia Y Futuro*, 2012. **4**(5): 33–42.
62. Manocha, L.M., et al., Microstructure of carbon/carbon composites reinforced with pitch-based ribbon-shape carbon fibers. *Carbon*, 2003. **41**(7): 1425–1436.
63. Zeng, S.M., et al., Preparation of isotropic pitch precursors for general-purpose carbon-fibers (Gpcf) by air blowing.3. air blowing of isotropic naphthalene and hydrogenated coal-tar pitches with addition of 1,8-dinitronaphthalene. *Carbon*, 1993. **31**(3): 421–426.
64. Zeng, S.M., et al., Preparation of isotropic pitch precursors for general-purpose carbon-fibers (Gpcf) by air blowing.2. air blowing of coal-tar, hydrogenated coal-tar, and petroleum pitches. *Carbon*, 1993. **31**(3): 413–419.
65. Hacker, P.J., G.B. Neighbour, and B. McEnaney, The coefficient of thermal expansion of nuclear graphite with increasing thermal oxidation. *Journal of Physics D-Applied Physics*, 2000. **33**(8): 991–998.
66. Lavin, J.G., Mesophase precursors for advanced carbon fibers: Pitches, stabilization and carbonization. *Design and Control of Structure of Advanced Carbon Materials for Enhanced Performance*, 2001. **374**: 151–161.
67. Huson, M.G., High-performance pitch-based carbon fibers, in *Structure and Properties of High-Performance Fibers*. Woodhead Publishing: Cambridge, 2017, 31–78.
68. Johnson, D.J., Structure property relationships in carbon-fibers. *Journal of Physics D-Applied Physics*, 1987. **20**(3): 286–291.
69. Greene, M.L., R.W. Schwartz, and J.W. Treleaven, Short residence time graphitization of mesophase pitch-based carbon fibers. *Carbon*, 2002. **40**(8): 1217–1226.
70. Marshik, B. and T. Apple, Conduction esr of vapor grown carbon-fibers. *Colloids and Surfaces*, 1990. **45**: 225–233.
71. Uchiyama, S., et al., Measurement of peroxide in oils by controlled-potential coulometry using a vapor-grown carbon-fiber electrode. *Electroanalysis*, 1990. **2**(3): 259–261.
72. Ishioka, M., T. Okada, and K. Matsubara, Mechanical-properties of vapor-grown carbon-fibers prepared from benzene in linz-donawitz converter gas by floating catalyst method. *Journal of Materials Research*, 1992. **7**(11): 3019–3022.

73. Chen, J.H., et al., Torsional modulus of vapor-grown carbon-fibers. *Applied Physics Letters*, 1992. **60**(19): 2347–2349.
74. Kato, T., et al., Formation of vapor-grown carbon-fibers on a substrate. *Carbon*, 1992. **30**(7): 989–994.
75. Masuda, T., S.R. Mukai, and K. Hashimoto, The production of long vapor grown carbon-fibers at high growth-rates. *Carbon*, 1992. **30**(1): 124–126.
76. Kato, T., K. Kusakabe, and S. Morooka, Process of formation of vapor-grown carbon-fibers by gas-phase reaction using ultrafine iron catalyst particles. *Journal of Materials Science Letters*, 1992. **11**(10): 674–677.
77. Tibbetts, G.G., D.W. Gorkiewicz, and R.L. Alig, A new reactor for growing carbon-fibers from liquid-phase and vapor-phase hydrocarbons. *Carbon*, 1993. **31**(5): 809–814.
78. Deborah, D., and L. Chung, *Carbon Fiber Composites*. Butterworth-Heinemann: Oxford, 1994, 209.
79. Tibbetts, G.G., et al., Physical-properties of vapor-grown carbon-fibers. *Carbon*, 1993. **31**(7): 1039–1047.
80. Masuda, T., S.R. Mukai, and K. Hashimoto, The liquid pulse injection technique: A new method to obtain long vapor-grown carbon-fibers at high growth-rates. *Carbon*, 1993. **31**(5): 783–787.
81. Kato, T., et al., Effect of carbon source on formation of vapor-grown carbon-fiber. *Carbon*, 1993. **31**(6): 937–940.
82. Ishioka, M., et al., Electrical-resistivity, magnetoresistance, and morphology of vapor-grown carbon-fibers prepared in a mixture of benzene and linz-donawitz converter gas by floating catalyst method. *Journal of Materials Research*, 1993. **8**(8): 1866–1874.
83. Gupta, S.K., N. Gupta, and D. Kunzru, Vapor-grown carbon-fibers from pyrolysis of hydrocarbons: Modeling of filament growth and poisoning. *Journal of Analytical and Applied Pyrolysis*, 1993. **26**(3): 131–144.
84. Ishioka, M., et al., Formation of vapor-grown carbon-fibers in Co-Co_2-H_2 mixtures, Ii influence of catalyst. *Carbon*, 1993. **31**(1): 245–245.
85. Ishioka, M., T. Okada, and K. Matsubara, Preparation of vapor-grown carbon-fibers by floating catalyst method in linz-donawitz converter gas: Influence of catalyst size. *Carbon*, 1993. **31**(5): 699–703.
86. Andrade, J.S., et al., Percolation conduction in vapour grown carbon fibre. *Physica A*, 1998. **248**(3–4): 227–234.
87. Carneiro, O.S., et al., Production and assessment of polycarbonate composites reinforced with vapour-grown carbon fibres. *Composites Science and Technology*, 1998. **58**(3–4): 401–407.
88. Darmstadt, H., et al., Surface spectroscopic analysis of vapour grown carbon fibres prepared under various conditions. *Carbon*, 1998. **36**(7–8): 1183–1190.
89. Madronero, A. and C. Merino, Some geometrical singularities in the characterization of vapor grown carbon fibers using laser diffraction technique. *Materials Research Bulletin*, 1998. **33**(10): 1503–1515.
90. Madronero, A., et al., Study by a modified scanning electron microscope fractography of hydrogenation process in vapour grown carbon fibres. *Journal of Materials Science*, 1998. **33**(8): 2079–2085.
91. Mukai, S.R., et al., The influence of catalyst particle size distribution on the yield of vapor-grown carbon fibers produced using the liquid pulse injection technique. *Chemical Engineering Science*, 1998. **53**(3): 439–448.
92. Hall, I.W. and F. Manrique, Surface-treatment of carbon-fibers for aluminum-alloy matrix composites. *Scripta Metallurgica Et Materialia*, 1995. **33**(12): 2037–2043.

93. Wu, G.M., et al., Effects of treatment on the surface-composition and energy of carbon-fibers. *Polymer Composites*, 1995. **16**(4): 284–287.
94. Vaidyanathan, N.P., et al., Surface-treatment of carbon-fibers using low-temperature plasma. *Journal of Adhesion*, 1995. **48**(1–4): 1–24.
95. Rotel, M., et al., Preadhesion laser surface treatment of carbon fiber reinforced PEEK composite. *Journal of Adhesion*, 1995. **55**(1–2): 77.
96. Bogoevagaceva, G., et al., The effect of surface-treatment on the interfacial properties in carbon fibre/epoxy matrix composites. *Journal of Materials Science*, 1995. **30**(13): 3543–3546.
97. Brown, P.R., D.R. Buchanan, and T.G. Clapp, Large-deflection bending of woven fabric for automated material-handling. *Journal of the Textile Institute*, 1990. **81**(1): 1–14.
98. Warren, W.E., The elastic properties of woven polymeric fabric. *Polymer Engineering and Science*, 1990. **30**(20): 1309–1313.
99. Cookson, P.G., Hygral expansion behavior of woven wool fabrics. *Textile Research Journal*, 1990. **60**(10): 579–590.
100. Ko, S., et al., Preparation of petroleum-based mesophase pitch toward cost-competitive high-performance carbon fibers. *Carbon Letters*, 2019. **30**(1): 35–44.
101. Lee, S., et al., The thermotropic liquid crystalline behavior of mesophase pitches with different chemical structures. *Carbon*, 2015. **81**: 694–701.
102. Mochida, I., K. Shimizu, and Y. Korai, Structure and carbonization properties of pitches produced catalytically from aromatic-hydrocarbons with Hf/Bf3. *Carbon*, 1988. **26**(6): 843–852.
103. Figueiredo, J.L., et al. *Carbon Fibers, Filaments and Composites*. Kluwer Academic Publishers: Netherlands, 1990, 37.
104. Waterbury, M.C. and L.T. Drzal, On the determination of fiber strengths by in-situ fiber strength testing. *Journal of Composites Technology and Research*, 1991. **13**(1): 22—28.
105. Chernin, S.M., High-temperature miniature blackbody radiation sources. *Applied Optics*, 1997. **36**(7): 1580–1591.
106. Tagawa, M., et al., Surface characterization of carbon fibers exposed to 5 eV energetic atomic oxygen beam studied by wetting force measurements. *Japanese Journal of Applied Physics*, 1991. **30**(9): 2134–2138.
107. Johnson, D., Structure-property relationships in carbon fibres. *Journal of Physics D: Applied Physics*, 1987. **20**(3): 286.
108. Janicki, G., V. Bailey, and H. Schjelderup (Eds). *Advanced Materials: Challenge Next Decade*. Society for the Advancement of Material and Process Engineering: Diamond Bar, CA, 1990, 139.
109. Nysten, B., et al., Determination of lattice defects in carbon fibers by means of thermal-conductivity measurements. *Physical Review B*, 1991. **44**(5): 2142.
110. Northolt, M.G., L.H. Veldhuizen, and H. Jansen, Tensile deformation of carbon fibers and the relationship with the modulus for shear between the basal planes. *Carbon*, 1991. **29**(8): 1267–1279.
111. Kumar, A. and T.E. Helminiak. Compressive strength of high performance fibers. *Materials Research Society Symposium Proceedings*, 1989. **134**: 363.
112. Janicki, G., V. Bailey, and H. Schjelderup (Eds), *Advanced Materials: Challenge Next Decade*. Society for the Advancement of Material and Process Engineering, Diamond Bar, CA, 1990, 52.
113. Krucinska, I. and T. Stypka, Evaluation of the intrinsic mechanical properties of carbon fibres. *Composites Science and Technology*, 1991. **4**(3): 287–301.
114. Khan, T., M. Dauzvardis, and S. Sayers, Carbon filament implants promote axonal growth across the transected rat spinal cord. *Brain Research*, 1991. **541**(1): 139–145.
115. Wegman, R.F., et al., *Advanced Materials: The Challenge for the Next Decade*. Society for Advancement of Materials and Process Engineering, Covina, CA, 1989, 35: 24.

116. Ohsawa, T., et al., Axial compressive strength of carbon fiber. *Journal of Applied Polymer Science*, 1990. **39**(8): 1733–1743.
117. Ismail, I.M.K. and W.C. Hurley, Modelling carbon fibers oxidation in air at constant heating rates. *Carbon*, 1992. **30**(3): 419–427.
118. Ho, I.H. and C.H. Chung, Bromination of graphitic pitch-based carbon fibers. *Carbon*, 1990. **28**(6): 831–837.
119. Fitzer, E.A. Gkogkidis, and M. Heine. Carbon fibers and their composites (a review). *High Temperatures–High Pressures*, 1984. **16**: 363–392.
120. Krucinska, I. and T. Stypka, Direct measurement of the axial Poisson's ratio of single carbon fibres. *Composites Science and Technology*, 1991. **41**(1): 1–12.
121. Fitzer, E. and W. Frohs, Modern carbon fibres from polyacrylonitrile (PAN)-polyheteroaromatics with preferred orientation. *Chemical Engineering & Technology*, 1990. **13**(1): 41–49.
122. Donnet, J.-B. and R.C. Bansal, *International Fiber Science and Technology (Carbon Fibers)*. Marcel Dekker: New York, 1990, 83.
123. Dhami, T.L., L.M. Manocha, and O.P. Bahl, Oxidation behaviour of pitch based carbon fibers. *Carbon*, 1991. **29**(1): 51–60.
124. John A. Preform based metal matrix composite fabrication for electronic carrier application. *16th International SAMPE Electronics Conference*. 1992. 6: 22–25.
125. Bunsell, A.R., S. Joannès, and A. Thionnet, *Fundamentals of Fibre Reinforced Composite Materials*. CRC Press: Boca Raton, FL, 2021, 193.
126. Parmar, H., et al., Metallisation of polymers and polymer matrix composites by cold spray: State of the art and research perspectives. *International Materials Reviews*, 2022. **44**: 1–25.
127. Abdelhak, B., M. Noureddine, and M. Hacen, Improvement of the interfacial adhesion between fiber and matrix. *Mechanics and Mechanical Engineering*, 2018. **22**(4): 885–894.
128. Van de Werken, N., et al., Investigating the effects of fiber surface treatment and alignment on mechanical properties of recycled carbon fiber composites. *Composites Part A: Applied Science and Manufacturing*, 2019. **119**: 38–47.
129. Navaneethakrishnan, G., et al., Structural analysis of natural fiber reinforced polymer matrix composite. *Materials Today: Proceedings*, 2020. **21**: 7–9.
130. Perumal, A.B., et al., Carbon fiber composites, in *Handbook of Nanomaterials and Nanocomposites for Energy and Environmental Applications*, O.V. Kharissova, L.M.T. Martínez, and B.I. Kharisov, Editors. Springer International Publishing: Cham, 2020, 1–32.
131. Akhil, M., et al., Squeeze infiltration processing and structural characteristics of lightweight aluminum-carbon metal matrix composites. *Transactions of the Indian National Academy of Engineering*, 2021. **6**(1): 41–48.
132. Duan, J., et al., Tribological behavior and applications of carbon fiber reinforced ceramic composites as high-performance frictional materials. *Ceramics International*, 2021. **47**(14): 19271–19281.
133. McKee, D.W., *Oxidation Protection of Carbon Materials*. CRC Press: Boca Raton, FL, 2021.
134. Chung, D.D., *Composite Materials: Science and Applications*. Springer Science & Business Media, Berlin, Germany, 2010, 93.
135. Gu, H., et al., *Introducing Advanced Composites and Hybrid Materials*. Springer: Berlin, Germany, 2018, 1–5.
136. Cooper, A. Toray plant suspends operations in Spartanburg County. 2020; Available from: https://upstatebusinessjournal.com/manufacturing/toray-plant-suspends-operations-spartanburg-county/.

5 Laser Micro-Processing of Sapphire

Guojie Chen
Foshan University

Wang Zhongke
Singapore Institute of Manufacturing
Technology (SIMTech)

CONTENTS

DOI: 10.1201/9781003182146-5

5.1 INTRODUCTION

Sapphire is the main substrate in the manufacturing of light-emitting diodes (LEDs) as it is the most economically viable material [1]. Since diamonds are harder than sapphire, diamond saw cutting is commonly used for machining sapphire [2]. However, the mechanical method results in a substantial material loss that is dictated by the width of the diamond saw blade employed. Due to the heat generated by the contact method, cooling of the material is required. The mechanical and thermal stresses induced on the sapphire produce outputs containing chips [3]. Therefore, the yield, the efficiency of the process and the quality of the outputs are reduced.

Lasers have been explored as an alternative contactless method to dice the hard but brittle sapphire wafers [4]. Laser scribing forms an indentation on the top surface of a sapphire wafer as a result of the high laser energy employed. After which, the sapphire wafer is separated into individual chips by a mechanical cleaving process [5,6]. It is well known that material optical absorption plays a dominant role in determining the laser machining capability. The material optical absorption is influenced by its surface morphology of the sample. Indeed, the rough patterned surface as seen in patterned sapphire substrates is desirable as it boosts the light extraction in LEDs through an increase in light scattering [7–9]. However, its apparent rough surface poses a challenge for laser singulation of such sapphire wafers.

Although the transparency of sapphire allows stealth dicing [6,10–12], its "apparent" absorptivity might differ depending on how light interacts with samples with different surface morphology. As such, a proper understanding on the interaction between light and sapphire with different surface morphology during laser irradiation would be vital. It would be highly desirable if the material surface morphology can lead to a higher level of material optical absorption required for laser machining of the sample.

5.2 SAPPHIRE WAFER

5.2.1 Sapphire Crystal Microstructure

Industrial grade sapphires are scientifically known as single-crystal corundum. Corundum is a type of mineral. Sapphire is also referred to as the α phase of aluminum oxide with a chemical formula of Al_2O_3 [13], also known as α-Al_2O_3 [14].

Al^{3+} cations and O^{2-} anions are bonded together in a hexagonal crystal lattice. Al^{3+} cations have a smaller ionic radius than O^{2-} anions, i.e., 0.57 and 1.40 Å, respectively. The larger O^{2-} anions can be imagined as spheres and are closely packed together. Due to the shape of the anions, an octahedron hollow is formed between six spherical O^{2-} anions. Two-thirds of these octahedron hollows are filled with the smaller Al^{3+} cations and therefore there is no symmetry center for the cations. The Al^{3+} cations distort the crystal lattice minimally and the structure remains stable.

A schematic representation of the hexagonal crystal lattice of sapphire is shown in Figure 5.1. In the schematic, looking in the direction of the *c*-plane axis,

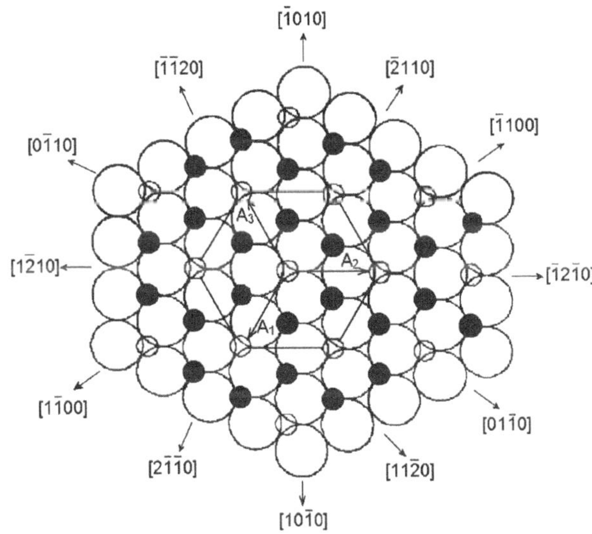

FIGURE 5.1 Schematic of crystal lattice of sapphire [14].

the larger O^{2-} anions are depicted as larger white circles, the smaller Al^{3+} cations as small black circles and the octahedron hollows as smaller white circles. Only a single lower basal plane of the O^{2-} anions were shown. Hence, the octahedron hollows appeared to be only surrounded by three O^{2-} anions. The <$11\bar{2}0$> translational vectors are depicted as A_1, A_2 and A_3.

Sapphire is grown upwards in the <0001> c-plane direction as illustrated in Figure 5.2. Since the Al^{3+} cations occupy two-thirds of the octahedron hollows,

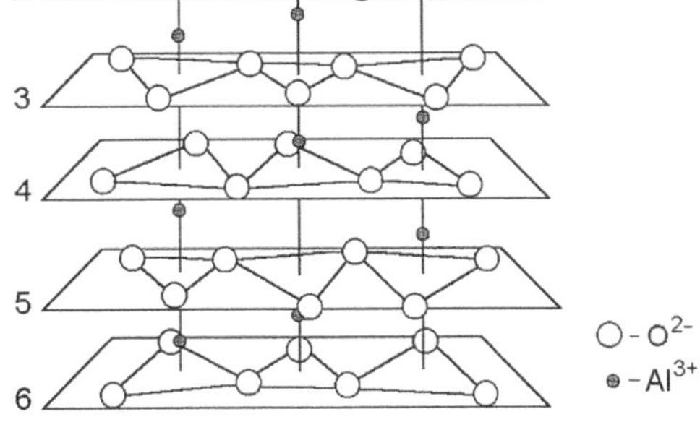

FIGURE 5.2 Schematic of the layering of Al^{3+} and O^{2-} ions in sapphire [14].

the positions of the unoccupied octahedron hollow changes in each layer and the patterns are repeated every three layers. Due to the layering of the larger O^{2-} anion spheres, the entire structure with both Al^{3+} and O^{2-} ions are repeated every six layers of each ion (i.e., Layers 1 and 7 are the same) [14].

5.2.2 Sapphire Properties

Sapphire is used as a substrate in many applications. Industrial sapphire is grown as single crystals. On the Mohs scale for crystal hardness, diamond is ranked the hardest crystal, with the highest value of 10 Mohs, and sapphire is ranked second with a value of 9 Mohs [15]. To quantify hardness in terms of Pascals, the sapphire has a high hardness of 22–25 GPa [16,17], which is second to diamond with a hardness value of 80–92 GPa [18,19]. Sapphire is chemically inert and optically transparent (for wavelengths ranging from ultraviolet to infrared) [19,20]. These properties enable sapphire to be a substrate in the manufacturing of watch covers, phone displays, integrated circuits and optoelectronics [21].

More importantly, sapphire is the main substrate in the manufacturing of LEDs as it is economically viable. Although silicon and silicon carbide have been tested as a substrate for LED manufacturing, from an economic and manufacturing quality standpoint, these materials are not deemed to be suitable for large-scale production [20].

5.2.3 Singulation of Sapphire

5.2.3.1 Diamond Wire Cutting

The dicing of sapphire wafers using diamond saw is the standard method and has been extensively studied. Kim et al. studied the effect of the initial deflection of a diamond wire on the variation of thickness of sapphire wafers [22]. A single diamond wire saw machine that was equipped with a core wire of 0.18 mm in diameter, with diamond abrasives of diameter 30–40 μm, was used. The sapphire ingot had a cross-sectional area of 50 mm × 50 mm. The study concluded that the variation in wafer thickness was highly dependent on the cutting speed and diamond wire wear, which can be controlled by fine-tuning the initial wire deflection. A consistent thickness was favored as post-processing time and production cost to remove the inconsistency could be minimized.

5.2.3.2 Laser Scribe and Break

Tamhankar et al. studied sapphire wafer singulation with the laser scribe and break technique [23]. A diode pumped solid state Q-switched laser with a wavelength of 355 nm was used. The laser had a pulse width of 18–42 ns and a repetition rate of 150 kHz. A 400 μm thick, single-crystal, single-side polished sapphire sample was used in the experiments. Laser was incident on the unpolished surface. Fluence, pulse width and repetition rate were varied in the parametric study. The cutting depth and kerf width were measured subsequently. The laser was focused within the material which resulted in deeper cuts. It was concluded that shorter pulse widths lowered the threshold for material removal, which resulted in deeper, cleaner and narrower scribes with a reduced heat-affected zone at lower fluence levels (see Figure 5.3). Higher repetition rates allowed for faster scribing speeds with deeper scribes (see Figure 5.4). A more noteworthy conclusion in the ablation of sapphire was that for higher repetition rates, it was better to split the excess energy over the optimal value into several beams at optimal energy.

Although Tamhankar et al. employed a single-side polished sapphire wafer with the laser incident on the unpolished surface, no reason was given for choosing this configuration. However, as surface roughness has an effect on light propagation direction, a better understanding can be gained through a study on the light-material interaction for different surface finishes, i.e., the difference between a smooth polished surface and a rough unpolished surface. With such knowledge, different mechanisms can be exploited for different machining purposes.

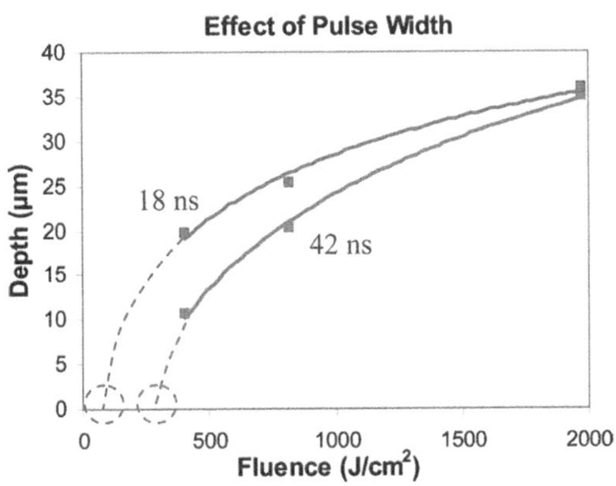

FIGURE 5.3 Graph of scribe depth against laser fluence on sapphire sample. A shorter pulse width resulted in deeper scribes with the same laser fluence [23].

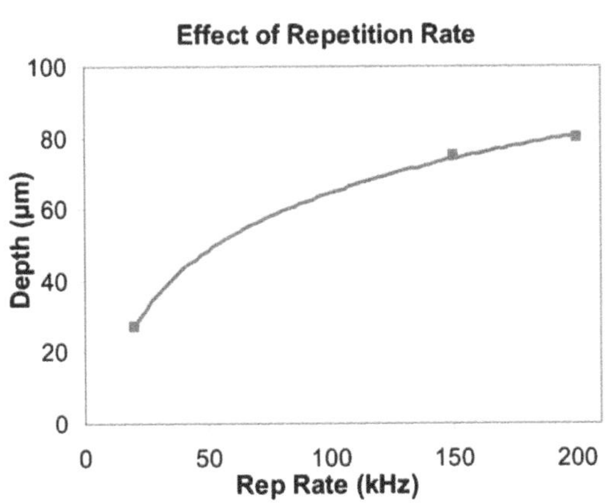

FIGURE 5.4 Graph of scribe depth against pulse repetition rate [23].

5.2.3.3 Stealth Dicing

Yadav et al. studied the stealth dicing of sapphire using near infrared (NIR) femtosecond lasers [24]. They employed an 800 nm (NIR range) Ti:Sapphire with a pulse width of 140 fs and a repetition rate of 1 kHz. Microscope objectives, with 100×, 20× and 5× magnifications, were chosen as focusing optics. The single-side polished sapphire wafer had a thickness of 350 μm. Even with a low absorption coefficient at a wavelength of 800 nm, it enabled stealth dicing to occur. Within the thickness of the transparent sapphire sample, a single focus spot was placed below the middle of the thickness. At high intensity levels, multiphoton absorption occurred. It was suggested that once the threshold for nonlinear absorption was surpassed, nonlinear absorption will occur at the laser focus. The damage will be localized and will not affect the other regions within the structure. The single focus within the sample will induce a line of stress points within the sapphire as a result of the moving laser beam. The sapphire can then be easily cleaved mechanically for a high-quality separation even for thick sapphire samples. The results from scanning electron microscope (SEM) and atomic force microscopy (AFM) are presented in Figure 5.5. Each spot received a hundred laser pulses. Even though ripples were observed around the focal spots due to nonlinear absorption, the mechanically cleaved regions were observed to be smooth and have a low surface roughness of about 20–50 nm (Figure 5.5b–d). In addition, chipping was observed on the top surface at the beam side (Figure 5.5a).

With a high pulse energy and high repetition rate, only one scan was required to weaken the internal structure. This enabled the process to be carried out quickly. Furthermore, this method was applicable for the dicing of blue or violet

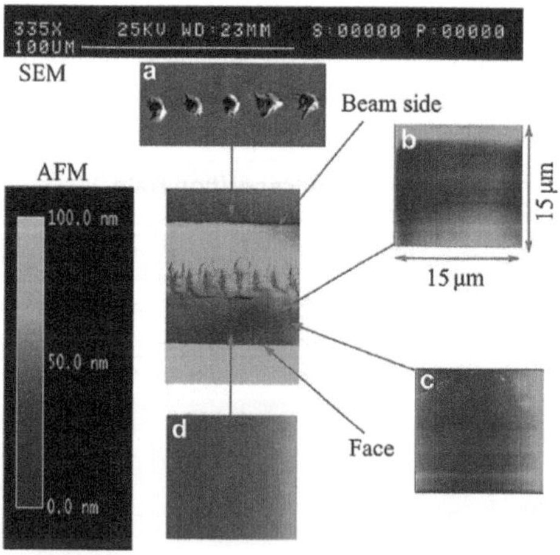

FIGURE 5.5 SEM and AFM results from stealth dicing of sapphire [24]. (a) damage of the edges on th beam side. (b) The roughness ditribution in the area near ripples. (c) The roughness ditribution in the middle area between ripples and bottom face. (d) The roughness ditribution in the area near the bottom face.

LEDs as the quality of the edge at the beam exit was good with reduced thermal damage and debris formation.

For stealth dicing, instead of using a single focus spot as in Yadav et al. [24], employing multiple focal spots simultaneously through the thickness of a sapphire sample could well be explored. With multiple foci, there will be more even distribution of laser energy through the thickness of a sample, with the potential of a better controlled dicing of the sample.

5.2.3.4 CO_2 Laser Thermal Cracking

Xu et al. controlled the thermal cleavage of sapphire wafers using laser irradiation [25]. A CO_2 laser with a wavelength of 10.6 μm was employed. The sapphire wafer used was 120 μm thick that housed a 5 μm-thick gallium nitride film. Their study concluded that a guide groove formed by laser ablation prior to wafer cleavage would guide the crack extension without deviation. The thermal cleavage resulted in a smooth cross-sectional surface of the sample without damage (see Figure 5.6). When processing successive scans, ample time between scans must be allowed for temperature and mechanical stresses to dissipate. Otherwise, the asymmetrical thermal and mechanical stresses from adjacent scans due to heat conduction will cause the crack to deviate from the initial scribed lines.

5.2.3.5 Laser-Induced Plasma-Assisted Cutting

Lee et al. devised a method to separate blue LED wafer via the laser-induced plasma-assisted ablation (LIPAA) technique [26]. The laser used was a Nd:YAG, Q-switched laser with a wavelength of 1064 nm. It had a pulse width of 150 ns and a power of 5 W. A galvanometer equipped with a f-theta lens with a focal length of 163 mm was employed for laser scribing. The experiment was carried out at atmospheric pressure without a vacuum. In LIPAA, the transparent sapphire was placed on a metal plate. The laser was focused through the sapphire onto the metal plate. After which, the high intensity laser generated plasma from the metal plate was then used to scribe the underside of the sapphire (see Figure 5.7). It was concluded that a nickel coated steel was the best choice for plasma generation.

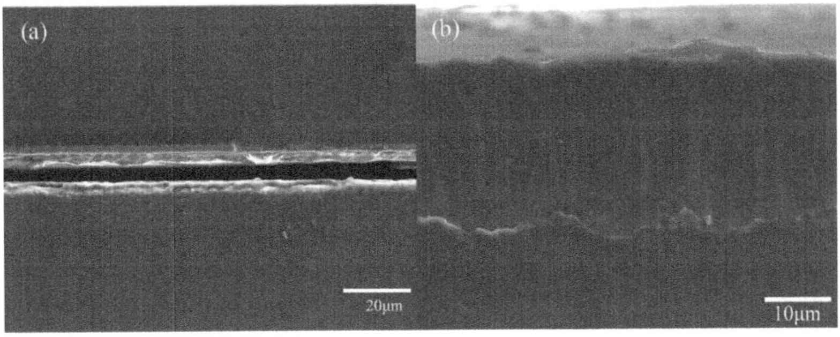

FIGURE 5.6 Optical images of sapphire after controllable laser thermal cleavage of (a) surface and (b) cross section [25].

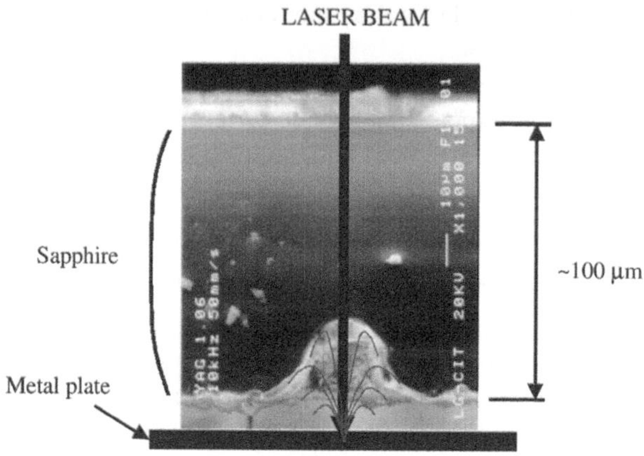

FIGURE 5.7 Laser irradiation through sapphire onto metal plate using the LIPAA technique [26].

Although no explanations or details were provided, it was observed that placing the wafer sample closer to the metal plate resulted in easier scribing, but it was more difficult to separate the wafer from the metal plate. To avoid the necessity of separating the wafer from the metal plate, the gap between the metal plate and wafer was kept at 100 μm. When the scanning speed was decreased, the depth of scribing increased. However, when the scanning speed was excessively slow, uncontrollable thermal cracks were observed.

5.2.3.6 Bessel Beam Cutting

Tsai et al. investigated the use of Bessel beams to separate fused silica [27]. Femtosecond Bessel beams were formed by a Ti:Sapphire laser with a wavelength of 800 nm and a regenerative laser amplifier source. The laser had a pulse width of 120 fs, a repetition rate of 1 kHz and a power of 250 mW. The Bessel beam was produced using an axicon lens with a wedge angle of 5° and a refractive index of 1.453. After the laser beam propagated through the axicon lens, the outspread laser intensity distribution of a Gaussian beam profile was transformed into a highly concentrated Bessel beam profile (see Figure 5.8a and b).

Two thicknesses of fused silica were used. The thicker glass was 45 mm thick, while the thinner glass was a flexible glass display with a thickness of 100 μm. On the thicker glass, the pulse number was increased to a maximum of 1000 shots at a single point and only a slight change in the diameter and depth of modification was observed (see Figure 5.9). Hence, the diameter and depth modification depended more on the geometry of the beam than the number of shots. The modified zone had an average width of 6.7 μm and a depth of 3 mm, with a variation in width of <9%. Therefore, the upper portion of the Bessel beam was selected for thin glass cutting for a more uniform cutting width and an even distribution of lateral residual stress. A scanning speed of 1 mm/s, which translated to 1 μm/pulse, was used for laser irradiation on the thinner glass. The thin glass

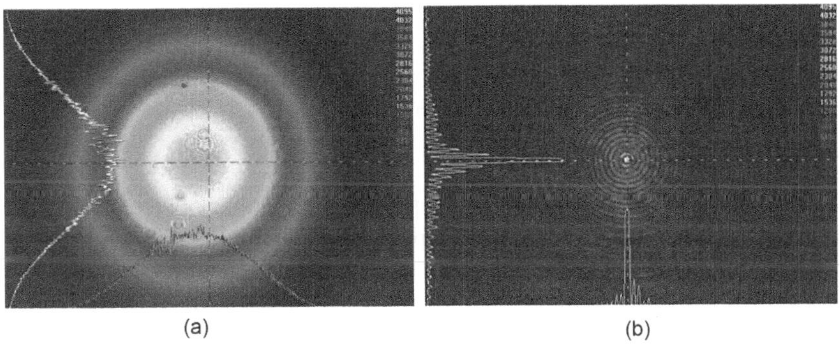

FIGURE 5.8 Beam profiles of (a) a Gaussian beam and (b) a Bessel beam [27].

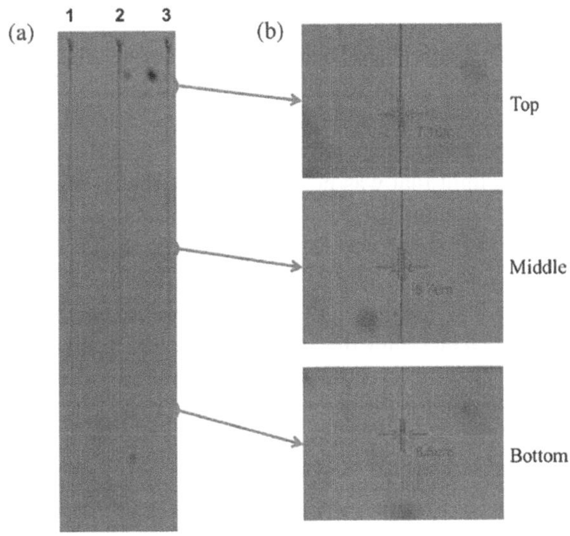

FIGURE 5.9 Cross-sectional view of laser irradiation within sapphire using Bessel beams of 1000 pulses for (a) overall view and (b) at the top, middle and bottom regions [27].

was well separated through internal modification and breakage, with chipping of <1 μm along the separation edge.

Rapp et al. utilized picosecond and femtosecond Bessel beams for stealth dicing of sapphire wafers through controlled crack formation and propagation [28]. The nature of the Bessel beams allowed uniform damage in the longitudinal direction. With a high repetition rate and a high energy level, shots were spaced out a few micrometers away from each other. This resulted in long vertical damages in the bulk sapphire thickness. The cracks then joined and thus formed a cleavage plane with less laser pulses. The transverse morphology of the top surface was vastly different for 3 ps and 140 fs pulse durations. For the picosecond pulses, three-fold cracks were observed that corresponded to the hexagonal lattice of oxygen in Al_2O_3 (sapphire). While for the femtosecond pulses, cracks were only formed in one direction. Therefore, femtosecond pulses enabled a fast separation of sapphire without tapering or chipping.

5.2.3.7 Multiple Foci Laser Cutting

Hitherto, there are no reports on laser cutting using multiple foci technology that has been implemented in the dicing of sapphire wafers. However, such technology has been studied intensively in the singulation of crystals and glasses.

Liu et al. explored using laser multi-focus separation (LMFS) technology in the separation of thick transparent potassium dihydrogen phosphate (KDP) crystals that were four times thicker than existing techniques [29]. Two lasers were employed for LMFS. Firstly, a femtosecond laser with a wavelength of 1030 nm was used to form pretreated lines within the crystal to increase the absorptivity and weaken the bonds of the material. Next, a second continuous wave fiber laser of 1070 nm wavelength was utilized to form the multiple foci within the crystal. The 50 mm thick crystal was placed between two concave reflectors, one of which housed a small pore that was sufficiently large for the laser beam to pass through while maximizing the reflection of the laser (see Figure 5.10). The laser beam was first focused by the focusing lens within the crystal. Since the transparent crystal has low absorptivity, light that was not absorbed by the crystal will be reflected by the rear concave reflector back into the crystal again at another focus. The cycle repeated until the energy was fully absorbed by the crystal through either linear or nonlinear absorption. With this setup, three foci were formed within the crystal. As a result of heating within the crystal, the temperature difference between the crystal surface and the interior induced a higher tensile stress to form at the surfaces. Thus, microcracks formed on the surfaces quickly extended to the middle of the crystal, forming a penetration crack. The resulting sidewall profile was smooth with some ripples observed near the top and bottom regions due to a tearing effect from the higher tensile stresses induced. Although the MFLS technique succeeded in producing several foci within the crystal, but the equipment employed did not appear to be adaptable for other transparent materials with different thicknesses. The number of foci formed was highly dependent on the laser power and the absorptivity of the material. For thinner materials, the

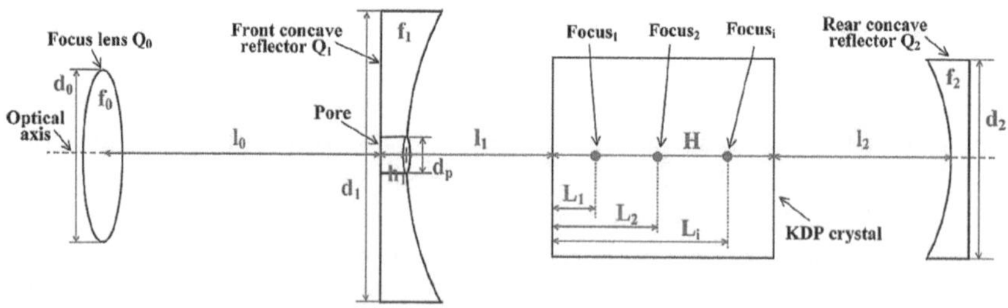

FIGURE 5.10 Schematic of MFLS setup with two concave reflectors [29].

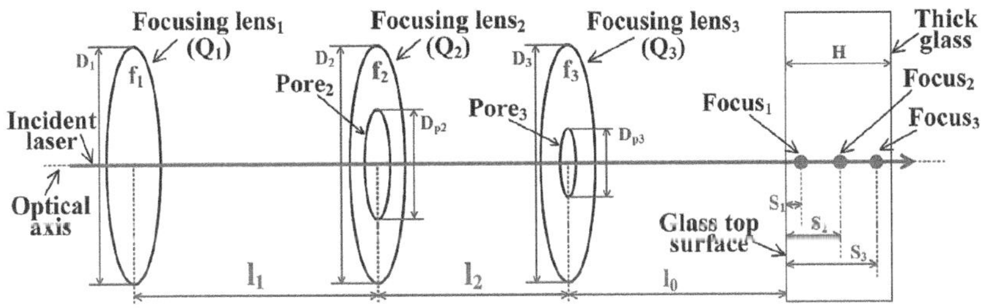

FIGURE 5.11 Schematic of FMFLS setup of three coaxial focusing lenses [30].

secondary and tertiary foci may be entirely out of the material. If the absorptivity of the material was too high, it might be impossible to form a secondary focus with enough energy to induce the penetration crack.

Liu et al. developed the flexible multi-focus laser separation (FMFLS) technology to separate a thick glass sample that is two to four times thicker than conventional separation techniques using lasers [30]. A fiber laser with a wavelength of 1070 nm, a power of 80 W and a scanning speed of 1 mm/s was used. A soda-lime plate glass that was 20 mm thick was employed in the experiments. The optics included three coaxial focusing lenses, with each housed a hole of different diameter as seen in Figure 5.11. The outermost part of the beam would be focused by all three focusing lens and had the shortest focal length. While the center part of the beam would be focused only by the first lens and had the longest focal length. This allowed the focusing of each focus with high precision and at the same time controlling the power density of each focus. Laser power at 80 W with a scanning speed of 1000 μm/s were determined to be the optimal parameters for cutting the sample. Increasing the power level or decreasing the scanning speed away from the optimal levels resulted in the formation of slant ripples on the separated sidewall. When the parameters were set further away from the optimal values, deeper ripples, microcracks and even uncontrolled crack deviation were observed. This was attributed to the difference in cooling rates of the separated surface and the interior of the glass. The FMFLS technology was especially suited for cutting glass with varying thickness for aerospace, military and optics applications. To separate an even thicker glass, additional focusing lenses could be introduced, thereby increasing the number of foci, and generating a uniform thermal stress distribution, which was an important factor in the separation of a thick glass.

5.3 EFFECT OF SURFACE MORPHOLOGY ON THE OPTICAL PROPERTIES OF SAPPHIRE

5.3.1 Surface Roughness

The 50.8 mm diameter and 0.43 mm thick single-crystal Al_2O_3 sapphire wafers from Latech Scientific Supply Pte. Ltd. (Singapore) were employed in this study. The wafers were either double-side polished or single-side polished. Sapphire has

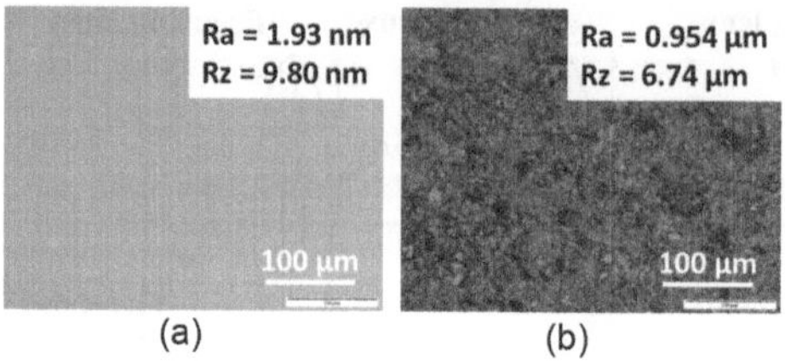

FIGURE 5.12 Optical images of (a) a polished surface and (b) a grinded (unpolished) surface.

a refractive index of 1.768 and band gap energy of 9.9 eV. A stylus profilometer Form Talysurf Series 2 by Taylor Hobson measured the surface roughness of the sample's surfaces, i.e., polished and unpolished (grinded rough) surfaces.

Figure 5.12 shows the optical images of a polished and grinded (unpolished) surface at 20 × magnification. The stylus moved in a 10 mm line on the sample surface to measure the surface roughness of the polished or grinded surface. The roughness values of each surface finish were measured 12 times in two directions (6 times in each direction), i.e., 0° and 90°, respectively.

The polished surface had average roughness values of Ra=1.93 nm and Rz=9.80 nm; the standard deviation for Ra is 0.00015 µm and Rz is 0.00061 µm. The unpolished surface had surface average roughness values that were significantly higher at Ra=0.954 µm and Rz=6.74 µm; the standard deviation for Ra is 0.059 µm and Rz is 0.544 µm. For both surface finishes, the surface roughness had no orientation effect as the roughness values were independent of the direction of measurements (i.e., 0° and 90° directions).

5.3.2 Linear Optical Property

Linear optical properties of sapphire wafer were characterized with a PerkinElmer Lambda 1050 UV/Vis/NIR Spectrometer. Optical transmittance and reflectance of each sample were measured for wavelengths ranging from 300 to 1200 nm, with 5 nm intervals. Three orientations of the sapphire samples were tested to analyze the surface roughness effect on light transmission. Sample *A* had two polished surfaces; Sample *B* was a single polished sample with its unpolished (rough) surface facing toward the light source; and Sample *C* was a single polished sample with its polished surface facing toward the light source.

When a parallel white light was incident on the sapphire samples, the transparent nature of the samples meant that most of the light were transmitted through the samples. Transmittance was always the largest proportion, followed by reflectance and in some cases, absorbance. However, light interaction differed for each sample and is highly dependent on the surface morphology.

The spectrometer used ambient light and was highly sensitive to the fluctuations in water vapor and carbon dioxide content. At very low absorbance region and high sensitivity of the experiments, some of the total values measured for reflectance and transmittance were greater than 100% due to experimental inaccuracies; this resulted in negative values calculated for absorbance which is physically impossible. As such, the absorbance value is set to zero indicating a very small or negligible absorption component at the measured wavelength.

Figure 5.13a shows the optical properties of transmittance and reflectance for double-side polished Sample *A* with a surface roughness *Ra* of 2 nm. The sample displayed an optical transmittance of about 85% with no or negligible absorbance.

Figure 5.13b shows the optical properties of transmittance, reflectance and absorbance for a single-side polished Sample *B* with its rough surface facing the spectrometer. The single grinded rough surface had a surface roughness *Ra* of ~1 µm (see Figure 5.13b). The transmittance was decreased. This is likely caused by the enhanced total internal reflection of light at the bottom smooth surface, reflecting and trapping the light within the material. This occurred possibly due to the top rough surface diffusing the light, and thus resulting in some light having an incident angle larger than the critical angle for total internal reflection when the light reached the bottom surface. The sample displayed some absorbance at shorter wavelengths (i.e., from 325 to 900 nm). With this rough entry surface, the transmittance was decreased to ~80% and absorption of 1% within the range of visible light.

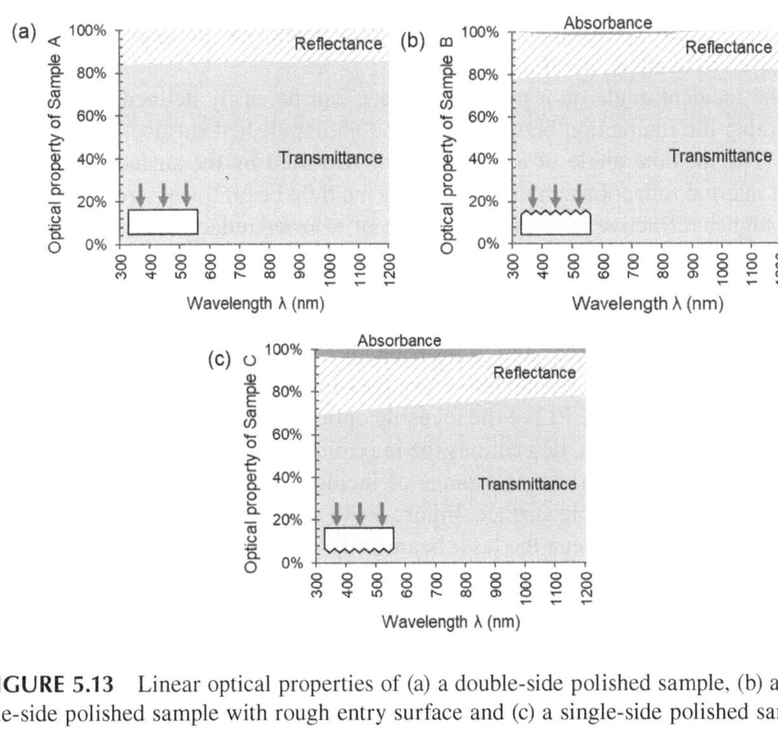

FIGURE 5.13 Linear optical properties of (a) a double-side polished sample, (b) a single-side polished sample with rough entry surface and (c) a single-side polished sample with rough exit surface.

Figure 5.13c shows the optical properties of transmittance, reflectance and more absorbance for a single-side polished Sample *C*, with its polished surface facing the spectrometer. The sample displayed more absorbance within the range of wavelengths tested (i.e., from 300 to 1200 nm). Compared to Sample *B*, at the bottom rough surface of Sample *C*, more light will have an incident angle greater than the critical angle when the light interacted with the irregular rough surface, resulting in total internal reflection. The presence of total internal reflection at the bottom rough surface increased the amount of light absorption as the length of a light beam path within the material was increased. As a result, there was less light transmission and reflection by the sample. With the introduction of a rough exit surface, the transmittance was decreased to ~70% and increased absorption to about 5% within the range of visible light.

As the wavelength increased, transmittance increased while reflectance decreased, regardless of the samples tested. Double-side polished Sample *A* had the largest amount of transmittance, followed by the single-side polished Samples *B* and *C*. In contrast, reflectance was the highest for the rough exit surface Sample *C*, followed by Sample *B* and then Sample *A*. Absorbance was only detected in samples with a rough surface (i.e., Samples *B* and *C*), with the highest absorbance detected in Sample *C*. This indicates the effect on linear absorbance by a grinded rough surface causing total internal reflection. This provided a plausible explanation for the observed increased absorbance, especially in Sample *C*.

Further investigations on the effect of the polished and grinded rough surfaces on the linear optical property was conducted. Instead of the parallel white light used in the spectrometer, a focused laser beam (i.e., an angled beam) was employed to irradiate the three samples.

The incident angle on a polished surface can be easily defined. In contrast, analyzing the interaction between light and an unpolished surface is more complex. The incident angle of a light beam is distorted by the surface roughness. Total internal reflection can only occur when a light beam travels from a medium of a higher refractive index to a medium of a lower reflective index when the critical angle of incidence is exceeded. Thus, the orientation of the unpolished surface, either the entry or exit surface, will affect the optical properties of the sample (i.e., light transmission, reflection and absorption).

Figure 5.14a illustrates how a laser beam can be refracted with the aid of a focusing optic and subsequently converges to a spot at the focal length. The effective focal length (EFL) of the focusing optics employed dictates the focused spot size, which in turn, determines the maximum divergence angle of the laser beam [31]. Therefore, there is a range of incident angles when the laser beam interacts with the sample surface. Figure 5.14b and c illustrates some examples on the interaction between the laser beam and a double-side polished Sample *A*. Figure 5.14d and e shows examples on the interaction between the laser beam and a single-side polished Samples *B* and *C* respectively. Subscripts 1 and 2 represent the entry and exit surfaces respectively and subscripts *I* and *r* represent the incident and refracted rays, respectively. In some cases, the subscript *r* also represents the internally reflected ray (e.g., in Figure 5.14e). α denotes

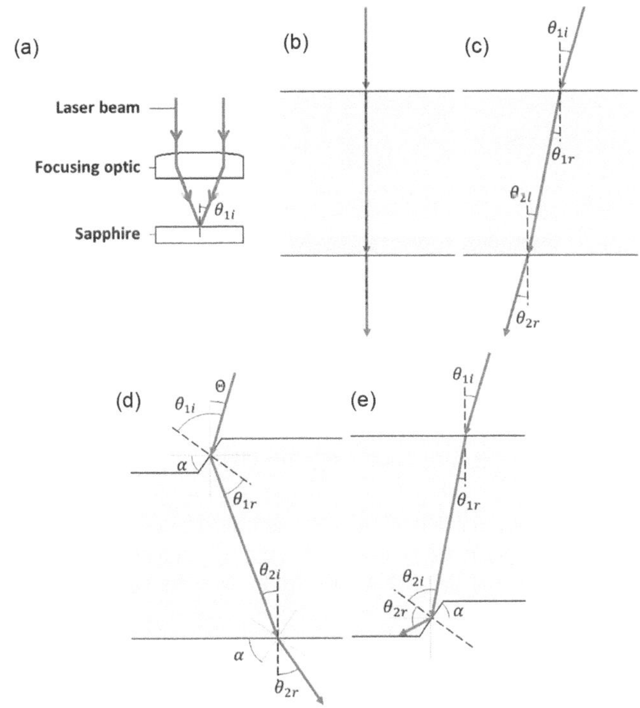

FIGURE 5.14 Illustration of a focused laser beam (a) incident on sample surfaces, interaction with (b and c) Sample A, (d) Sample B and (e) Sample C.

an idealization of the plane of surface roughness with respect to the reference polished surface.

Figure 5.15a and b illustrates the divergence half angle and the idealized plane of surface roughness with respect to the reference polished surface. As an example, when a focusing lens of EFL 4.38 mm was employed, the divergence half angle was calculated to be 32.7°. The critical angle for a sapphire sample, with a refractive index of 1.768, was calculated to be 34.4° [32]. Thus, at the exit surface, if the incident angle exceeds 34.4°, the light beam will be reflected into the sapphire sample. The light interaction at the exit surface can be assessed for all samples by varying α (see Figure 5.15c). It indicates the range of α when light will undergo total internal reflection or transmitted through the respective samples as dictated by Snell's law [32].

According to Figure 5.15c, the interactions between light beams and samples in Figure 5.14 are further discussed below. Figure 5.14b and c illustrates the interaction between light beams and Sample A. The two parallel polished surfaces of Sample *A* do not facilitate the occurrence of total internal reflection within the sample as it is impossible for the incident angle at the exit surface to be greater than the critical angle. Therefore, theoretically all light incident on Sample *A* can

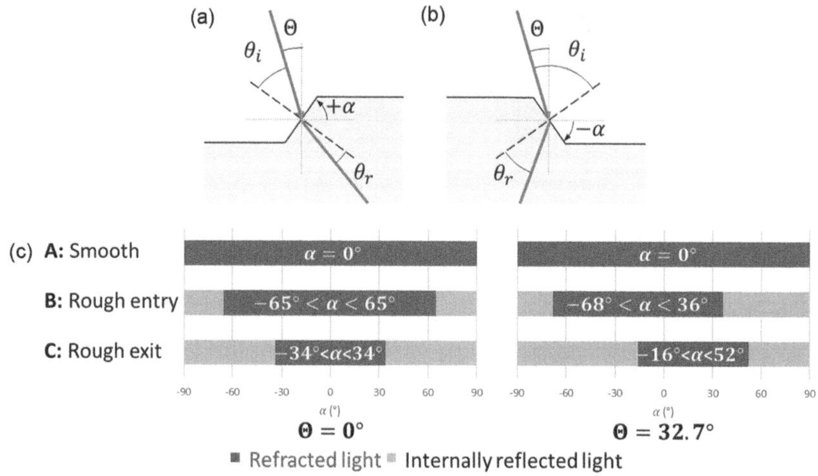

FIGURE 5.15 Schematic for definitions of α and divergence half angle θ: (a) where $0° \leq \alpha \leq 90°$ (clockwise: positive α) and (b) $-90 \leq \alpha \leq 0°$ (counter-clockwise: negative α). (c) Laser beam interaction at the exit surface for the various samples, where $\Theta = 0°$ or 32.7°.

be transmitted through the sample (see Figure 5.15c). However, in reality, a small portion of incident light is reflected by the shiny polished surface [33]. In contrast, the rough surface of Samples *B* and *C* in Figure 5.15c facilitates the occurrence of total internal reflection within the samples.

Figure 5.14d illustrates the interaction between light beams and Sample *B* in Figure 5.15c. A portion of the incident light diffuses on the top rough surface of Sample *B*. As a result of the top rough surface of Sample *B*, it causes light to be refracted into the sample at different angles. Consequently, these light beams will have different incident angles on the bottom surface. Since some light beams will have an incident angle greater than the critical angle, a small portion of light incident on the bottom surface will be internally reflected into the sample. In most cases, the incident angle at the bottom exit surface is smaller than the critical angle and light will exit out of the sample (see Figure 5.15c). With a focusing lens of EFL 4.38 mm as an example, when the laser beam with $\Theta = 0°$ is incident on the rough entry surface (i.e., the central light beam), light can be transmitted through the sample for the range of plane of surface roughness angles between $-65° < \alpha < 65°$. When the laser beam with $\Theta = 32.7°$ was incident on the rough entry surface (i.e., at maximum half divergence angle), the amount of light transmitted through the sample was reduced; light can only be transmitted when the range of plane of surface roughness angles is between $-68° < \alpha < 36°$.

Figure 5.14e illustrates the interaction between light beams and Sample *C*. At the top polished surface, similar to Sample *A*, most of the light beams will be refracted into Sample *C*. In contrast to Sample *A*'s polished bottom surface, light forms different angles of incidence with the rough exit surface of Sample *C*, with some of these angles exceeded the critical angle. When the incident angle exceeded the critical angle at any surface, the light beams will be totally internally reflected and thus trapped within the sample. Sample *C* would have the least amount of transmitted light due to a larger portion of light totally internally reflected within the sample as compared to Samples *A* and *B* (see Figure 5.15c). With a focusing lens of EFL 4.38 mm as an example, when $\Theta = 0°$, light can be transmitted through the sample for the range of the plane of surface roughness angles between $-34° < \alpha < 34°$. When $\Theta = 32.7°$, the range of plane of surface roughness angles for light transmission remains the same, but it has shifted to be $-16° < \alpha < 52°$. The range of plane of surface roughness angles for light transmission of Sample *C* is less than that of Sample *B*.

Smaller range of α angle for light refraction out (and thus transmission) of the sample indicates higher probability of total internal reflection and therefore higher light absorption in sapphire. This suggests that Sample *C* could be a better laser machining candidate as total internal reflection at the rough exit surface is the most efficient in trapping light within the sample. Therefore, with high intensity laser irradiation, more energy accumulated could lead to more nonlinear absorption within the material for machining applications. Indeed, these analyses indicate that the range of the α angle for transmission, which has a direct impact on total internal reflection, is not only a function of surface roughness, but also the orientation of the rough surface to the light source.

When a longer EFL is employed in the experiments with the single-side polished Sample *C*, the divergence half angle of the laser beam will be reduced. However, the amount of light transmitted through the sample would remain approximately the same as the range of α angle for light refraction out of the sample would remain the same but have merely shifted.

5.3.3 Nonlinear Optical Absorption

Characterization of nonlinear absorption properties through the measurement of the defocused laser power was performed with a PowerMax PM30 power sensor with a FieldMaxII-TO laser power meter from Coherent, Inc. (California, United States). Figure 5.16 illustrates the setup with a sample. A flat top beam shaper facilitated the analysis of the measured power transmission. The location reference height z was positioned at the sample's top surface.

The nonlinear absorption threshold for each sample was determined by adjusting the laser power output to approximately 5%, 20%, 40%, 60%, 80% and 95% of the total power output of 2.4 W. Once the nonlinear threshold was surpassed, additional data points were supplemented, at power level increments of 5%, between the two power levels to increase the precision of nonlinear threshold determination. The experiments were conducted three times, each time at a fresh, unmodified spot. There was no overlapping of laser spots when the power transmitted measurements were taken for each data point.

The effect of changing the focus position on the threshold for nonlinear absorption was examined. The focus spots tested were positioned at the top

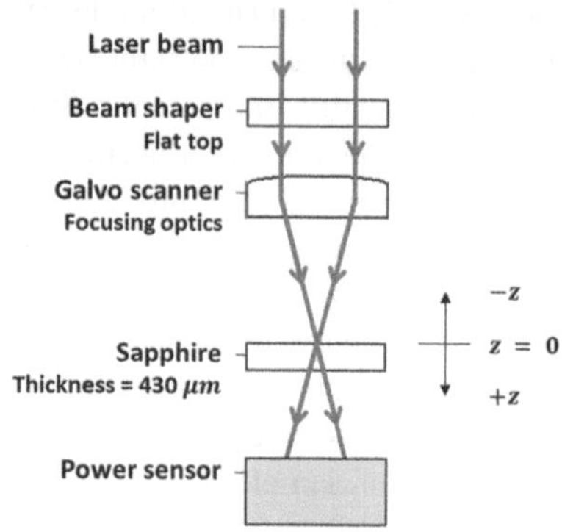

FIGURE 5.16 Schematic representation of nonlinear absorption power measurement setup.

surface (z=0 μm), midplane (z=+210 μm) and bottom surface (z=+430 μm) of the samples.

The laser power output was measured before and after all experiments, without the sapphire samples, and the stability of the laser could be deduced. The laser was stable regardless of the beam profile employed. At a laser power of 2.2 W (i.e., the highest laser power employed in the experiments), for a flat top beam profile, the standard deviation ranges from 0.005 to 0.009 W; for a Gaussian beam profile, the standard deviation ranges from 0.004 to 0.007 W. The maximum percentage change in the laser power output was 0.4% (i.e., the laser was stable). Under the assumption that if the laser power output was stable, then it could be implied that the power of a single pulse was also stable.

Laser processing of sapphire requires ultrashort laser pulses of high intensity to induce nonlinear absorption for internal modification of the material through the weakening of bonds. Therefore, surface morphology on the nonlinear absorption threshold is an important consideration for material processing.

For most practical applications, a Gaussian beam distribution is used for material processing. However, its spatial non-uniform power intensity distribution meant that the region of nonlinear absorption cannot be clearly distinguished; this would complicate nonlinear absorption analysis. Instead, this study employs a flat top laser beam distribution to facilitate ease of analysis. The advantage of a flat top distribution is that the intensity at the focus spot is rather constant. Hence, deviations from the expected result of linear absorption can be easily detected and identified for the determination of the nonlinear absorption threshold. A flat top distribution can be generated from a Gaussian distribution with a beam shaper.

For a constant laser power source, the nonlinear absorption threshold may be defined as the instant when the power absorbed by the sample was no longer constant with a change in the laser power intensity irradiated on the sample. Alternatively, with an increase in laser power, the nonlinear absorption threshold may be identified by detecting the deviation of linear proportionality between the absorbed laser power and the laser power source. Since it is challenging to directly measure the amount of power absorbed by the sample, the amount of power transmitted through the sample can be measured instead as there are similar characteristics exhibited between absorption and transmission. As such, this nonlinear absorption threshold could be determined by identifying the instant where the percentage of transmitted power through the sample has been reduced.

An experiment where only linear absorption was present in the samples was conducted as a reference for subsequent studies on nonlinear absorption. To provide the highest probability for nonlinear absorption, the power level of the laser was set at 95% of the maximum power output. To determine z height (see Figure 5.16) at which the sample only experienced linear absorption, the distance between the focal position and the top surface of the sample was increased incrementally (i.e., in the negative z direction) until nonlinear absorption was not observed. Some surface damage was observed on the sample's surface when a drop in the amount of laser power transmitted was detected. Thus, a previously untested location or sample was used for each experiment.

Figure 5.17 shows the power transmitted against the changing z height. Three measurements were carried out. The standard deviation ranges from 0.002 to 0.059 W for linear absorption and 0.004 to 0.139 W for nonlinear absorption. The highest standard deviations were mostly clustered around the nonlinear absorption threshold, with lower standard deviations for data points further away from the threshold. These low standard deviations indicate the repeatability of results.

The transmitted power measured remained relatively constant when the focal position was adjusted between 2 to 10 mm above the sample's top surface (i.e., from $z=-2$ mm to $z=-10$ mm). This indicates a rather constant absorbed power and that the samples experienced only linear absorption. Therefore, a focal position located at 5 mm above the top surface (i.e., $z=-5$ mm) could satisfactorily ensure that only linear absorption was present for all the three Samples *A*, *B* and *C*. The condition of focusing at 5 mm above the top surface of the sample where the power transmission was measured was used as a linear absorption reference (i.e., without nonlinear absorption).

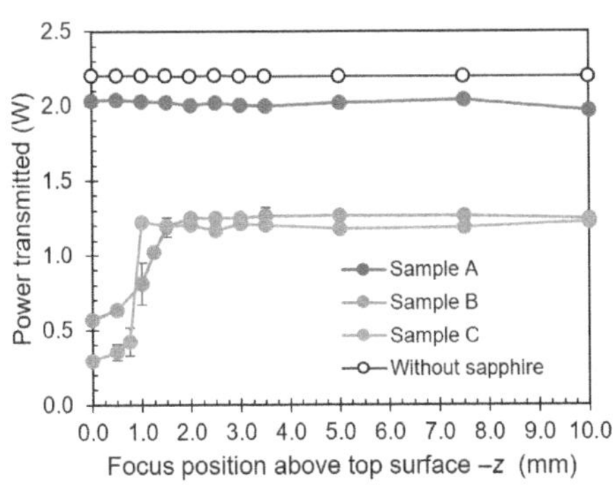

FIGURE 5.17 Power transmitted against focal position above top surface for all samples at input power of 2.3 W which was 95% of maximum power (2.4 W).

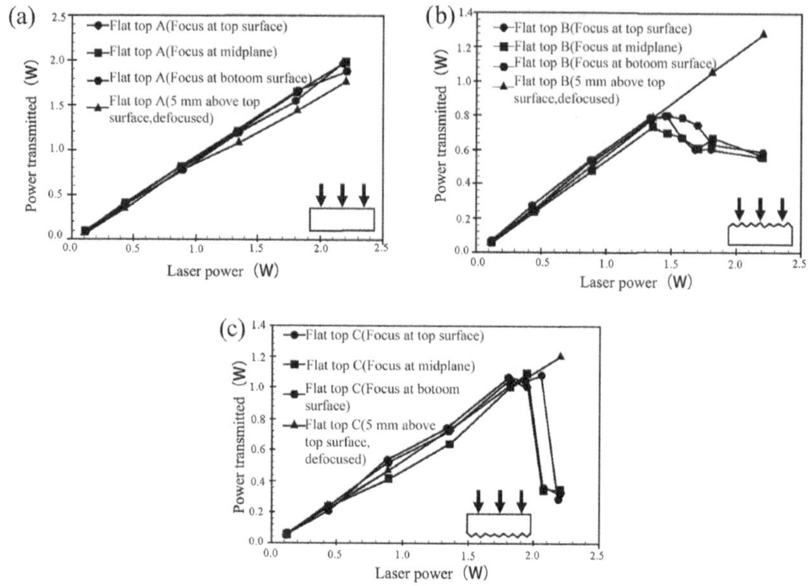

FIGURE 5.18 Power transmitted against the laser input power irradiated for (a) Sample A, (b) Sample B with a top rough surface and (c) Sample C with a top polished surface. Legend: dot: focus at top surface, square: focus at midplane, hexagon: focus at bottom surface, triangle: 5 mm above top surface (defocused). (Note: Error bars with values <0.045 W cannot be observed in the figures.)

Figure 5.18a shows that the power measured after the laser passed through Sample *A* was rather linearly proportional to the power irradiated on the sample. The percentage of power transmitted did not fluctuate significantly when the laser power was increased. The standard deviation ranges from 0.001 to 0.042 W. This indicates that only linear absorption was present within the double-side polished smooth sapphire wafer.

Figure 5.18b shows the power transmitted for various input laser powers for Sample *B* with a rough entry surface. As the incident irradiation power to the sample was increased, a drop in the transmitted power was distinctly observed. Figure 5.18b indicates that nonlinear absorption occurred at input laser power of ~1.4 W. The standard deviation ranges from 0.001 to 0.025 W for linear absorption and 0.002 to 0.100 W for nonlinear absorption. Intuitively, due to the increased surface roughness of Sample *B*'s top surface, a portion of the reflected light at the top surface could be reflected back onto the material. As a result, light was trapped near the top surface and accumulated energy within this region. When the level of energy accumulated was sufficiently high, nonlinear absorption within the sample could be induced.

Furthermore, it was observed that the nonlinear absorption threshold was insensitive to the laser focal position within the sample thickness. The calculated

Rayleigh length was 417 μm and that was comparable to the 430 μm sample thickness. Since the depth of focus of the laser is twice the Rayleigh length, this meant that the laser intensity did not vary significantly over the thickness of the sample. Therefore, the fluence and energy deposition onto the sample would be similar regardless of the focal position of the laser.

When the incident power exceeded the power threshold of 1.4 W for nonlinear absorption, the amount of power measured gradually decreased, even when the laser power output was steadily increased. A large portion of the reduction in power transmitted could be the result of nonlinear absorption within the sample, with the assumption that the reflectivity of the sample had little effect or remained relatively constant. When the laser power output was increased to 2.3 W (95% of maximum power of 2.4 W), the percentage in power transmitted through the sample fell to about 26%. This trend suggests that further increasing the power output could lead to a reduction in percentage of power transmission, and hence, a more pronounced nonlinear absorption.

A parallel beam of light was employed in the spectrometer. Hence, the extent of light scattering was low and most of the transmitted light could be detected by the sensors. In contrast, for results reported in Figure 5.18b, the convergent nature of a laser beam caused light to be incident on the sample surface at various angles; this would result in more light scattering as well as additional light absorption at the rough surface. As such, the amount of transmitted laser power measured would naturally be much lower.

Figure 5.18c shows the power transmitted for various input laser powers for Sample *C* with a top smooth surface and bottom rough surface. The standard deviation ranges from 0.001 to 0.031 W for linear absorption and 0.003 to 0.059 W for nonlinear absorption. The nonlinear absorption occurred at power level of about 1.9 W which was higher than the laser power for Sample *B*, which had a top rough surface and bottom smooth surface. The threshold was relatively constant regardless of the focal position within the sample. Interestingly, even though Sample *C* displayed the highest linear absorbance as discussed previously, Sample *C* did not have the lowest threshold of nonlinear absorption. Instead, to reach the nonlinear absorption threshold, Sample *C* required a larger amount of laser power of about 1.9 W as compared to Sample *B*'s 1.4 W. Even more counter intuitive, although Sample *C* had a higher threshold, once the nonlinear threshold was exceeded, the rate of decrease of transmitted power (and thus the rate of increase of absorbed power) was much higher for Sample *C* than Sample *B*.

These experimental observations indicate that Sample *B* had the lower nonlinear absorption threshold of about 1.4 W; however, it is likely that the volume of material affected by nonlinear absorption would be small. The study by Yadav et al. shows that once the laser power exceeded the nonlinear absorption threshold, strong multiphoton absorption will occur at the high intensity laser focus [6]. Nonlinear absorption only results in localized damage and will not affect the surrounding regions. Since Sample *B* has a rough top surface which is effective in trapping light, it is logical to deduce that most of the laser energy would be accumulated within this region. Nonlinear absorption would thus be mainly confined to the top surface. This strong absorption nonlinear region at the top surface acted as a barrier in preventing energy from transmitting downwards and out of the sample.

Comparing Sample *C* with Sample *B*, the observation of a higher nonlinear absorption threshold of Sample *C* (about 1.9 W) can be explained by the lack of a top rough surface to assist in the trapping of laser power. In addition, it can be expected that due to linear absorption, the incident laser intensity at the top surface would be higher than at the lower surface. This will be the case even with the focal position placed on the bottom surface of the sample as the Rayleigh length is in the same order as the sample thickness. This was a result of energy loss due to linear absorption within the bulk material.

However, with increasing laser power and total internal reflection at the bottom rough surface of Sample *C*, the resulting laser intensity as a sum of the incident and reflected laser intensity could well exceed the nonlinear absorption threshold at the bottom surface region before the top surface.

Comparing Sample *C* with Sample *A*, Sample *A* had two polished surfaces, while Sample *C* had a top polished surface and a bottom rough surface. Since both samples had top polished surfaces, for the same laser power, the state at the top surface and the sample interior would be similar. The only difference between the samples will be the surface roughness of the bottom surface. However, since no nonlinear absorption was detected in Sample *A* even at high power levels, this observation provides validation that the bottom rough surface of Sample *C* was the cause of the observed nonlinear absorption. This observation provides further evidence that the interaction between the totally internally reflected light at the bottom rough surface and the incoming laser beam had raised the laser power intensity beyond the threshold to induce nonlinear absorption in Sample *C* at the bottom surface.

When laser power exceeded the 1.9 W nonlinear threshold, the transmitted power sharply decreased in Figure 5.18c for Sample *C*. There was an ~40% decrease in transmitted power once nonlinear absorption was induced in Sample *C*. As the power was increased, due to total internal reflection at the rough exit surface, increasingly more energy was accumulated in the region near the bottom surface. The interaction between the total internal reflected laser beam and the incoming laser beam increases the intensity of the laser energy, triggering nonlinear absorption in this region. This region could expand from the very vicinity at the bottom surface to induce nonlinear absorption into the surrounding area, in particular along the optical axis. An increase in nonlinear absorption in the sample was observed through the drastic fall in power transmitted. This provides the explanation that once the nonlinear absorption threshold was reached, slightly increasing the laser power could result in a significant increase in nonlinear absorption. This is due to the increased effectiveness of trapping laser power through total internal reflection with increasing incident laser power. This explains the rather counter intuitive observation of the rapid decrease in transmitted laser power once the nonlinear threshold was exceeded for Sample *C* as compared to Sample *B*, despite Sample *C* having a higher threshold than Sample *B*.

5.3.4 Picosecond Laser Scanning of Sapphire

A study was conducted to explore if the effects of surface roughness and its orientation to the laser power source may be exploited for laser processing of transparent samples with a single roughened surface. The picosecond laser beam was employed to scan the sapphire wafers. For a practical consideration, a Gaussian beam profile was used instead of the flat top beam. The laser powers levels were set to be 25%, 50% and 75% of the maximum power output of 2.4 W. To be free of debris, it has been suggested that the optimal scanning speed for ablation scribes is to exceed 200 mm/s for glass related materials [34]. In this study, a speed of 240 mm/s was employed to scan the sapphire samples. A high peak power with high intensity for scanning was produced by a 50-kHz pulse repetition rate. It provided a higher probability of inducing nonlinear absorption within the sapphire samples at these higher powers. The number of pulse overlap was calculated to be 5.65 pulses at 240 mm/s [35]. The calculated fluence incident on a sample's surface for each laser power setting is presented in Table 5.1. To ensure consistent scanning results, three lines were scanned on each sample, with each line scanned with a single pass.

Figure 5.19 shows the optical images of the top and bottom surfaces for Sample *A* (double-side polished). With insufficient laser energy at 0.5 W, no scribes were observed on either the top or the bottom polished surface.

TABLE 5.1
Power Level and the Corresponding Fluence

Power Level (%)	Actual Power (W)	Fluence (J/cm^2)
25	0.5	1.73
50	1.1	3.82
75	1.7	5.90

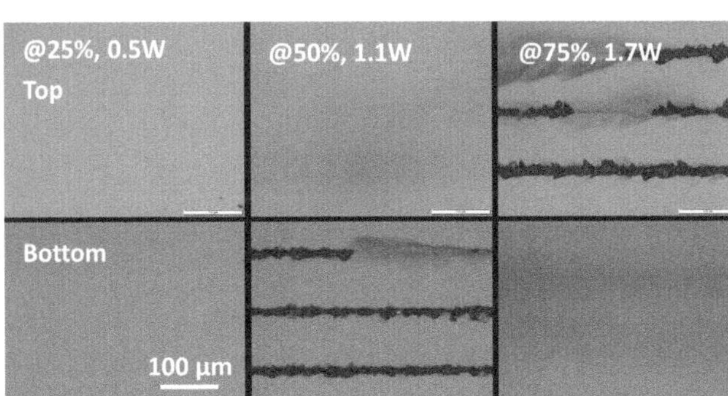

FIGURE 5.19 Optical images of double-side polished sapphire Sample A irradiated under different laser power levels.

At 1.1 W, crack-dominated scribes were observed on the bottom surface. As discussed previously, reflection naturally occurred at an interface between two materials. The interaction between the reflected light and the incoming laser light resulted in energy accumulation at the region near the bottom polished surface. When the accumulated energy exceeded a certain value, microcracks were produced.

When the laser power was increased to 1.7 W, crack-dominated scribes were observed on the top polished surface instead of the bottom surface. It indicates that a power of 1.7 W from the incoming laser beam alone was sufficient to induce crack-dominated scribes on the top surface of the sample. It was also observed that the cracks produced many micro-surfaces and each facing a different random direction. These micro-crack surfaces could reflect light onto other micro-surfaces when a laser irradiated the sample. Therefore, the micro-surfaces formed near the top surface acted as a rough surface shield preventing laser energy from propagating to the bottom surface. Insufficient energy at the bottom surface meant that there was no crack-scribe formation at the bottom surface.

If light can be trapped effectively to accumulate laser energy exceeding the nonlinear threshold and beyond the ablation threshold, ablation could occur to induce ablation scribes. At areas where there was insufficient energy for ablation, only cracks were observed. As such, the lack of nonlinear absorption in Sample *A* did not manage to produce ablation scribes and thus was not conducive for machining.

Figure 5.20 shows that the optical images of the top and bottom surfaces for Sample *B* (with a top rough surface and a bottom smooth surface). Scribes were only observed on the top rough surface for all three laser power levels employed. Ablation dominated scribes were observed even when the sample was irradiated with a low power of 0.5 W. This was unsurprising because the top rough surface has a high optical absorption characteristic as previously discussed. Since nonlinear absorption was mostly confined near the top surface of the sample, it prevented most of the laser beams from propagating to the bottom surface.

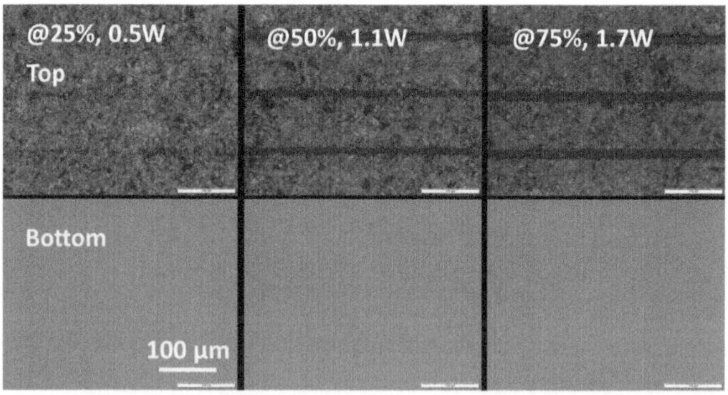

FIGURE 5.20 Optical images of Sample B with a top rough surface irradiated under different laser power levels.

Therefore, scribes were absent on the bottom polished surface of Sample *B* for all power levels. The top rough surface could effectively trap light and accumulate sufficient energy for ablation. Furthermore, the thickness of the scribes increased with increasing laser power from 0.5 to 1.7 W. With increasing laser power, and thus more energy accumulated, thicker and more pronounced ablation dominated scribes were observed at the top grinded rough surface. Hence, the presence of nonlinear absorption was detected through the observation of ablation scribes on the top rough surface, Sample *B* was deemed to be more suitable for machining as compared to Sample *A*.

Figure 5.21 shows the optical images of the top and bottom surfaces for Sample *C* (with a top smooth surface and a bottom rough surface). The ablation dominated scribes on the bottom rough surface were formed in a similar manner to that of the top rough surface of Sample *B* for all power levels tested.

At 0.5 W, there was insufficient energy on the top polished surface for the formation of scribes. When the incident laser power was increased to 1.1 W, the energy deposited was still mostly insufficient for any scribes to form on the top surface. However, occasionally, energy accumulated from the interaction between the incoming laser light and the internally reflected light at the top surface exceeded the threshold for crack formation. Hence, cracks were observed occasionally.

At a higher power of 1.7 W, the accumulated energy was just sufficient for ablation at the top polished surface and thin ablation scribes were mostly observed. However, at certain areas when there was insufficient accumulated energy for ablation, crack-dominated scribes were observed instead. Indeed, ablation scribes were mostly observed on both surfaces of Sample *C*. This is an indication of strong nonlinear absorption within the sample thickness that suggests that Sample *C* was the best candidate for machining.

The geometricity of a sample's surface affects how light interacts with a surface. This in turn determines if it is possible for the laser beam to be totally internally reflected at the exit surface, where the laser beam propagates from a medium

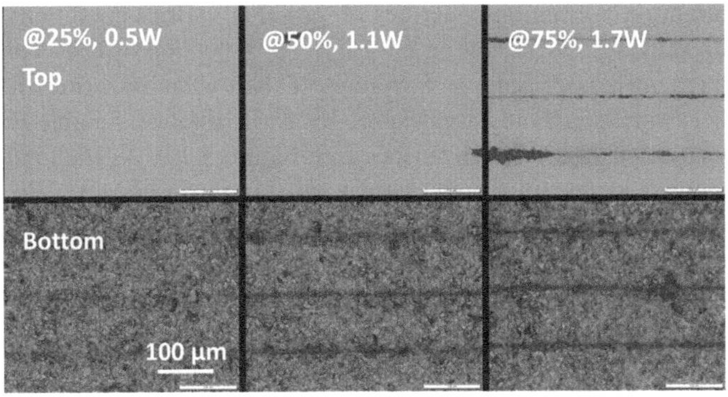

FIGURE 5.21 Optical images of Sample C with a top polished surface irradiated under different laser power levels.

of a higher refractive index to a medium of a lower refractive index. This has a significant effect on the distribution of the laser beam energy within the sample; with total internal reflection, the reflected laser beam will reinforce the incoming laser beam. As such, a larger amount of laser beam energy can be trapped and accumulated within the sample, until the nonlinear threshold is exceeded. Only when the nonlinear threshold and subsequently the ablation threshold were exceeded, ablation of the sample was possible.

Noting that for the same laser power input, the higher peak laser fluence in the center of the focal spot for a Gaussian beam as compared to a flat top beam would result in a reduced input laser power threshold required for nonlinear absorption. Indeed, for a 0.5 W Gaussian beam, it has a peak fluence of 12.8 J/cm^2 approximately at the center of the focus spot. This is much higher than the (peak) fluence of 6.60 J/cm^2 at 1.9 W of a flat top beam for the measured nonlinear absorption threshold of Sample *C*. As such, it is reasonable to deduce that for all the 3 power levels investigated here for the Gaussian beam, namely 0.5, 1.1 and 1.7 W, they have all exceeded the nonlinear absorption threshold. However, it has to be highlighted here that although the nonlinear absorption threshold fluence has been exceeded for a Gaussian beam, the area (or volume) of material affected by nonlinear absorption will be much more confined in a small region as compared to a flat top beam.

Taking the differences between a Gaussian beam and a flat top beam into consideration, the results of the different types of scribe formation in the picosecond laser irradiation experiments by a Gaussian beam were consistent with the observations of the study here on the nonlinear absorption threshold of samples with different surface morphologies irradiated by a flat top beam.

Firstly, although the power of 0.5 W has exceeded the nonlinear threshold, no scribe was observed for Sample *A* due to the small, confined area of nonlinear absorption. However, with increased nonlinear absorption by the rough surface, faint ablation scribes were observed on the rough surfaces of Samples *B* and *C*. This shows that the industrial standard of employing a Gaussian beam for material processing was preferred for its higher laser intensity at the laser focal spot.

In addition, the formation of scribes observed at various conditions are consistent with previous discussion on total internal reflection of laser beam at the bottom rough surface of Sample *C*. Sample *C* had ablation scribes formed on both surfaces when irradiated at power of 1.7 W. In contrast, Sample *B* only had ablation scribes on its top rough surface (see Figure 5.20). As the total incident laser power were identical for both samples, most of the laser energy was trapped at the top rough surface for Sample *B*. In contrast, the top smooth surface scribe of Sample *C* (see Figure 5.21) was formed by nonlinear absorption in the region at the vicinity of the focus spot. The region immediately outside this nonlinear absorption region would only have linear absorption. However, the laser beam at this linear absorption region would then experience total internal reflection at the bottom rough surface. As such, the totally internally reflected laser beam is trapped, and laser energy is accumulated in the region near the bottom surface. The interaction between the trapped laser beam and incoming laser beam would enhance nonlinear absorption and could result in scribe formation at the bottom rough surface (see Figure 5.21).

In addition, the nonlinear absorption of the laser energy was concentrated in the top rough surface for Sample *B* for scribe formation, and distributed for both smooth top surface and bottom rough surface for Sample *C*. As a result, the scribes on the top rough surface only of Sample *B* (see Figure 5.20) were much thicker than the scribes of Sample *C* at both surfaces (see Figure 5.21).

These results would suggest that for a practical singulation of a sapphire wafer, the sample should be orientated such that its polished smooth surface faced toward the incident laser beam and the rough surface away from the laser beam, i.e., Sample *C*. With this sample orientation, it promotes a highly effective use of the laser energy for nonlinear absorption throughout the thickness of the sample with scribe formation at both the top smooth surface and the bottom rough surface. For singulation, it is highly advantageous for scribes to occur at both surfaces.

5.4 MULTIPLE FOCI SCANNING OF SAPPHIRE

5.4.1 Multiple Foci Generation

The multiple foci were obtained through multi-focal optics (multi-focus, MF, lens) consisting of a combination of a diffractive optical element (DOE) and a conventional focusing lens (see Figure 5.22).

The distance between the DOE and the focusing lens can be arbitrary and is not a key parameter as the phase of the laser beam only changes after propagating through the diffractive elements. The EFL of the focusing lens dictates the average spacing between two adjacent foci (i.e., the foci spacing); the total foci spacing is defined as the distance between the focus closest to the laser source to the focus furthest from the laser source. A short EFL resulted in a shorter foci spacing, and a long EFL a longer foci spacing. The focus of diffraction order 0 (i.e., the middle focus) is located along the optical axis at the EFL. Foci closer to the laser source have positive diffraction orders and foci further from the laser source have

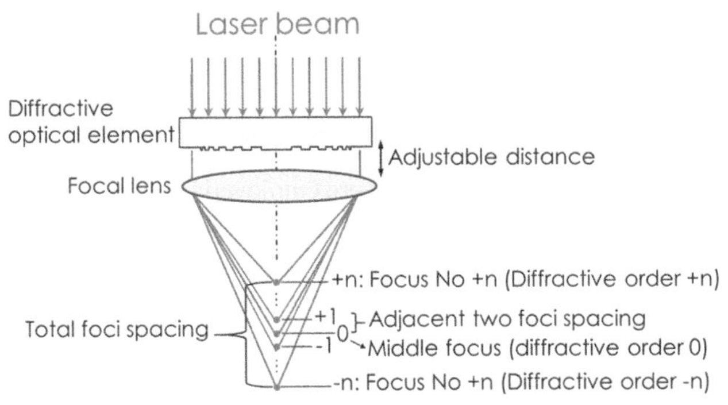

FIGURE 5.22 Schematic illustration of generation of multiple foci with a combination of optical diffractive element and focal lens.

negative diffraction orders, with diffraction orders numerically increasing away from diffractive order 0.

To investigate the influence of foci number, the spacing between two adjacent foci and the foci spot diameter, two types of DOEs for 27 foci and 9 foci from HOLO/OR Ltd, Israel, were employed in the singulation experiments. Two conventional focal lenses with EFL 12 and 7.5 mm from Edmund Optics, Singapore, were selected to produce a total foci spacing close to the entire thickness of the sapphire wafer under the combination of DOE 27 foci with EFL 7.5 mm and DOE 9 foci with EFL 12 mm, respectively. This will result in different adjacent two foci spacing, with larger adjacent two foci spacing for DOE 9 foci and shorter adjacent two foci spacing for DOE 27 foci. To further reduce the two foci spacing of DOE 27, OB40× objective lens from Meiji Techno, USA, was employed as it had a much shorter EFL. In addition, the different optical setups would result in different foci spot size. The details of the optics combination with the associated foci diameter and laser processing parameters are shown in Table 5.2.

In the experiments, all foci were placed within the interior of the sample thickness under a single pass laser beam scanning with different speed of 0.1, 1 or 10 mm/s. Figures 5.23 and 5.24 show the schematics of the scanning sections for full section scanning (i.e., Experiments F9 and F27) and partial one-third section scanning experiments (i.e., Experiments L27, M27 and U27) respectively, with the vertical red line signifying the locations of the foci in each experiment. The respective z value (i.e., the position of diffraction order 0) with respect to the top surface of the sample (with the positive direction into the sample) indicates the location of the scanning section within the thickness of the sapphire wafer.

5.4.2 Multiple Foci Scanning of Sapphire

5.4.2.1 Effect of Focal Length on Multiple Focus

This investigation aims to understand how the EFL affects the scribe characteristics of the samples. The EFL of the focusing optic determines the foci separation between the multiple foci. By changing the EFL of the objective lenses and laser scanning heights, the effects of foci separation, number of foci focused within the sample and the power distribution among foci can be studied.

The study investigates the effect of distribution of laser power across the entire thickness of the sample by employing 9 foci and 27 foci MF lenses, for Experiments F9S-50 and F27S-50 respectively (see Table 5.2). Single-side polished samples with a rough exit surface were employed. The laser power employed was 1.13 W at a 50 kHz pulse repetition rate, as it was determined to be critical to have a high peak intensity to induce nonlinear absorption.

The laser produced 22.6 μJ for each pulse and the amount of pulse energy allocated to a single focus was dependent on the number of foci formed by the MF lens. Each focus formed by the 9 foci lens had approximately triple the amount energy of a single focus formed by the 27 foci lens, with pulse energies of 2.52 and 0.839 μJ, respectively.

TABLE 5.2
Detailed Optics Combination for Producing Multi-Foci and Laser Processing Parameters

	Shorter Foci Spacing over Entire Thickness	Larger Foci Spacing over Entire Thickness	Further Shorter Foci Spacing over One-Third Thickness		
Experiment No.	F27	F9	L27	M27	U27
z value (μm) (position of middle focus), (total foci covering thickness section)	+210 Middle (M), Full section (F)	+210 Middle (M), Full section (F)	+360 Lower (L), 1/3 section	+210 Middle (M), 1/3 section	+70 Upper (U), 1/3 section
DOE (Foci No)	27	9	27	27	27
Focusing optic EFL (mm)	7.5	12	4.38 (OB40×)	4.38 (OB40×)	4.38 (OB40×)
Foci spot diameter (μm)	2.20	3.52	1.29	1.29	1.29
Ave foci spacing (μm)	15.4	43.6	5.2	5.2	5.2
Total foci spacing (μm)	399.5	348.9	136.3	136.3	136.3
Samples tested (S or D)	Single and double polished (S&D)	Single polished (S)	Single polished (S)	Single polished (S)	Single polished (S)
Laser power (W)	1.13	1.13	0.80	0.80	0.80
Pulse repetition rates (kHz)	50, 500, 5120	50	50	50	50

Notes:
1) "F", "L", "M" or "U" indicates the scanning position, meaning that the sample was scanned across the full thickness, lower, middle or upper one-third sample section, respectively.
2) The number "9" or "27" indicates the number of foci of the laser beam.
3) "S" or "D" indicates the sample is either single-side polished or double-side polished, respectively.
4) The number after "S" or "D" indicates the pulse repetition rate in kHz.
5) For example, Experiment F27D-50 indicates that the experiment was carried out on a double-side polished sample with 27 foci laser beam distributed across the entire section with a pulse repetition rate of 50 kHz.

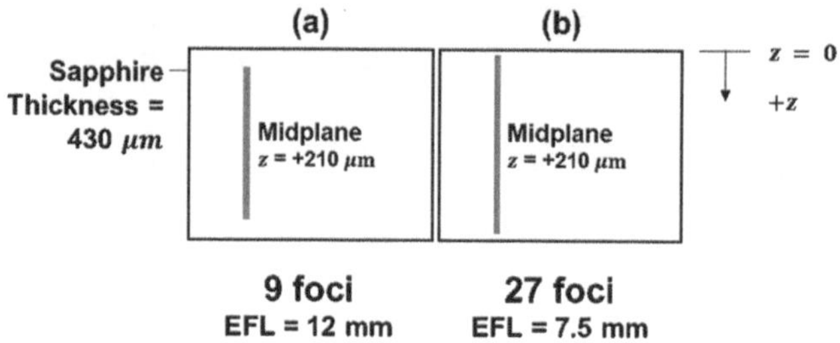

FIGURE 5.23 Schematic of scanning over the full thickness with diffractive order 0 positioned at z value=+210 μm for (a) 9 foci setup (Experiment F9) and (b) 27 foci setup (Experiment F27), respectively.

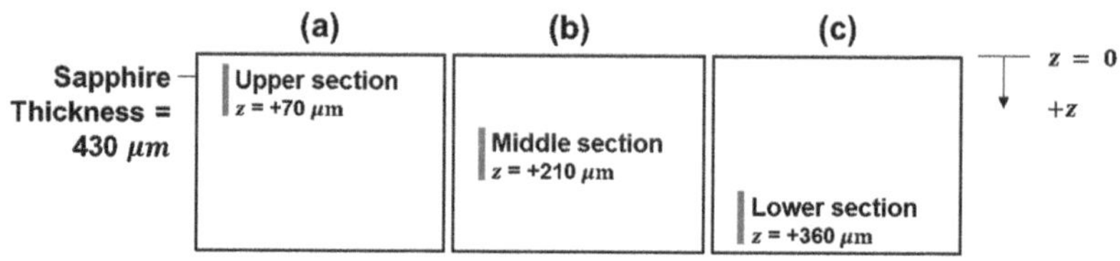

FIGURE 5.24 Schematic of the three scanning sections over one-third thickness of the sample with diffractive order 0 positioned at the z values for (a) lower (Experiment L27), (b) middle (Experiment M27) and (c) upper (Experiment U27) sections, respectively.

However, since focusing optics of different EFL were employed for the respective MF lenses, the focused spot diameter for the 9 foci lens was larger than the 27 foci lens, i.e., 3.52 μm for 9 foci lens and 2.20 μm for 27 foci lens. The spot size has a direct impact on power density. Taking into the account the difference in the laser power per pulse per focus and the focused spot size, and despite the larger focus spot size, each focus in Experiment F9S-50 has a higher peak power density per pulse per focus as compared to Experiment F27S-50, with peak power densities (or intensity) per pulse per focus of 2.51 and 2.14 TW/cm^2, respectively, for Experiments F9S-50 and F27S-50. Although the difference in peak power densities per pulse per focus was only ~17%, this could result in a significant difference in nonlinear absorption behavior as nonlinear multiphoton absorption is expected to be proportional to the laser intensity to the power of number of photons simultaneously absorbed [36,37]:

$$dI/dz = -\alpha I^n \tag{5.1}$$

where I is the intensity (or power density), z is the distance along the optical axis, α is the nonlinear absorption coefficient and n is the number of photons simultaneously absorbed in a multiphoton absorption. For the current investigation, the irradiation of sapphire under a 1064 nm laser resulted in a nine-photon absorption process (i.e., $n=9$) [37].

The intensity at a focus spot is proportional to the laser power, and inversely proportional to the number of foci and the focus spot area. As such, for the same laser power applied, the ratio of intensity I and dI/dz for a single focus in Experiment F9S-50 (with 9 foci and a focus spot diameter of 3.52 μm) to that of F27S-50 (with 27 foci and a focus spot diameter of 2.20 μm) may be calculated to be approximately 1.17 and 4.2 times, respectively. This implies that significant amount of energy would be absorbed over a much shorter distance for F9S-50 as compared to F27S-50. This higher energy absorption over a shorter distance would result in higher heating and thus thermal stress for F9S-50. In addition, the larger focus spot diameter of F9S-50 as compared to F27S-50 would also means that a larger area in the x–y plane parallel to the surface of the sample would also experience the higher heating and thermal stress for F9S-50.

On closer examination, the distribution of the energy absorbed was rather different for these two experimental setups. With a large foci spot diameter and area, F9S-50 would have a larger proportion of the absorbed energy distributed in the x–y plane parallel the surface of the sample than that of F27S-50. In contrast, with more foci, F27S-50 would have higher proportion of absorbed energy more evenly distributed in the through thickness direction of the sample as compared to F9S-50. As singulation is a "through thickness" operation, intuitively one would expect that a higher proportion and more evenly distribution of energy in the through thickness direction will be more desirable.

As such, this study attempts to identify if it is more effective (i) to split the laser energy over fewer foci, each with higher intensity and focus spot area (and overall more energy absorbed per focus), but with a larger distance between two adjacent foci (i.e., Experiment F9S-50) or (ii) to have more foci with closer proximity to one another and with lower intensity and energy absorbed for each focus but with higher proportion of the absorbed energy more evenly distributed along the desirable through thickness direction (i.e., Experiment F27S-50).

The effects of energy deposition within the samples on scribe formation were further investigated by changing the laser scanning speeds (i.e., 0.1, 1 and 10 mm/s). The amount of energy deposited can be derived as [37,38]:

$$\begin{aligned} E_d &= QN \\ &= \frac{P}{R_p} \times \frac{R_p \phi}{v} \end{aligned} \tag{5.2}$$

$$E_d = \frac{P\phi}{v} \tag{5.3}$$

where E_d is the energy deposited within the sample, Q is the pulse energy, N is the number of pulses overlap, P is the average power, R_p is the pulse repetition rate, ϕ is the focused spot diameter and v is the laser scanning speed. For the same laser power and focused spot diameter, Equation (5.3) indicates that the energy deposited into the sample decreases with increasing scanning speed, and it does not depend on the pulse repetition rate.

5.4.2.2 Multiple Foci Scanning

5.4.2.2.1 Top Polished Entry Surface after Full Section Scanning

Figure 5.25 shows the optical images of the top polished entry surface after laser irradiation at various scanning speed for the full section scanning experiments. The widths and their standard deviations of the top surface scribes (if any) are indicated in the figure.

Samples of Experiment F9S-50 (i.e., 9 foci scanning) had scribes on the top polished entry surface for all scanning speeds (see Figure 5.6). This is somewhat unexpected as the total foci spacing only occupied ~80% of the thickness of the sample, with no foci in close proximity to the surfaces. Indeed, this illustrated the importance to have sufficient laser power for scribe formation even when the laser foci are not near the surface.

Crack-dominated scribes were formed at the slower scanning speeds of 0.1 and 1 mm/s, while ablation scribes were formed at the faster scanning speed of 10 mm/s. When the scanning speed increased from 0.1 to 10 mm/s, less energy was deposited into the samples and the scribe widths decreased. The different types of scribe formation were dependent on the amount of energy deposition.

At the slower scanning speeds (i.e., 0.1 and 1 mm/s), the energy deposited internally modified the sample and the excess energy was channeled to the formation and propagation of internal cracks. Since crack propagation could not be controlled, deviated cracks were formed on the top surface. The internal cracks acted as micro-surfaces that reflected the incoming laser beam onto other micro-surfaces, effectively trapping light within the region. In addition, since the laser energy was split only between 9 foci, the energy allocated to a single focus was higher (as compared to 27 foci). The combination of the two effects accumulated energy near the top surface by the unfocused laser. With sufficient energy

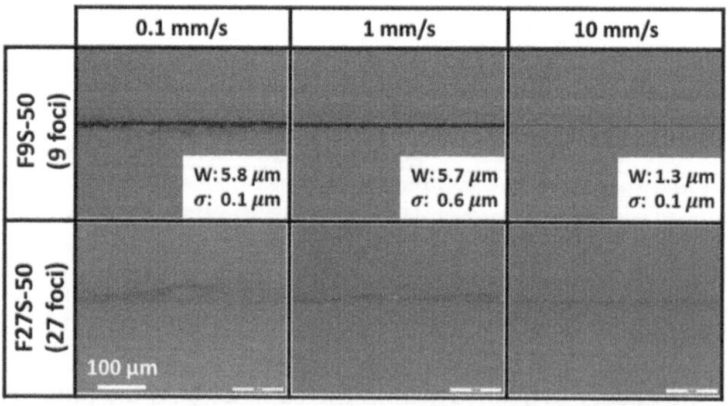

FIGURE 5.25 Optical images of top polished entry surface after full section scanning at 1.13 W with the respective widths (W) and standard deviations (σ).

accumulated, ablation occurred that resulted in a crack-dominated scribe to form on the top surface.

At the 10 mm/s scanning speed, the energy deposited was just sufficient for the formation of a thin ablation dominated scribe on the top polished surface of the sample. The amount of energy deposited was not in excess to form cracks seen in the samples irradiated with slower scanning speeds.

In contrast, the total foci spacing occupied ~90% of the sample thickness in Experiment F27S-50 (i.e., 27 foci scanning) (see Figure 5.25). Although some foci were close to the top polished entry surface, no scribes were observed. This could be attributed to the lower energy allocated to a single focus which could not induce scribe formation. Instead, cracks were observed on the top surface with the occasional subsurface damage for all scanning speeds. As the scanning speed was increased, less excess energy was deposited into the samples to form and propagate the internal cracks to the top surface. The amount of excess energy deposited to the samples dictated the length of cracks formed and the extent of crack deviation from the scanning direction.

5.4.2.2.2 Bottom Unpolished Exit Surface after Full Section Scanning

Figure 5.26 shows the optical images of the bottom unpolished exit surface after laser irradiation at various scanning speeds for full section scanning experiments. The widths and their standard deviations of the bottom surface scribes (if any) are indicated in the figure.

Ablation dominated scribes were formed on the bottom unpolished rough surface for all scanning speeds in both Experiments F9S-50 and F27S-50 (see Figure 5.26). Figure 5.26 indicates that the bottom surface scribes were always thicker than the top surface scribes shown in Figure 5.6 regardless of the scanning speed employed. Diffusion of light at the bottom rough surface caused the laser beam to be incident on a larger surface area. With total internal reflection and accumulation of energy in this region, a thicker scribe could be formed on the bottom surface.

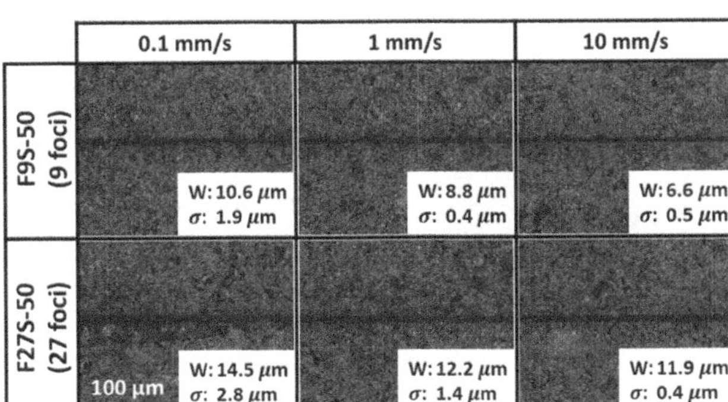

FIGURE 5.26 Optical images of bottom unpolished exit surface after full section scanning at 1.13 W with the respective widths (W) and standard deviations (σ).

With increasing scanning speed, both Experiment F9S-50 (i.e., 9 foci scanning) and Experiment F27S-50 (i.e., 27 foci scanning) produced scribes with decreasing widths. As the energy deposited decreases with increasing scanning speed, this indicates the width of the scribes is a function of energy deposition.

The greater the distance the laser beam of the last focus had to travel to the bottom surface, the greater the energy loss through linear absorption. As such, it is expected that that the width of the scribe should also be a function of proximity of the foci to the surface. The last focus of the 9 foci setup in Experiment F9S was further from the bottom surface than the last focus of the 27 foci setup in Experiment F27S. Indeed, consistently, at the same scanning speed, the widths of the ablation scribes of Experiment F9S-50 were narrower than that of F27S-50.

It should also be noted that only at the slowest 0.1 mm/s scanning speed in Experiment F27S-50, there was sufficient excess energy to propagate the internal cracks to the bottom surface. Cracks that largely deviated from the scanning direction were observed adjacent to the scribe.

5.4.2.2.3 Sidewall Profiles after Full Section Scanning

Figure 5.27 shows the optical images of the sidewall profiles of the successfully cleaved samples after full section scanning with different scanning speeds.

The sample irradiated at the 0.1 mm/s scanning speed in Experiment F9S-50 (i.e., 9 foci scanning) was mechanically cleaved (see Figure 5.8). The sidewall profile revealed wave-like structures in the region closest to the top surface of the sample, justifying the cracks observed on the top surface. The region below this comprised of laser modified areas with small wave-like cracks. The formation of the two distinct regions was dictated by the amount of energy each focus had at its focus position. The closer a focus was to the top surface, the shorter the distance it had to travel and the amount of energy loss through linear absorption by the sample was reduced. Therefore, there was more excess energy to form both laser modified areas and larger internal cracks. The converse was true, whereby focus further from the top surface of the sample would have less excess energy to form large internal cracks after modification.

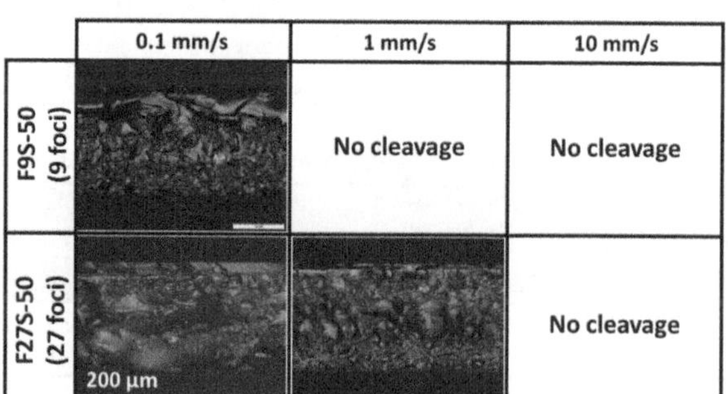

FIGURE 5.27 Optical images of sidewall profiles after full section scanning at 1.13 W.

Samples irradiated with a scanning speed of either 0.1 or 1 mm/s in Experiment F27S-50 (i.e., 27 foci scanning) were successfully cleaved (see Figure 5.8). At the 0.1 mm/s scanning speed, the sidewall profile shows that the plane of cleavage was not perfectly perpendicular to the top and bottom surfaces of the sample. A plausible explanation could be due to the randomness in internal cracks formation from the excess energy deposition to the sample. As such, mechanical cleavage would tend to propagate along the weakened paths of the internal cracks instead of along the plane of laser modification. Since the internal cracks were formed in a random manner, the plane of cleavage was not perfectly perpendicular to the surfaces of the sample.

At the 1 mm/s scanning speed, as a result of less excess energy deposited, the optical image of the sidewall profile displayed a plane of cleavage that seemed to be along, or closer to, the plane of laser modification. The non-uniformly roughened sidewall profile seemed to contain smaller cracks in the plane of laser modification. Therefore, the sidewall profile observed could contain both the laser modified areas and small internal cracks.

For samples irradiated by a 1 mm/s scanning speed, the 9 foci setup in Experiment F9S-50 provided a larger amount of energy deposited to the samples than the 27 foci setup in Experiment F27S-50. Yet, mechanical cleavage of the samples was only successful for Experiment F27S-50 samples but not for Experiment F9S-50 samples. This illustrates the importance of the role of energy distribution and foci spacing. It was more favorable to split the laser energy over more foci along the desirable through thickness direction as in Experiment F27S-50.

Although each focus in Experiment F27S-50 had a lower pulse energy, smaller spot size and less absorption, larger proportion of the energy was distributed in the through thickness direction with the foci closely spaced to each other and effectively interacted with the adjacent foci as compared to Experiment F9S-50. As demonstrated, although there was less energy absorption at each focus, the absorbed energy more evenly distributed along the desired direction was more ideal as there was potentially less undesired damage (such as undesired cracking) made to the sample. With a close proximity of the foci, the desirable separation crack in the through thickness direction induced by one focus could extend to another crack induced by an adjacent focus. Since the 27 foci were more evenly spaced over the thickness of the sample, it effectively formed a line of cracks in the plane of the laser optical axis in the through thickness direction of the sample. These crack interactions sufficiently weakened the sample for mechanical cleavage. In contrast, in Experiment F9S-50, with a higher pulse energy, larger spot size and more energy absorption over a shorter through thickness distance, it might cause greater undesirable damage to a larger area as a much larger proportion of energy absorbed was distributed along the x–y plane parallel to the sample surface. The larger spacing between foci meant that interaction between the separation cracks induced by the foci was minimal along the optical axis in the through thickness direction of the sample. As such, mechanical cleavage could well not be successfully completed at higher scanning speed. This conclusion was only valid if each focus had an appropriately high intensity for nonlinear absorption.

Hence, employing the 27 foci setup in Experiment F27S was a more effective use of laser energy as sample cleavage was possible with less energy deposition.

Mechanical cleavage was not successful for samples irradiated at a scanning speed of 1 or 10 mm/s in Experiment F9S-50 or 10 mm/s scanning speed in Experiment F27S-50. Due to insufficient energy deposited, the samples could not be effectively weakened for cleavage.

5.5 SUMMARY AND PERSPECTIVE

Nonlinear absorption within the transparent sapphire samples is crucial in its machining with ultrashort pulse laser technology. Hitherto, there is no investigation on the existence of total internal reflection caused by surface roughness, and its significance on laser energy absorption in laser-material interaction. This investigation demonstrated that the orientation of the rough surface with respect to a laser beam source is a key parameter on total internal reflection, and thus nonlinear absorption. By comparing samples with double-side polished surfaces, a rough entry or exit surface, this investigation demonstrated vividly that only the samples with a rough exit surface had significant total internal reflection. This phenomenon was significant and effective in trapping the laser energy within the sample enhancing nonlinear absorption required for laser machining. Due to this unique characteristic of the exit surface roughness, ablation scribes could be formed simultaneously at both the entry and exit surfaces; this is not possible for samples with either double-side polished surfaces or a rough entry surface.

The investigation adapts the multi-foci ultrashort pulse laser technology of transparent glass cutting to sapphire singulation with unique recommendations. These include (i) the exploitation of total internal reflection of laser beam exiting from an unpolished rough surface to enhance nonlinear absorption for laser processing and (ii) to only scan the middle one-third section of the wafer thickness with much smaller foci spacing instead of evenly distributing the foci over the entire thickness of the wafer resulting in laser energy optimization. This study also elucidates that splitting the laser energy over more foci with closer proximity to each other and a smaller focus spot in the wafer thickness direction is more effective for wafer separation than employing fewer foci with more energy and a larger focus spot but spaced further apart. To produce a cleavable sample, sufficiently high laser intensity to induce nonlinear absorption generated at low pulse repetition rates within the wafer is critical. In addition, a rough sidewall profile desirable for improving the light extraction of LEDs could be attained through multi-foci laser irradiation of sapphire wafers.

ACKNOWLEDGMENTS

The research work was supported by Research Fund of Guangdong-Hong Kong-Macao Joint Laboratory for Intelligent Micro-Nano Optoelectronic Technology (No. 2020B1212030010) and A-Star Research Agency, Singapore Institute of Manufacturing Technology (SIMTech) under SIMTech-NTU Joint Laboratory with project No. U12-M-007JL. Deep appreciation also goes to master student Ms Celescia Siew Mun Lye and Professor Yee Cheong Lam for their strong help in the project research.

REFERENCES

1. W.E. Baylies, C.J. Moore, The use of sapphire in mobile device and LED industries | semiconductor digest (2015). https://sst.semiconductor-digest.com/2015/11/the-use-of-sapphire-in-mobile-device-and-led-industries/ (accessed October 3, 2018).
2. D.M. Allen, R. Redondo, M. Dany, Fabrication methods for the manufacture of sapphire microparts, *Microsyst. Technol.* 18 (2012) 1835–1841. doi: 10.1007/s00542-012-1484-2.
3. G. Savriama, V. Jarry, L. Barreau, C. Boulmer-Leborgne, N. Semmar, Laser micro-cutting of wide band gap materials (2012). doi: 10.1063/1.4739871.
4. R. Mayerhofer, Innovative laser technologies for cutting, drilling and structuring of glass and other transparent, brittle materials. In: *Advanced Solid State Lasers*, Optical Society of America, Berlin (2015), p. AF3A.6. doi: 10.1364/ASSL.2015.AF3A.6.
5. C. Rüttimann, sapphire cutting with pulsed fiber lasers, *Laser Tech. J.* 11 (2014). doi: 10.1002/latj.201400034.
6. A. Yadav, H. Kbashi, S. Kolpakov, N. Gordon, K. Zhou, E.U. Rafailov, Stealth dicing of sapphire wafers with near infra-red femtosecond pulses, *Appl. Phys. A.* 123 (2017) 369. doi: 10.1007/s00339-017-0927-0.
7. M. Yamada, T. Mitani, Y. Narukawa, S. Shioji, I. Niki, S. Sonobe, K. Deguchi, M. Sano, T. Mukai, InGaN-based near-ultraviolet and blue-light-emitting diodes with high external quantum efficiency using a patterned sapphire substrate and a mesh electrode, *Jpn. J. Appl. Phys.* 41 (2002). doi: 10.1143/JJAP.41.L1431.
8. D.-S. Wuu, W.K. Wang, W.C. Shih, R.-H. Horng, C. Lee, W.-Y. Lin, J.-S. Fang, Enhanced output power of near-ultraviolet InGaN-GaN LEDs grown on patterned sapphire substrates, *Photonics Technol. Lett. IEEE.* 17 (2005) 288–290. doi: 10.1109/LPT.2004.839012.
9. G. Yang, J. Chang, J. Zhao, Y. Tong, F. Xie, J. Wang, Q. Zhang, H. Huang, Y. Da-Wei, Investigation of light output performance for gallium nitride-based light-emitting diodes grown on different shapes of patterned sapphire substrate, *Mater. Sci. Semicond. Process.* 33 (2015). doi: 10.1016/j.mssp.2015.02.002.
10. S.C. Chang, L.M. Chang, J.Y. Chen, C.S. Hsu, D.S. Kuo, C.F. Shen, W. Chen, T.K. Ko, GaN-based light-emitting diodes prepared with shifted laser stealth dicing, *J. Disp. Technol.* 12 (2016) 195–199. doi: 10.1109/JDT.2015.2478598.
11. R. Ma, L. Ji, T. Yan, Laser multi-focus precision cutting of thick sapphire by spherical aberration rectification, *Opt. Lasers Eng.* 126 (2020) 105876. doi: doi: 10.1016/j.optlaseng.2019.105876.
12. Y. Zhang, H. Xie, H. Zheng, T. Wei, H. Yang, L. Jing, X. Yi, X. Song, G. Wang, J. Li, Light extraction efficiency improvement by multiple laser stealth dicing in InGaN-based blue light-emitting diodes, *Opt. Express.* 20 (2012) 6808–6815. doi: 10.1364/OE.20.006808
13. L. Lytvynov, Aluminum oxide. In: R. Fornari (Ed.) *Single Crystals of Electronic Materials: Growth and Properties.* Woodhead Publishing, Cambridge (2019), pp. 447–485.
14. V.N. Kurlov, Sapphire: Properties, growth, and applications. In: Saleem Hashmi (Ed.) *Reference Module in Materials Science and Materials Engineering.* Elsevier Ltd., Pergamon (2001), pp. 8259–8264.

15. D.M. Allen and R. Redondo, Fabrication methods for the manufacture of sapphire microparts, *Microsyst. Technol.* 18(11) (2012) 1835–1841.
16. SJC Advanced Ceramics, Physical properties of Aluminia, Zirconia and Sapphire, SJC Advanced Ceramics, n.d.. [Online]. Available: http://www.swiss-jewel.com/property.htm. [Accessed 22 November 2018].
17. Kyocera, Single Crystal Sapphire, (2018). [Online]. Available: https://global.kyocera.com/prdct/fc/product/pdf/s_c_sapphire.pdf. [Accessed 22 November 2018].
18. S. Dub, P. Lytvyn, V. Strelchuk, A. Nikolenko, Y. Stubrov, I. Petrusha, T. Taniguchi, S. Ivakhenko, Vickers hardness of diamond and cBN single crystals: AFM approach, *Crystals* 7(369) (2017) 1–13.
19. C. Rüttimann, Sapphire cutting with pulsed fiber lasers, *Laser Technik J.* 11(3) (2014) 48–50.
20. W.A. Baylies, C.J. Moore, The use of sapphire in mobile device and LED industries, (November 2015). [Online]. Available: https://electroiq.com/2015/11/the-use-of-sapphire-in-mobile-device-and-led-industries/. [Accessed 3 October 2018].
21. S.N. Dahal, J.A. LeBeau, S. Bowden, C. Honsberg, Sub-surface laser damage in sapphire and silicon: A path towards laser wafering, *In 2016 IEEE 43rd Photovoltaic Specialists Conference*, Portland (2016).
22. D. Kim, H. Kim, S. Lee, H. Jeong, Effect of initial deflection of diamond wire on thickness variation of sapphire wafer in multi-wire saw, *Int. J. Precis. Eng. Manuf. Green Technol.* 2(2) (2015) 117–121.
23. A. Tamhankar, R. Patel, Optimization of UV laser scribing process for light emitting diode sapphire wafers, *J. Laser Appl.* 23(3) (2011) **032001**.
24. A. Yadav, H. Kbashi, S. Kolpakov, N. Gordon, K. Zhou, E.U. Rafailov, Stealth dicing of sapphire wafers with near infrared femtosecond pulses, *Appl. Phys. A Mater. Sci. Proc.* 123(5) (2017) 1–7.
25. J. Xu, H. Hu, C. Zhuang, G. Ma, J. Han, Y. Lei, Controllable laser thermal cleavage of sapphire wafers, *Opt. Lasers Eng.* 102 (2018) 26–33.
26. J.-M. Lee, J.-H. Jang, T.-K. Yoo, Scribing and cutting a blue LED wafer using a Q-switched Nd:YAG laser, *Appl. Phys. A Mater. Sci. Proc.* 70(5) (2000) 561–564.
27. W.-J. Tsai, C.-J. Gu, C.-W. Cheng, J.-B. Horng, Internal modification for cutting transparent glass using femtosecond Bessel beams, *Opt. Eng.* 53(3) (2014) 11.
28. L. Rapp, R. Meyer, L. Furfaro, C. Billet, R. Giust, F. Courvoisier, Crack formation and cleaving of sapphire with ultrafast Bessel beams, *In 2017 Conference on Lasers and Electro-Optics Europe & European Quantum Electronics Conference*, Munich (2017).
29. P. Liu, L. Deng, J. Duan, B. Wu, X. Zeng, S. Ying, X. Wang, A study on laser multi-focus separation technology of thick KDP crystal, *Int. J. Mach. Tools Manuf.* 118–119 (2017) 26–36.
30. P. Liu, J. Duan, B. Wu, L. Deng, S. Ying, X. Zeng, X. Wang, A flexible multi-focus laser separation technology for thick glass, *Int. J. Mach. Tools Manuf.* 135 (2018) 12–23.
31. W.M. Steen, J. Mazumder, *Laser Material Processing*, Fourth edition. Springer, London (2010). doi: 10.1007/978-1-84996-062-5.
32. A. Ghatak, *Optics*. McGraw-Hill, New York (2005).
33. D.W. Ball, Field guide to spectroscopy, SPIE (2006). doi: 10.1117/3.682726.
34. Z. Wang, H. Zheng, W.L. Seow, X.C. Wang, Investigation on material removal efficiency in debris-free laser ablation of brittle substrates, *J. Mater. Process. Technol.* 219 (2015). doi: 10.1016/j.jmatprotec.2014.12.013.

35. R. Geremia, D. Karnakis, D.P. Hand, The role of laser pulse overlap in ultrafast thin film structuring applications, *Appl. Phys. A.* 124 (2018) 641. doi: 10.1007/s00339-018-2045-z.
36. M. Gedvilas, J. Mikšys, J. Berzinš, V. Stankevič, G. Račiukaitis, Multi-photon absorption enhancement by dual-wavelength double-pulse laser irradiation for efficient dicing of sapphire wafers, *Sci. Rep.* 7 (2017) 5218. doi: 10.1038/s41598-017-05548-x.
37. R. Paschotta, Field guide to laser pulse generation, SPIE (2009). doi: 10.1117/3.800629.
38. R. Geremia, D. Karnakis, D.P. Hand, The role of laser pulse overlap in ultrafast thin film structuring applications, *Appl. Phys. A.* 124 (2018) 641. doi: 10.1007/s00339-018-2045-z.

6 Metal Additive Manufacturing

Hao Zhao and Garrison Zong
3D Printing Technology, Inc.

CONTENTS

DOI: 10.1201/9781003182146-6

6.1 INTRODUCTION

6.1.1 Definition of Additive Manufacturing

Additive manufacturing (AM) is a near-net-shape technology that directly fabricates three-dimensional (3D) objects from 3D drawings. The term 'additive' represents its nature of adding, consolidating or joining material accurately to a predefined 3D space to form a 3D object usually in a layer-by-layer fashion. AM processes are in contrast to the conventional subtractive processes, such as CNC machining, grinding and drilling, besides the various forming processes like casting, forging and injection molding. In the past decades, AM technology was also referred to as 'rapid prototyping' due to its capability of rapidly creating a physical model or a verification part before the final product is derived. Based on the additive concept, to date a class of various approaches has been developed that can deposit a powder, wire, sheet or liquid material in a layer-by-layer fashion, using a heat source, UV activator or glue to consolidate the raw material into once piece [1,2].

The AM technologies could not have emerged without the earlier developments and maturation of computers, lasers, controllers, etc. before the outbreak of AM in the 1980s. These technological advancements together brought AM system from concept to reality. In 1984, the first generally recognized patent in the field of AM was filed by Charles Hull, who founded 3D systems and commercialized the first light curing (LC) apparatus called stereolithography, or SLA. In 1986, DTM was founded with their patented selective laser sintering (SLS) technology. SLS is the precursor of the mainstream AM class named powder bed fusion (PBF), which now includes selective laser melting (SLM), selective electron beam melting (SEBM) and SLS itself. In 1989, two more technologies were commercialized in the United States, which are fused deposition modeling (FDM) by Scott Crump at the Stratasys company and 3D printing (3DP) process by the MIT groups. The FDM process found its success in the low-cost AM market and is often known as desktop 3D printing. The process employs a wire or granular feedstock to provide an easy-to-operate solution for amateurs, educational users and some industrial tool makers. The 3DP process is alternatively defined as a binder jetting (BJ) process. Both of the above 1989 processes are still being extensively used today, and both have been successfully extended to the production of metal. Another important process is the material jetting (MJ) developed by Sanders in 1994.

Most AM sub-classes have arisen and started to be commercialized in the 1990s. Since 2000, the later innovations and new developments were mostly updating the previously established methods, improving the material diversity, accuracy, efficiency and reliability or creating hybrid processes. It is important to note that the definition of AM today has been largely broadened and sometimes diluted as it fuses in other technologies and becomes part of a comprehensive process, i.e., the discussions in the following section on 'indirect metal AM processes' will be a good example of this situation.

6.1.2 Classification of AM

According to ASTM F42 and ISO TC 261 terminologies, AM techniques can be classified into seven sub-classes, that are:

- *Light curing*: processes that utilize a liquid photopolymer or a UN-resin-based slurry that is placed in a container, or delivered using a UV-transparent foil, and selectively exposed to a UV light source to form a part cross-section.
- *PBF*: processes that utilize a powder evenly distributed on a platform, or a powder bed, that is processed selectively using an energy source, most commonly a scanning laser or an electron beam.
- *Material extrusion*: processes that deposit a material by extruding it through a nozzle, typically while scanning the nozzle in a given pattern that produces a part cross-section.
- *MJ*: processes that deposit a fluid material or a slurry by jetting it through a print head and ink-jet prints each layer cross-section.
- *BJ*: processes where a binder is selectively jetted into a powder bed to form part cross-sections.
- *Sheet lamination*: processes that deposit a layer of sheet material at a time and the layers are joined by adhesive force or diffusion-induced metallic bonding.
- *Directed energy deposition (DED)*: processes that simultaneously deposit a material (usually powder or wire) and provide energy to melt the material through a deposition device.

The AM processes described above are summarized in Table 6.1, where their general properties are rated comparatively. Among the seven AM classes, several processes are capable for manufacturing metals. These can be classified

TABLE 6.1
Comparative Comments on the Seven Classes of Additive Manufacturing

Process Name	Resolution	Freedom of Design	Productivity	Cost	Material Variety	Material Performance
Light curing	+++	+++	+	++	++	–
Powder bed fusion	++	++	—	—	+++	+++
Material extrusion	–	++	+	+	+	++
Material jetting	+++	++	+	–	–	++
Binder jetting	++	++	+++	+++	+++	—
Sheet lamination	++	–	+	++	–	+
Directed energy deposition	—	–	++	++	+	++

into direct metal AM processes and indirect metal AM processes. In the former case, i.e., PBF and DED, metals are heated to above their melting temperature and solidify to form a near-net-shape metal part. The resolution of the heat source, as well as the form of raw material, determines how much additional machining will be required to reach the desired dimensional accuracy. Direct metal AM, as its name indicates, is yet the most versatile process that requires minimal preparation, tooling and post-treatment steps. However, direct metal AM utilizes a focused heat source to melt the metal, and is usually costly and slow. In contrast, indirect metal AM processes create molds or metal-glue composites rather than the metal alone. The intermediate products are then forwarded to a casting process, or respectively, to debinding and sintering processes, and the final metal part is derived. The indirect processes can empower conventional metallurgical processes by adding geometric flexibility to manufacturing flow without altering the metallurgical nature. As such, indirect metal AM generally offers lower cost, larger part sizes and higher production rate than direct processes.

6.1.3 Role of AM in the Manufacturing Industry

As a class of versatile, flexible and highly customizable production techniques, AM provides great opportunities for manufacturing industries. In the early 1990s, AM technologies were first adopted in creating physical prototypes to assist new product developments. For example, in 1994 an SLS system invented by Longyuan China was named 'rapid prototyping machine'. However, the use of metal AM has been limited until the 2010s due to its limited material availability and high cost. In the 2010s, a number of successful PBF materials and applications were developed and fulfilled the demand from several key industries in the high-end, such as aerospace, medical, marine and energy industries [3]. Typical aerospace parts include fuel nozzles, hinge brackets, lightweight frameworks, housings and oil valves. In the medical sector, there is an intrinsic demand for penalized manufacturing. Metal AM successfully entered medical areas, such as dentistry, bone surgery and wearable medical devices.

As the technologies advance, a number of mass-production industries, such as automobile, casting, metal injection molding (MIM), sportswear, fashion and consumer electronics, have started to adopted AM as a supplementary technology to make their production more flexible, swift and accurate. The public recognition of metal AM also motivated many production service centers and online shops to be founded, which further added to the availability and convenience of using metal AM.

Today, AM has become an important advanced manufacturing technology to overcome many limits and barriers that could not be solved by conventional processes. For example, topological optimized and lightweight structures can now be fabricated at a fraction of cost and many extremely complex structures that could never be produced before are now made possible (Figure 6.1).

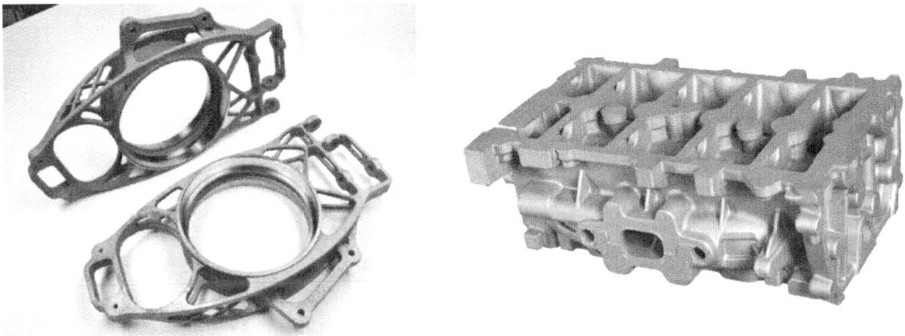

(a) (b)

FIGURE 6.1 Typical applications of AM. (a) A direct metal printing topological optimized pair of uprights for racing cars. (b) An indirect metal AM - SLS wax and lost wax casting aluminum alloy cylinder head.

In addition, the unique capabilities of metal AM and its advantages can be summarized as follows:

- Reducing time and cost of launching the product into production due to the absence of need for specialized tooling;
- Possibility and economic feasibility of the small-scale output;
- Operational changes in the project at the production stage;
- Functional optimization of products (e.g., the implementation of the optimal shape of cooling channels);
- Economic feasibility of the production of customized products;
- Reduction of production losses and waste;
- Opportunities to simplify logistics, reduce delivery time and reduce inventory volumes; and
- Personalization of the design.

6.2 POWDER BED FUSION SYSTEMS

6.2.1 Introduction

PBF processes were among the first commercialized AM processes. Metal PBF emerged soon after the successful adoption of high-energy lasers and electron beams. The schematic PBF setup is shown in Figure 6.2. A PBF process typically includes a powder spreading, or recoating mechanism, which adds and smooths the powder at the beginning of each layer. The system may include one or more heat sources for inducing fusion between powder particles. The heat source is focused at the powder bed and is controlled to scan a prescribed pattern of each layer. The recoating and scanning process repeats until a 3D object is completely fabricated.

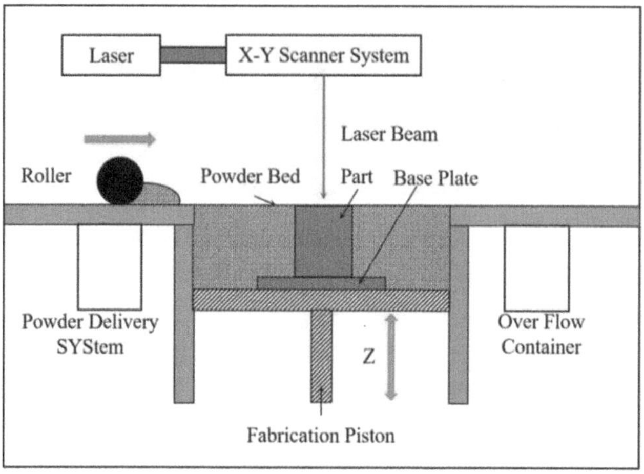

FIGURE 6.2 Schematic layout of a laser PBF system.

PBF systems are of particular interest to the industry because of their relatively higher material performance as a result of the high-temperature fusion, as well as the wide range of materials that can be used. In the case of metal PBF, the selective laser or electron beam fully melts the metal powder, in a similar way to laser welding or electron beam welding. As such, a strong and consistent metallic bond is formed throughout the printed bulk. The high-temperature fusion of a PBF system is comparable to that of a DED or an FDM system, but PBF also takes advantages of its much higher resolution. These advantages together make PBF an ideal approach for producing high accuracy and high-performance parts, despite its relatively low production rate and high cost.

6.2.2 PBF Processes

The most commonly used heat source in the PBF class is laser. Before direct metal PBF was available, polymer laser PBF, or SLS, was first invented and commercialized by the University of Texas at Austin, USA. An SLS machine utilizes an infrared point-wise laser to sinter a polymer, a thermoplastic material or their mixture, wherein the coated metal was found a possible medium to indirectly fabricate metal parts [3]. This indirect approach to fabricating metal parts is hereafter defined as mSLS, namely metallic SLS. mSLS-printed polymer-metal mixture requires additional debinding, furnace sintering and optional infiltration treatments, to finally become a dense metal part. Although SLS of nonmetal was successfully commercialized and are still being used today, the mSLS has somehow lost its competitiveness and is rare to see.

Finally, the developments of SLS led to the successes of direct metal PBF in the late 1990s, after successfully adopting higher energy focused beams, e.g., fiber lasers and electron beam systems. The fiber laser and electron beam PBF processes directly melt metallic particles instead of sintering them through the polymer and are, respectively, called SLM and SEBM. Both processes have become mainstream metallic AM technologies today with well-known brands like EOS, SLM Solutions, Arcam, Renishaw, BLT, etc.

The SLM was first enabled by the successful adoption of an Nd–YAG laser instead of the prior CO_2 laser, because of its much better absorptivity when working with metallic powders. Subsequently, the more recent fiber lasers became the dominant SLM powder source as they were found more compact, energy-efficient and provide better beam quality than the former. An SLM system requires a low oxygen atmosphere to prevent oxidation. The airflow inside the print camber is also important as it must expel the molten pool smoke effectively, and to reduce contamination to the powder bed. A preheating base plate is typically integrated to help reduce residual stress. The other enablers to make SLM processes successful include the use of optimized scanning themes, optimized processing parameters for each material, online monitoring and the enhanced optical systems including f-theta lenses or automatic focusing lenses to minimize beam distortion during scanning [2].

In the mid-2010s, further variations in SLM were invented to improve the production speed and to exploit the potentials that the high-flexibility SLM may imply. A most common attempt is that many SLM companies installed multiple laser sources and multiple laser scanning systems inside the same chamber to multiply productivity. This enabled the successful printing of much larger parts within an acceptable length of time than the prior single-laser solution. However, the addition of optical systems does not change the nature of the process or lower the overall cost. So yet the largest number of laser systems within one machine is around 10. In the multi-laser systems, the scanning strategy and handling of the overlapped areas are critical to guarantee the part consistency. Another attempt to speed up the printing is by adopting different laser powers and different focus settings during the laser scanning. Furthermore, to improve the internal surfaces during a SLM process (Figure 6.3), some SLM companies combined the used of fiber laser with a picosecond laser or a femtosecond laser. The latter pulsed lasers act as a fine-machining or a polishing tool, can remove the adhered semi-detached powder on the part surface and thus improve the surface quality. The additional surface treatment is particularly meaningful for bio-implant applications, as the dissociated metal particles could cause damage to the recipient.

Another technical trend is that several machine manufacturers introduced multi-material SLM (Figure 6.4) that can dispense multiple materials on the same platform and form a multi-material part. Long Yuan, China, invented a continuous-gradient-material SLM system that can customize the compositional gradient curve for up to three different participants. The mechanism involves the use of an integrated powder mixer and a programmable processing parameter controller. The gradient SLM systems have been adopted as a high throughput material preparation technique in the Material Genome Programme, as well as a means to fabricate Functionally Graded Material (FGM) parts.

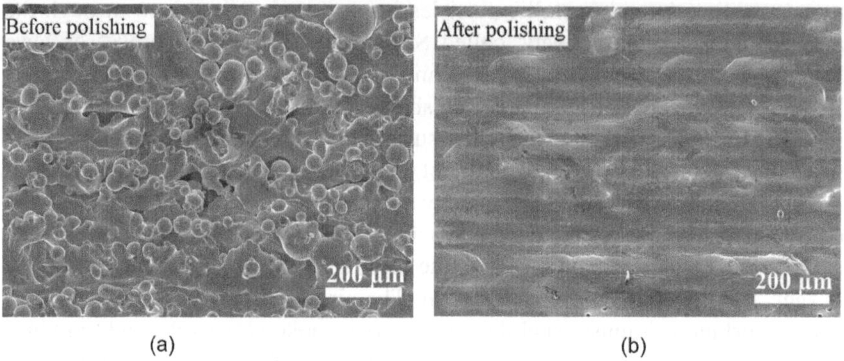

FIGURE 6.3 Medical grade Ti-6Al-4V lattice structure surfaces with (a) and without (b) femtosecond laser polishing.

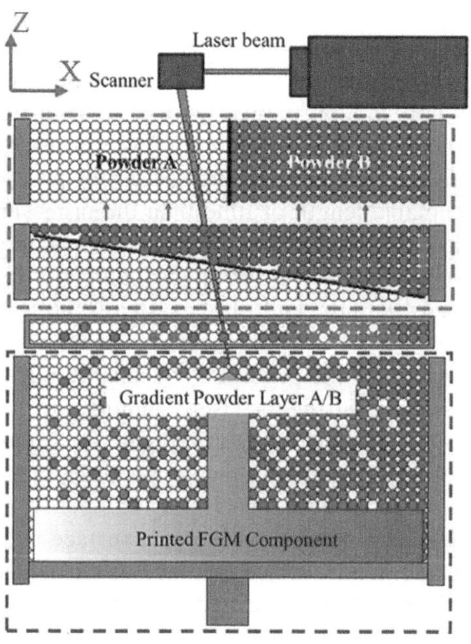

FIGURE 6.4 Layout of a typical multi-material SLM machine, where the compositional gradient is formed along the horizontal direction.

An alternative PBF heat source to laser is the electron beam. The electron beam PBF, or SEBM (Figure 6.5), was first developed at Chalmers University of Technology, Sweden, and was commercialized by Arcam AB, Sweden, 2001. An SEBM process employs a focused electron beam instead of lasers for SLM. The transmission of an electron beam requires high vacuum instead of an inert atmosphere. The melt pool behavior is also different, as the penetration ability of an

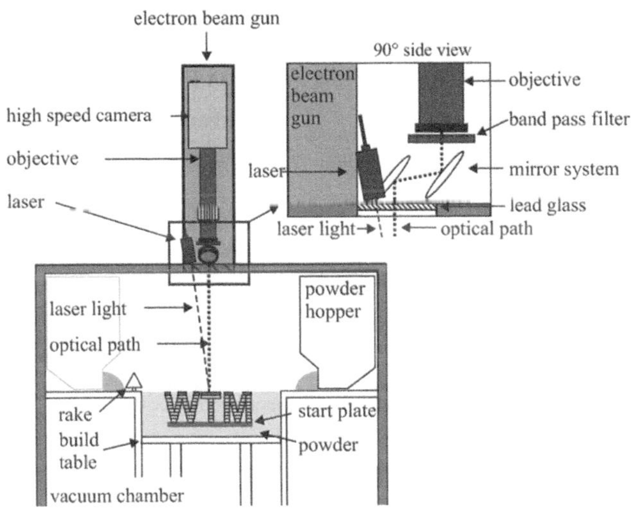

FIGURE 6.5 A schematic SEBM system integrated with an online monitoring module.

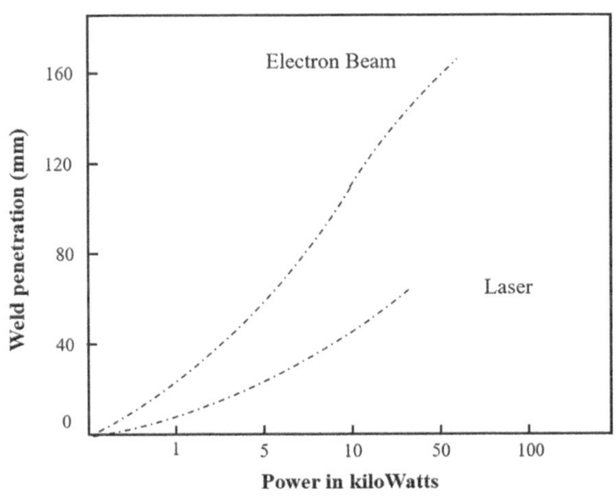

FIGURE 6.6 Comparing the penetration depths between the laser and electron beam welding processes.

E-beam is much higher than a laser (Figure 6.6). This causes the SEBM to have a larger remelting region and to form a coarser but more dense structure than SLM. The energy efficiency for SEBM is usually higher than lasers as most electrical energy is converted into electron beam and the absorption and diffusion inside the powder bed are also high. In addition, the electron beam is controlled by digitally controlled electromagnetic fields, so the scan can be extremely agile and flexible compared to SLM's scanning mirrors.

TABLE 6.2
Comparison of SEBM and SLM Processes

Properties	SEBM	SLM
Minimum feature size (mm)	0.25	0.15
Porosity	↑	–
Tensile strength	–	↑
Ductility	↑	–
Residual stress	↑	–
Productivity	↑	–
Cost	–	↑

The upper arrow represents a better performance than the other.

It's important to note that the electric force can cause metallic particles to repel and fail to melt. This is the reason why only coarser powders are used in SEBM than those for SLM. The SLM powder diameter is typically between 15 and 53 μm, whereas SEBM powders are between 40 and 80 μm. To avoid the repelling effect, a pre-sintering scan is performed using a lower energy density (e.g., by defocusing the E-beam) and a much faster raster speed. This preheating step is also a means to maintain a uniformly high powder bed temperature before the melting scan. As a consequence, the residual stress after printing is usually much smaller than for SLM and thus requires much less support structure.

Overall, the advantages and disadvantages of an SEBM process over SLM can be summarized in the following form (Table 6.2).

6.2.3 PBF Materials

In the 1990s, several important mSLS materials were successfully developed and commercialized after the names: RapidSteel, RapidSteel 2.0, LaserForm ST-100, etc. Some successful results of H13, A6 tool steel, Copper alloys, Tungsten, Nickel-based alloys and hard alloys were also reported as possible candidates for commercial mSLS. Although the mSLS processes are more or less obsolete in recent years, the technology still has the potential to be re-used if next-generation line-wise or layer-wise laser systems will be successful and regain competitiveness for mSLS.

The more recent developments of SLM and SEBM are now capable of processing a number of engineering-grade alloys (Table 6.3). Generally, any metal that can be welded is considered to be a good candidate for PBF processing. Several types of steel, typically stainless and tooling steels, titanium and its alloys, nickel-based alloys, some aluminum alloys, magnesium alloy, copper alloy and cobalt-chrome have been processed and are commercially available. In addition, PBF of precious metals, such as gold, silver, tantalum and rare earth, are commercially available.

TABLE 6.3
Benchmark PBF Metals and Their Mechanical Properties

Machine Type	Alloy	Condition	Specimen Orientation	E (GPa)	Yield Strength (MPa)	Ultimate Tensile Strength (MPa)	Elongation (%)	Hardness
SLM250[HL]	TNM(Ti-Al-Nb-Mo)	As built (compression)	NA	NA	1620–1651	1816–1903	NA	NA
		Heat treated (compression)			886–1070	1428–1671		
EBM	TiAl	Heat treated (compression)		54	544	1800	40	NA
SLM	316L	As built	NA	NA	NA	600.2±2.2	55±2.5	NA
		Heat treated				617.9±1.4	54.1±1.6	
		HIP				586.6±2.4	64.4±2.9	
SLM concept laser M1	316L	As built	NA	NA	640	760	30	NA
EOS M270	15 to 5 PH (precipitation hardening stainless steel)				1100	1470	15	
EOS M250	4340	Stress relieved	XY	213.7	1303	1372	16–17	462–495(Hv)
SLM 250 realizer	316L	As built	XY	NA	220–270	520–680	40–45	NA
Arcam S12	IN625	As built	Z	NA	410	750	44	14(HRC)
		HIP			330	770	69	8(HRC)
EOS M280	IN718	Heat treated	Z	NA	1034	1309	27	NA

(Continued)

TABLE 6.3 (*Continued*)
Benchmark PBF Metals and Their Mechanical Properties

Machine Type	Alloy	Condition	Specimen Orientation	E (GPa)	Yield Strength (MPa)	Ultimate Tensile Strength (MPa)	Elongation (%)	Hardness
			XY		1068	1344	27	
SLM	IN718	As built	XY	NA	816±24	1085±11	19.1±0.7	35(HRC)
			Z		737±4	1010±10	20.6±2.1	30(HRC)
		Heat treated	XY		1227±1	1447±10	10.1±0.6	45(HRC)
			Z		1136±16	1357±5	13.6±0.2	45(HRC)
SLM	$AlSi_{10}Mg$	As built	NA	68±3	NA	396±8	NA	136±9(HV)
		Heat treated		66±5		399±7		152±5(HV)
SLM250^{HL}	$Al_{12}Si$	As built	NA	NA	220.5±9.4	418.9±9.3	3.91±0.3	NA
		Heat treated			202.2±4.3	369.3±3.4	4.38±0.16	
SLM	$Al_{12}Si$	As built	NA	NA	223.5	355.1	4.2	NA
SLM	$AlSi_{10}Mg$	As built	Z	70.7±1.3	168.8±1.3	272.8±2.9	8.2±0.3	NA
			XY	70.2±1.2	169±1	267	9.1±0.5	
SLM	$AlSi_{10}Mg$	As built	NA	NA	207.8	367.7	4	NA
SLM	$Al_{12}Si$	As built	XY	NA	224±7	368±11	4.8±0.6	NA
SLM Realizer	$Al_{12}Si$	As built	XY	NA	240	360	4	NA
		Heat treated			110	190	5	
SLM concept laser M2	$AlSi_{10}Mg$	As built	XY	NA	~250	~330	~1.2	NA

(*Continued*)

TABLE 6.3 (*Continued*)
Benchmark PBF Metals and Their Mechanical Properties

Machine Type	Alloy	Condition	Specimen Orientation	E (GPa)	Yield Strength (MPa)	Ultimate Tensile Strength (MPa)	Elongation (%)	Hardness
			Z		~240	~320	~1	
EOS M270	$AlSi_{10}Mg$	As built	XY	73±1	243±7	330±3	6.2±0.3	105±2(HV)
			Z	72±1	231±3	329±2	4.1±0.2	108±3(HV)
SLM concept laser M1	$AlSi_{10}Mg$	As built	XY	68±3	NA	391±6	5.55±0.4	127(HV)
			Z	NA		398±8	3.47±0.6	
SLM	CoCrMo	As built	XY	NA	738±9.9	1050±12.2	5.2±0.3	39.7±1.2(HRC)
			Z		685.3±10.5	970±9.8	3.9±0.2	34.3±1.2(HRC)
EBM	CoCrMo	As built	XY	NA	717	1110	5	NA
			Z		786	869	0.8	
		HIP+heat treated	XY		586	1145	30	
			Z		586	1151	30	
SLM	CoCrMo	As built	NA	NA	1050±150	1300±150	>6	42±4(HRC)
EBM	Niobium	As built	Z	NA	141	225	34.25	NA
SLM	AZ91D	As built	XY	NA	254	296	~1	100 (HV)
Arcam	Ti_6Al_4V	As built	XY	NA	870±8.1	971±3.1	12.1±0.9	NA
			Z		812±12	953±8.8	13.8±0.9	
		Heat treated	XY		866±6.4	959±8.2	13.6±0.6	
			Z		868±2.9	942±2.6	12.9±0.8	

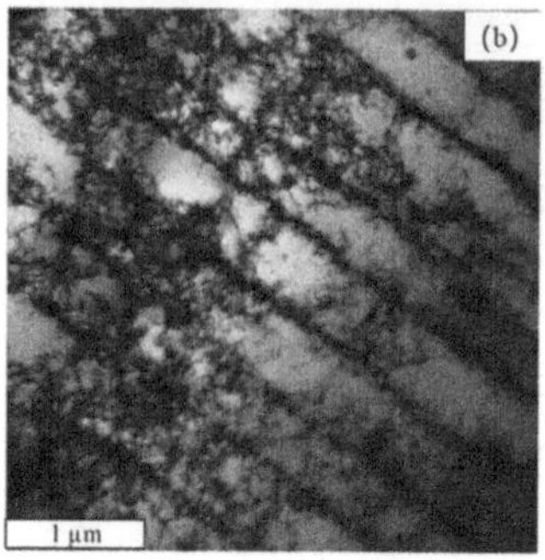

FIGURE 6.7 FEM photo of an SLM SS316L steel exhibiting a strong metastable martensitic phase and high density of dislocations.

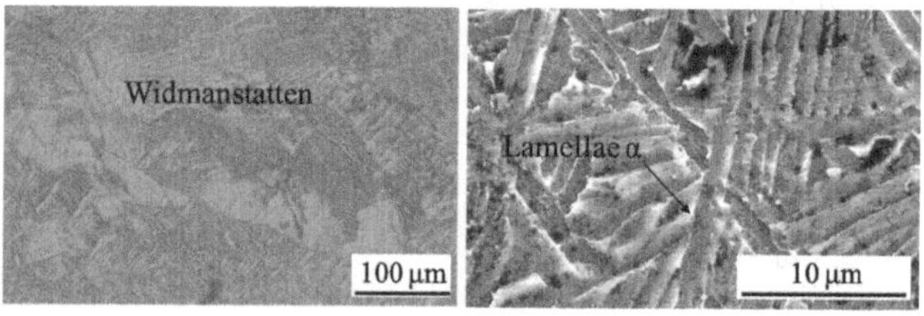

FIGURE 6.8 Typical transformation induced Widminstatten microstructure observed in SLM Ti-6Al-4V.

Questions are often asked 'what the PBF metal material performance is like?' A most typical answer is to compare it with the same alloy in its cast or wrought form. PBF metals are advantageous over casting due to their much higher solidification rates, and higher cooling rates during solid-state phase transformation (SSPT) when it takes place. The high cooling rates are induced by the rapid moving heat source and the steep temperature gradient inside and around the traveling weld pool. Such thermal conditions result in an extremely fine microstructure with supersaturated and non-equilibrium phases, high density of dislocations and dramatically reduced segregation (Figure 6.7). The high cooling rate during SSPT can result in an ultra-fine transformation microstructure in multi-phase alloys like $\alpha+\beta$ Titanium alloy and Austenitic steel (Figure 6.8). After deposition, the consequent heat passes remelt the last few layers for several times. The repeated heating and cooling have similar influences to the material except for the decreasing peak temperature. When the peak temperature falls below the melting temperature, the consequent non-melting thermal passes effectively act as a short time heat treatment to the deposited layers that will lower the chemical energy stored in the rapid cooled non-equilibrium structures. For those passes that peaked below the melting point but above the SSPT temperature, an SSPT microstructure can be formed in addition to the

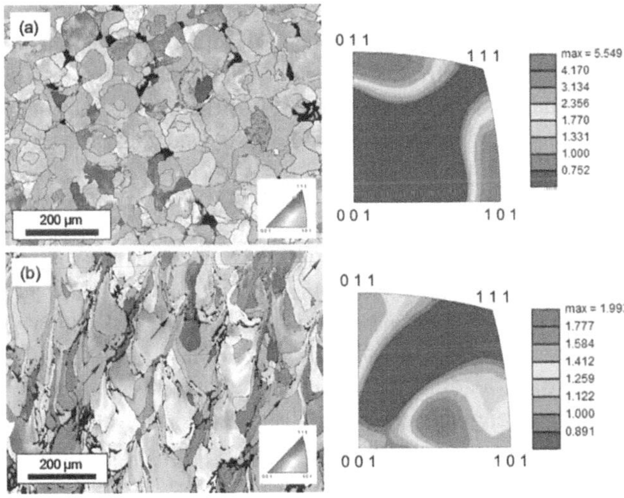

FIGURE 6.9 Strong crystallographic texture observed on (a) X-Y plane and (b) X-Z plane in additive manufactured pure nickel.

FIGURE 6.10 The banding effect seen in SEBM of Ti-6Al-4V.

prior solidification microstructure. These features together make the successful development of several benchmark PBF alloys, including Ti-6Al-4V, the Airbus Scamalloy (AlMgScZr), SS316L and 1.2709 tooling steel [4–7].

The directional and steep temperature field can also cause problems. First, the aforementioned thermal stress issues are almost inevitable. Second, the PBF metals normally present a strong crystallographic texture that is parallel to the build direction (Figure 6.9). This causes anisotropic mechanical properties that make the use of PBF metal less trivial to end users. Third, the repeated heating and cooling thermal history causes a cyclic microstructure variation along the build direction. In SLM and SEBM Ti-6Al-4V, consecutive sets of bands are seen perpendicular to the build direction. The effect is called the banding effect in some literature (Figure 6.10). This effect is more severe in DED processes, where the layer thickness is larger [5,6]. Lastly, gas porosity is inevitable because of the nature of the powder melting process. In safety-critical applications, hot isostatic pressing (HIP) may be required to reduce porosity and improve fatigue performances [7]. In some cases, when the material is difficult to melt or the parameters are not optimized, lack-of-fusion defects may exist and cause the part to crack at much lower fatigue cycles than usual.

6.2.4 Process Advantages and Drawbacks

Overall, PBF metal processes are advantageous over other metal AM classes by the better material performance and material availability. PBF can process a wide variety of metals as has been mentioned above. The printed parts are usually dense enough for static-loading applications, even though small gas pores cannot be eliminated. Many engineering-grade PBF metals have been developed and can be used with good confidence. The spot size of a laser or an E-beam is small enough to fabricate very complex geometries, such as those used in aerospace, bio-implants and conformal cooling mold.

On the other hand, PBF processes are slow, and the parts are difficult to post-process. For instance, a mainstream single-laser SLM machine only deposit $<20\,cm^3$ metal per hour. This causes the printing cost to be too high for most industries, except for a small number of high-added value applications. Multiple laser systems also increase machine costs so cannot help much with the cost per part. In addition, the powder-based process requires plenty of support structures both to constrain the part against thermal stress and to supply cooling channels. This means post-processing of metal parts after PBF can also be expensive and time-consuming.

6.3 DIRECTED ENERGY DEPOSITION SYSTEMS

6.3.1 Introduction

DED is a relatively large-scale direct metal AM class when compared to PBF. DED processes are very much like metal welding that the material is melt as it is being deposited. Laser, electron beam and electric arc have all been used as DED heat sources to melt a metal from its powder or wire form. The system setup can be either multi-axis or robotic (Figure 6.11) and can be operated in an vacuum chamber, or in an open space with a moving atmosphere control, similar to that of an welding system. The build envelop can be as large as a few meters, which is yet the largest possible metal AM technology within all AM classes.

6.3.2 DED Processes

A DED process feeds a wire or a powder material and uses a laser, electron beam or electric arc that is directed to weld the material to the substrate or to the previous layer. These processes can be classified by the various combinations of the feed material and the heat sources [8,9].

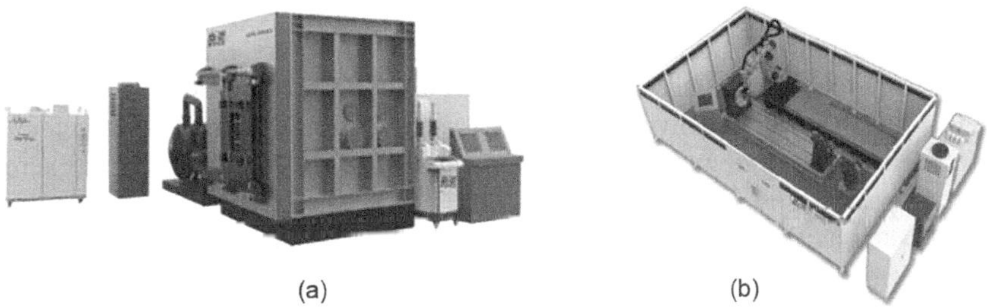

FIGURE 6.11 Examples of (a) multi-axis and (b) robotic DED systems.

FIGURE 6.12 Photo of a LENS machine.

The first well-known DED process is the laser metal deposition process, or LMD, which was initially commercialized by Optomec, USA, following the name LENS (Figure 6.12). The LMD process combines a laser heat source and a blown-powder feeding system. The most characteristic feature in an LMD machine is the 'deposition head', which mainly consists laser optics and powder-gas delivery tubes. The deposition head is clearly distinguished from the other DED setups and can be further classified according to the different layouts. Most deposition heads place the powder nozzle beside the laser aperture, and usually four nozzles are installed to guarantee symmetry regardless the moving direction. If the powder feed and the laser beam are positioned along the same axis, the deposition head is called 'coaxial feeding head'. The coaxial setup provides better material utilization and better gas shield in place.

The inert gas not only protects the weld from oxidation but also acts as the carrying gas to move the powder from the powder tank to the weld spot. Simultaneously, the laser beam is directed to melt the fed powder to form a layer of dense metal. The first layer is deposited on an substrate or on an existing part for repairing purposes. The following layers stack up layer by layer as any other AM process does. The print path is accomplished by the motion of the print head and/or the motion of the substrate. In some cases, the deposition is onto a cylindrical substrate that is placed across a rotary axis. This setup can efficiently fabricate axially symmetric parts. In addition, many companies use LMD together with a conventional CNC system. This novel hybrid setup found its use in build large metallic frames and housings for some marine and aerospace applications. The system is also used in repairing the high-value metal parts, such as the steel rollers, hydroposts and heavy chain wheels, to extend their service times.

In the case of wire feeding, the machine setup is almost the same as LMD except for changing the powder feeding into wire feeding. The deposition head is identical to a weld head. Coaxial wire-laser setup is also available where that the wire feed module is integrated within the weld head. In addition, the metal wire can also interact with electron beam and electric arc, so the corresponding variations of the wire-based DED were developed after the names: electron beam freeform fabrication and wire-arc additive manufacturing, respectively. Similar to the case of welding, the use of a wire is easier to handle than powder, and there is nearly 100% feedstock capture efficiency, but the wire feeding is not as geometrically flexible as powder feeding, so it is mainly used to fabricate simple geometries without too many transitions or overhangs. Another major benefit from a wire feeding DED is the extremely low porosity, in other words, nearly 100% density. This is yet the only AM process that can offer full-dense metal, and it is ideal for the fatigue resistant applications, e.g., safety-critical applications, or structural parts under intense cyclic load.

Some recent work attempted to integrate the DED process with inter-layer micro-forging steps. This could be a ground-breaking technology as it will combine the wrought material property with the DED geometric flexibility. The micro-forging devices, operating at high temperatures, vary from rolling, vibrational pressing, to ultrasonic pressing. All the attempts aim to break the directional grain growth and the strong crystallographic texture, as well as to reduce the residual stress. It has been proven that such a micro-forging mechanism can enhance the material performance significantly (Figure 6.13).

6.3.3 DED Materials

DED processes have the capability of producing fully dense functional parts. Any powder material or wire material that can form a stable molten pool, namely with good weldability, is considered suitable for DED. The powder DED processes are much coarser than PBF processes, so coarser powder is used for lower oxygen pick up and at lower costs. Typical DED powder diameters are between 20 and 150 μm. In both wire-based and powder-based DED processes, it is possible to deposit multiple materials by putting together multi-material feeding and controlling the feeding speed of each.

Similar to metallic PBF processes, the DED processes also offer very high cooling rates, in the order of 10^2–10^4 °C/s. This causes a very fine microstructure and very good yield and tensile strengths in most cases, given the fact that most

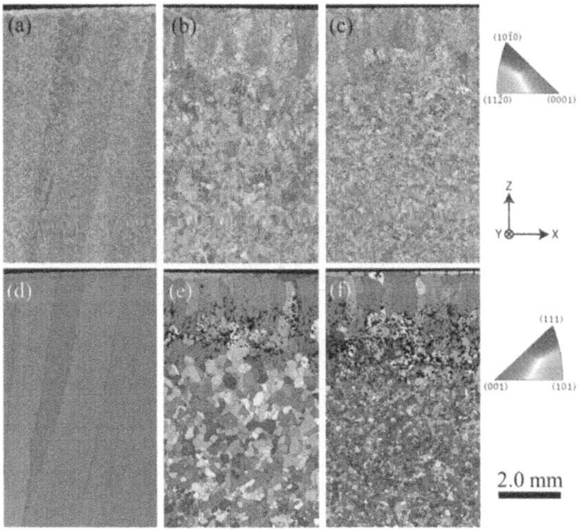

FIGURE 6.13 The room temperature $\alpha+\beta$ microstructure (a–c) and re-constructed beta microstructure (d–f) before (a, d) and after (50 kN: b, e; 75 kN: c, f) the inter-pass rolling on a hybrid DED system.

commercially available DED metal prints are nearly full-dense. However, the ductility is generally considered to be inferior to wrought or cast equivalents. The poor ductility has several possible causes including the heterogeneous inter-layer microstructure with the presence of detrimental phases, the presence of oxidation layers, possible lack-of-fusion defects and the vertical columnar grain boundaries. A most frequently used means to improve ductility is heat treating. The anisotropy, heterogeneity and weak phases can generally be improved, although not completely, after proper heat treatment.

The other important concern for DED metal is the residual stress. Similar to a welding process, a large heat-affected zone (HAZ) results in large distortions. This is most apparent in the wire-arc processes where the HAZ is large. To improve this, the process may be conducted at high temperature, or adopt in situ thermal treatment, to minimize the thermal distortion on site. Stress relief heat treatment is also useful, but the geometric accuracy may be affected.

6.3.4 Advantages and Drawbacks of DED

The DED processes are capable of producing fully dense parts with a relatively wide processing window. This enables a higher degree of process control and microstructure control. Compositional control is also possible simply by feeding multiple materials at different rates. The rapid-cooling DED processes offer desirable fine microstructure, and moderate-to-high mechanical performance. Lastly, the DED systems can be made very large and flexible, so it can be used both to fabricate large parts and to repair existing parts on site.

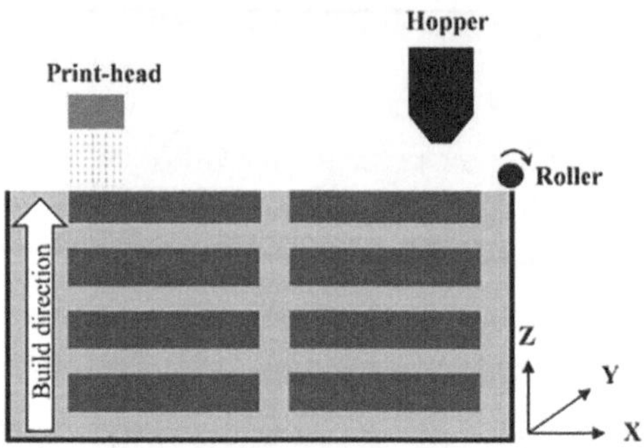

FIGURE 6.14 Schematic BJ system.

The main limitations of DED processes are the poor resolution and surface finish, such that nearly all DED parts must be CNC machined to get the desired final accuracy. Nevertheless, in many cases, DED still offers much lower buy-to-fly ratio compared to CNC only. In the future, the combined DED+CNC approach, as well as the optional micro-forging, will become a competitive means of metal AM.

6.4 METAL BINDER JETTING SYSTEMS

6.4.1 Introduction

BJ methods were developed in the early 1990s, primarily at MIT. Researchers developed the so-called 3DP process in which a binder is printed onto a powder bed to form a part cross-section (Figure 6.14). A wide range of polymers, ceramics and metals have been validated using BJ, and many have become commercially available [10].

The process joins the target powder by jetting binder droplets that are as small as ~50 µm in diameter. This sounds like a slow process, but when thousands of jetting nozzles work simultaneously, the deposition speed can become hundreds of times faster than other point-wise AM processes. This is exactly why BJ processes are of special interest today. To date, the 256–2048 nozzle industrial grade print heads are widely available. The typical resolution of each print head is 200–600 dpi (dots per inch), which is sufficient for most AM applications. For even higher-resolution applications, print heads can be interlaced to multiply the resolution.

The other motivation to develop BJ is the much lower cost than PBF, as the process operates without any high-powered energy, or particular protective atmosphere, and the portion of binder consumed is very low compared to those processes that directly deposit the target material.

6.4.2 BJ Process and Materials

The use of BJ in metal is a relatively new technology field. The process and material are closely interlinked, so this section introduces various BJ processes together with the material systems.

To produce a metal part via BJ, two key steps are usually taken. The first step is to BJ print the green part, namely the binder-metal mixture. At this stage, the liquid-binding agent is dispensed on metal powder, permeates and solidifies in the predefined area of the layer. Then, the build platform is lowered to make room for the next layer and the process is repeated until the 3D object is formed. Second, in some cases, the as-printed green part has to be further cured using a microwave furnace or equivalent heat source, to obtain sufficient strength for the following operations. Third, the fragile green part will be carefully removed from the powder bed, and depowdering process applied to remove the metal powder adhered to the green part. In order not to damage the fragile green part, the operator can choose to use soft brushes, compressed air, vibration or liquid bath to assist the depowdering process. The final step is binder and sintering that are to remove the binder from the object and densify the metal particle matrix. Differing from the separate debinding and sintering processes in MIM, metal BJ usually uses the same furnace unit or furnace array for both steps. This is because the BJ green part consists much smaller volume fraction of binder than for MIM. After sintering, the density of the sintered metal part can reach 95%–98% in most cases. To pursue even higher density, finer powder, compaction to the powder bed or additional HIP process may apply.

There are a number of requirements for the properties and quality of the powder material. First of all, they are related to the quality of the powder layer application (recoating process) and its suitable saturation with a liquid binder. The metal powder must be coarse enough to be dispensed efficiently, and fine enough to be sintered to high density. When intermediate-diameter metal powders are used, the BJ metal machines can utilize a similar PBF recoater, but more recent systems tend to use finer powders, or a combination of both finer and coarser powders, that require more complex powder dispensers to smooth and compact the powder bed. Some companies adopted vibrational compactors, or rollers, to densify the powder bed. When the particle size becomes too small (<1 μm), the surface forces and electrical forces will dominate over gravity, such that it can no longer be used for the BJ, nor for other powder bed processes. In this extreme case, the use of ultra-fine metal powders can only be accomplished by converting the dry powder into a slurry. However, the slurry-based metal BJ system has been proven but not yet commercialized.

A BJ process only joins metal powders together to form a 'green part', and whether a metal is suitable or not largely depends on its debinding and sintering behaviors and result. So generally those metals suitable for MIM, or powder metallurgy, are considered suitable for a metal BJ process, e.g., 316L stainless steel, 420 stainless steel, bronze, Inconel 625 and Tungsten alloys from metal BJ processes have become commercially available.

Lastly, the formulation of a binding agent is also determinant for any metal BJ systems. The binder, in the first place, must be printable. An ideal binder would have low viscosity, which allows the stream of individual droplet beads to form and then break off from the nozzles rapidly. Also, the binder must have stability against the significant shear stress induced by printing. There are many different ways of binding the powder, but some standard selection criteria include binder location (in liquid vs. in bed). The various metal BJ binders can be broadly classified into in-bed or in-liquid types. The in-bed type of binder means a two-component binding system is used, and one of the two components is added to the powder bed. The other component is delivered by ink jetting. The two components interact to form adhesive forces to join the metal. In this case, the ink composition is simple and the ink-jet system is easy to maintain. In the case of in-liquid organic binder, the adhesive force is created upon the polymerization of binding agent or upon the evaporation of the binder solution, the remaining component can adhere the metal particles with sufficient green strength. The post-print curing process is usually required to further polymerize (curing) the organic binder, dry the solvent crosslink polymers [11] and improve the structural integrity of the component when still surrounded by loose powder [12].

However, the formulation of the in-liquid type binder can be complicated and results in more risks to the ink-jet system. The product quality depends on the nature of the interaction between the printed droplets in the powder. The simulating of the powder-ink processes is a very challenging computational problem due to the complex fluid/powder interactions. Most work is still based primarily on experimental methods [13].

Many process parameters can impact the binder efficient application and/or final properties of a fabricated part. These include layer thickness, printing saturation, drying power level, drying time, spreading speed, powder size, powder size distribution, shrinkage, etc. Moreover, binder formulation should be sufficiently stable, and the cured binder product should show strong bonding forces with sufficient properties of elasticity and durability to ensure reasonable deformation of the resulting product. In addition, such factors as temperature, time and its rates used in binder curing, debinding and sintering play key roles. Either binding systems must meet several requirements from the ink-jetting machine, and the metallurgical process. These include

- Surface tension and viscosity to adapt the print heads operation;
- Efficient wettability of powder layer by the liquid binder to ensure a good interaction behavior and uniform saturation of powder material;
- PH value and corrosive performance to adapt the ink circulation system;
- Curing speed must be high enough to ensure productivity and slow enough to guarantee the penetration depth;
- Debinding performance, burnout completeness and environmental impacts; and
- Possible diffusion or contamination of the target metal.

So far, a variety of materials have been printed by the BJ method, including stainless steel, titanium, graphene oxide, copper, Inconel, lead zirconate titanate, calcium polyphosphate, etc. These printed parts could find applications in various industries, such as biomedical [14–16], aerospace [17], solid-oxide fuel cell [18] and ceramics scaffolds for bone tissue [19].

6.4.3 Advantages and Drawbacks

The main advantages of BJ metal are the superior deposition speed and high resolution. The BJ metal systems produce several liters per hour, which is hundreds of times faster than a usual metal PBF. In addition, the indisputable advantage of the technology is the possibility of its application for metal prototypes and small-scale production: the BJ method does not require the design of special forming tool equipment and the shape of the future product does not affect the settings of printing parameters. Printing of details in a powder bed does not require the creation of supports and provides high-dimensional accuracy and printing stability.

Metal BJ technology has several significant limitations yet to be improved. The fact that a binder-metal mixture is used is the main disadvantage. Long-time debinding and sintering must be applied to remove the binder and get the final part in a dense and pure form. Moreover, the geometric stability is found difficult to control, as a significant shrinkage can occur upon debinding and sintering. Similar to the MIM process, this shrinkage effect constrains the process of fabricating parts as large as a mainstream PBF machine normally offers. Another limiting factor for the fast-growing development of this technology is the high cost of equipment and consumables. The development of progress requires optimization of technological processes of the BJ method, increasing efficiency and reducing costs to improve profitability indicators.

6.5 INDIRECT METAL ADDITIVE MANUFACTURING

The previous sections have introduced the mainstream metal AM approaches. However, the manufacturing of a metallic part can also be achieved by combining AM modeling with casting or powder metallurgy processes. In these cases, a casting mold or a 'green part' is prepared by AM, and the printing process does not involve any metallurgical treatment. The indirect metal AM has many variations and involves a range of different AM processes; thus, this section is written to summarize the important facts about indirect metal AM.

6.5.1 AM and Casting

Casting is one of the major metal forming technologies, where hot liquid metal is poured or injected to a mold to form the desired bulk shape. Clearly molding is a critical step for all casting processes. Conventional molding involves intense labor and CNC machining and is usually costly and time-consuming. By utilizing AM, the pattern can be fabricated at a fraction of time, and in some cases, at a fraction of cost [20,21].

The first successful application of AM in the casting industry is SLS. In the 1990s, several SLS companies succeeded in developing a phenolic resin-coated sand for SLS. This type of sand is equivalent to the conventional injection molding sand; thus, users can use the SLS sand mold without changes to their sand-casting lines. These products quickly found their application in the high-end sand-casting industries and are still being used today. The phenolic resin-coated sand can be sintered in an oven to reach 4–6 MPa tensile strength. The relatively high strength makes the SLS sand a desirable sand core material for complex thin pipes and narrowly channeled structures.

Another competing sand mold printing technology is the BJ or 3DP. As has been discussed in Section 6.4, BJ is yet the most productive and low-cost approach for industrial applications. BJ sand is just a good example. BJ companies, such as Exone, Voxeljet and LongYuan, offer large sand BJ machines that can print 100 L/hour sand mold, over 20 tons per month. The BJ sand systems utilize several types of binders to match the casting requirements that include furan resin, phenolic resin and inorganic binders.

In addition, the ShengQuan company developed a resin pattern to replace the conventional wood pattern used in sand molding. The resin pattern is printed with an SLA system for the highest resolution and surface quality available among the AM classes, but at relatively lower deposition speed. Nonetheless, the resin mold is announced to be at least two times more durable than wood mold and is more cost-effective than an aluminum mold or a cast iron mold.

Lost wax casting, or precision casting, is another widely adopted casting technology. The shape of a wax pattern is identical to the casting, so it must be converted to a female shell mold to contain the molten metal. The shell mold is made by coating the wax pattern with several layers of refractory substances. The wax is then removed by heat and steam, leaving the shell mold in place. The shell mold is then sintered in a furnace to the desired density and size. Finally, the liquid metal is poured into the shell mold to complete the casting. When high-grade refractories are used, the shell mold can have excellent casting performance, such as low degassing and good stability at high temperatures. Thus, it has been successfully used in casting steel, titanium alloy and superalloys. The conventional method for making a wax mold is by first producing an aluminum or silicone master pattern. Wax is then injected into the master pattern to form the shape of the final product. By adopting AM technologies, foundries can directly 3D print wax molds without master patterns. This is particularly beneficial for small-batch production, as well as for new product developments. Lost wax casting molds can be additively manufactured by SLS, BJ, LC and MJ.

First, laser sintering (SLS) of polystyrene powder is yet the most well-established process for lost wax casting. Companies like EOS and LongYuan offer specialized wax SLS systems. The as-printed 'wax' patterns may have distortions and undesirable surface roughness, so the parts are usually sunk in a tank of liquid wax and manually polished according to 3D scanning results. A more productive wax printing approach is BJ. The printing quality and post-processing steps are similar to that for SLS wax. In the BJ wax approach, PMMA is the most developed and commercially available wax material. Neither SLS nor BJ can provide sufficient accuracy and surface finish without extensive manual work, so novel wax-like resins were also developed for LC and MJ systems.

Another possibility to create a wax pattern is by making a master mold for it. LC resin, as well as FDM polymers, is sometimes used to print such master molds. The master mold is then converted to a silicone female mold and finally

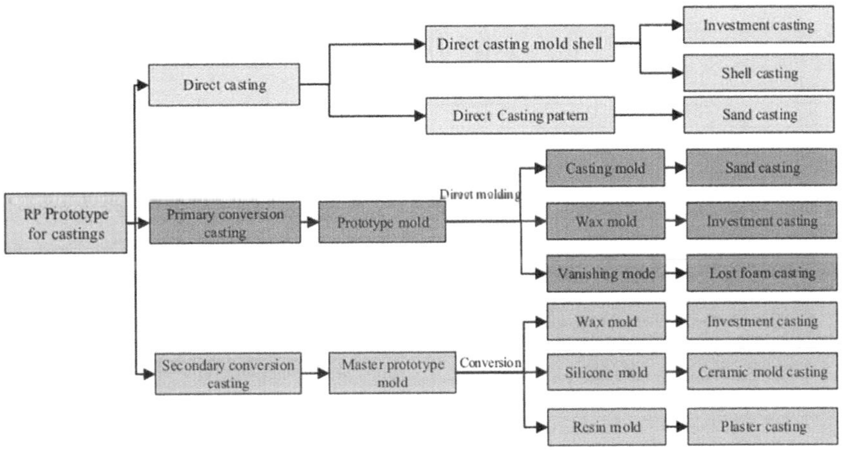

FIGURE 6.15 A summary of processes and materials used for AM-Casting.

in wax. In addition, Admatec, Netherlands, prints equivalent ceramic shell molds directly from a novel slurry-based LC system. However, the LC ceramic process is constraint by the part size and costs. In the future, BJ ceramics may be a promising means to fabricate ceramic shell mold for such applications.

The various combinations of AM and casting processes are summarized in Figure 6.15.

6.5.2 AM and Powder Metallurgy

Sections 6.2 and 6.4 have already introduced the indirect metal AM processes, e.g., mSLS and metal BJ process to print the green parts for powder metallurgy, but there are other possible AM processes used for the same purpose. For example, some companies developed metal FDM systems using premixed composite wire, bars or grains (Figure 6.16). The mixture of binder and metal powder is heated to a fluid state and extruded along the 3D paths to form a green part. Slurry-based feed stock, a mixture of UV resin and metal powder, can be used in an LC machine or an MJ machine (Figure 6.17). The UV resin is selectively cured by the UV light source and is removed during the debinding process. The X-Jet company, Israel, created a low-viscosity ink for metal MJ. The ink is a combination of binder and nanometer-size metal particles. The printing quality is very high, and the main drawback is the high cost.

6.5.3 Other Systems and Future Expectations

Many successful metal AM processes have been developed, but the technology is still evolving by nature. Many new heat sources, novel line-wise and layer-wise scanning devices (Figure 6.18) and new joining mechanisms are still being exploited. Meanwhile, the concept of AM overlaps with a wide range of

FIGURE 6.16 A commercial metal FDM machine and stainless steel feedstock.

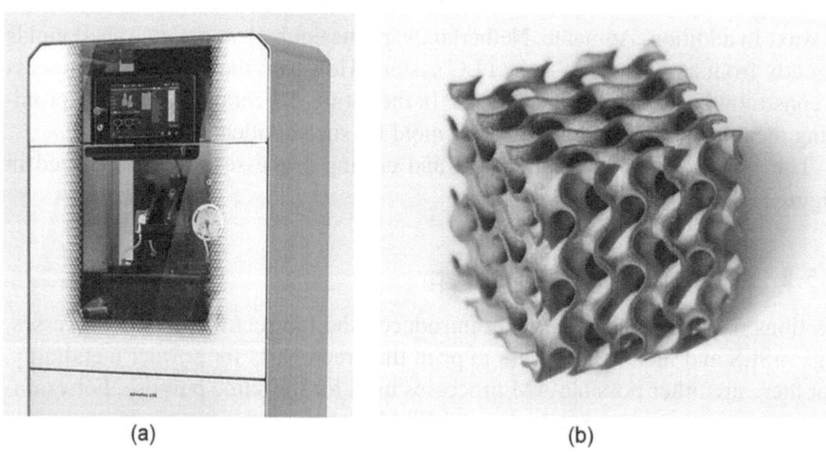

FIGURE 6.17 (a) The Admatec Admaflex-130 LC-based slurry metal printing system and (b) the as-sintered copper part.

manufacturing technologies that join materials together. For example, friction stir welding and diffusion welding have been proven to form parts in an additive fashion.

In the past 30 years, the AM of metal has evolved from modeling to direct metal manufacturing, from very few successful metal types to a wide range of commercially available metals and from high-end applications to more generalized use. The technology is becoming more diverse to adapt the different industries and applications and more competitive that the productivity per cost continues to increase.

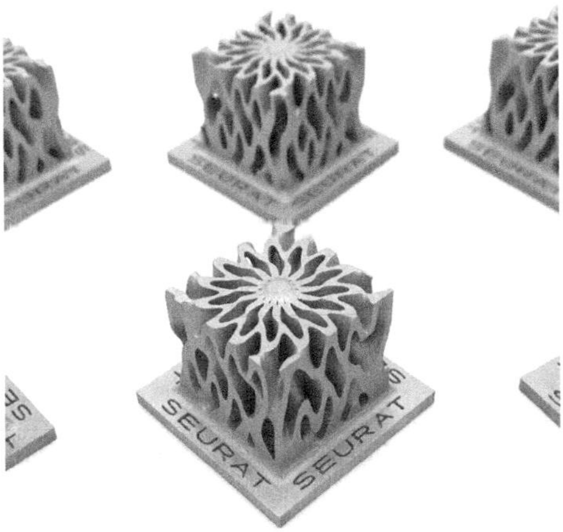

FIGURE 6.18 An areal laser melting fabricated heat sink part, credits to Seurat Technologies.

6.6 ADDITIVE THINKING

AM has the potential to revolutionize the global parts manufacturing and logistics landscape. It enables distributed manufacturing and the productions of parts-on-demand while offering the potential to reduce cost and energy consumption. In contrast with the conventional mass-production methodologies, these new possibilities offered by AM require new ways of additive thinking, new design principles, new simulation technologies, new serviced-based manufacturing modes, new data platforms, etc. [22–24]. This section discusses the additive thinking principles to help understand how to implement AM to its most potential.

6.6.1 Design for Additive Manufacturing

The first important point in additive thinking is to understand how AM can change the design for manufacturing. AM has the capabilities of producing shape complexity, hierarchical structures, material complexity, functional complexity and metallurgical complexity for metals. The unique capabilities of AM provide opportunities to rethink the design for manufacturing and assembly.

- *Shape complexity*: the layer-by-layer fashion of AM makes it possible to build virtually any shape. This offers freedom of design that is nonparallel to any other manufacturing technologies. A mechanical engineer no longer needs to be constrained by the CNC machining principles and can now focus on the functionality of the part itself. Typical examples include the topologically optimized structural parts (Figure 6.19) that were previously difficult to machine and the conformal cooling mold with complex cooling channels inside.

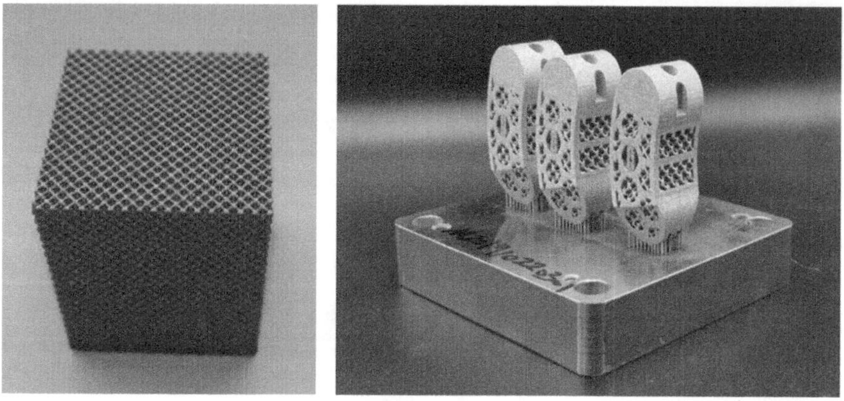

FIGURE 6.19 A porous structure and titanium implants fabricated by SLM machine.

- *Hierarchical complexity*: features can be designed with shape complexity across multiple size scales. A typical example is the lattice structures being used in surgical implants (Figure 6.20). The lattice can be designed to fit Young's modulus of the human bone structure and is helpful for regrowth and recovery.
- *Functional complexity*: functional devices can be produced in one build by integrated design. GE additive manufactured fuel nozzles that integrate 18 parts into 1 piece. The integrated manufacturing avoids welding and assembly errors. GE announced over 35,000 pieces per year. In auto engines, the compressor shell and vent pipe are cast together by using AM sand cores. The printed sand core has better dimensional accuracy and integrated the multiple molds to be one piece.
- *Material complexity*: material can be processed one point, or one layer, at a time a single material or a combination of multiple materials.
- *Metallurgical complexity*: metals can be heated and cooled at adjustable rates and multiple times, so the temperature profile can be tailored to customize the microstructure and mechanical properties both locally and globally.

6.6.2 Process Simulation

AM is used to fabricate parts swiftly on demand. Conventional trial and error efforts are very much disliked. Metal AM is a multi-scale process that can usually take dozens of hours per print job. In many cases, the AM metal part is complex, difficult and expensive to make. So the other additive thinking principle is to have the capability of predicting and optimizing an AM process before a failure could occur.

The PBF process involves micron-size metal particles that fiercely interact with the high-energy laser at a fraction of a second. The interaction causes a moving melt pool in the size of millimeters. The process can last for more than

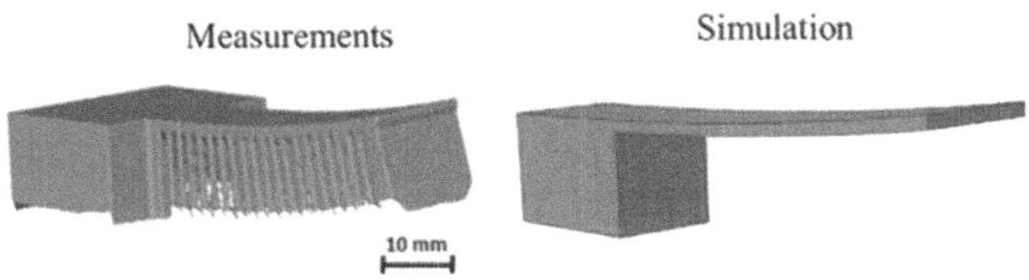

FIGURE 6.20 An SLM simulation that successfully predicted thermal distortion that agrees with the printing result.

10 hours to form a part, given the production rate is as low as 3–10 mL/hour. Successful modeling for PBF was accomplished by simplifying the microscopic melt pool process mathematically. The mathematical function is derived using a high-resolution FEM model, and the algorithm is then substituted into the macroscopic FEM model to accomplish a relatively swift multi-scale simulation. This type of AM simulation software is now commercially available from companies like 3D System, Materialise and Siemens. Many production features, such as the recoating risks, residual stress, distortion, material hardness and lack-of-fusion porosity, can now be simulated with reasonable accuracy (Figure 6.20).

The DED process is a relatively larger-scale process. So the finite element modeling is more or less the same to the equivalent welding processes. Usually the heat source is described as a point-wise moving heat source centered to the weld track. The simulation is particularly useful to prediction the distortion and potential cracking during the process.

The simulation of BJ metal includes modeling of the BJ forming process, as well as the high-temperature debinding and sintering processes. Commercial BJ metal simulation software is not yet available. This is one of the reasons why BJ metal is yet less used commercially than PBF processes.

6.6.3 Service-Based Manufacturing

An AM machine can virtually print any possible product at any time. So all the AM machines, or 3D printers, effectively formed a global network of productivity. Never in the human history had a comparable production technology with the same flexibility and availability to the public. This means that AM has the potential to become an important service-based manufacturing for everyone.

Many online platforms, in the forms of a website or an mobile phone APP, were established to connect the various market demands to the widely distributed AM productivity. For example, if someone places an order to 3D print a metal phone holder, the order may be directed to the nearest metal printing workshop. The next morning, an neighbor knocks at the door and brings in the finished product. Another example is that an metal sculpture in Wellington city center was 3D printed using more than ten different SLS wax machines in China, casted and assembled in a sculpture workshop, before being shipped to New Zealand. It is important to note that distributed manufacturing using AM is much more agile and beneficial for consumers, R&D users and innovators.

6.7 SUMMARY AND PERSPECTIVE

AM provides high flexibility, short lead time, near-net-shaping and material diversity for the modern manufacturing industry. Despite its relatively higher costs, AM technologies were first adopted in high-added-value applications, such as aerospace and medical care. In recent years, the technological maturity and supply chain maturity rapidly arises. AM facilities started to be seen in production line in industries such as rapid prototyping, dentistry, art and casting sectors, and the concept of mass customization starts to realize.

Various AM technologies have been adopted to fabricate metal parts, either directly or indirectly. With PBF and DED processes, fully dense metal parts are directly produced through the highly automated 3D printing processes. With LC, SLS, BJ and others, an intermediate product is fabricated and then converted to the metal via casting or sintering. Each of the processes has found its particular context of use, ranging from micron-scale fine structures to meter-scale large castings. Almost all the mainstream metals and alloys are candidates for AM. In particular, weldable metals are good candidates to direct metal AM. Casting alloys and powder metallurgy metals are straightforward options for indirect metal AM, because their metallurgy is identical to that of conventional casting, respectively, sintering processes.

The tendency to develop more efficient and economic AM systems is critical to generalizing the use of metal AM. Line-wise and area-wise printing systems have already been developed by several companies to replace the slow point-wise process. Another important effort is to further reduce the cost of raw materials. For instance, conventional powder metallurgy feedstock is used in BJ metal process to replace the expensive spherical gas-atomized metal powder. As such, the overall cost of AM will continue to decrease until getting to the threshold for serial production. This turning point has already been exceeded in some high-end complex geometry applications in aerospace, military defense, energy and marine and will continue to take place in more industries. The landscape of conventional manufacturing will potentially be converted into a highly flexible, fast response and distributed form and finally revolutionizes human creativity and productivity.

REFERENCES

1. Gibson, I., Rosen, D., & Stuker, B. (2015). *Additive Manufacturing Technologies.* Springer: Berlin/Heidelberg, Germany.
2. Gu, D. (2015). Laser additive manufacturing (AM): Classification, processing philosophy, and metallurgical mechanisms. In: *Laser Additive Manufacturing of High-Performance Materials SE-2* (pp. 15–23). Springer: Berlin/Heidelberg, Germany.
3. Kruth, J. P., Wang, X., Laoui, T., & Froyen, L. (2003). Lasers and materials in selective laser sintering. *Assembly Automation*, 23(4), 357–371.

4. Blakey-Milner, B., Gradl, P., Snedden, G., Brooks, M., Pitot, J., Lopez, E., Leary, M., Berto, F., & du Plessis, A. (2021). Metal additive manufacturing in aerospace: A review. *Materials & Design*, 209, 110008.
5. Brandl, E., Leyens, C., & Palm, F. (2011). Mechanical properties of additive manufactured Ti-6Al-4V using wire and powder based processes. In *IOP Conference Series: Materials Science and Engineering*, vol. 26, December 2011, p. 012004, Sheffield.
6. Al-Bermani, S. S., Blackmore, M. L., Zhang, W., & Todd, I. (2010). The origin of microstructural diversity, texture, and mechanical properties in electron beam melted Ti-6Al-4V. *Metallurgical and Materials Transactions A: Physical Metallurgy and Materials Science*, 41(13), 3422–3434.
7. Tammas-Williams, S., Zhao, H., Léonard, F., Derguti, F., Todd, I., & Prangnell, P. B. (2015). XCT analysis of the influence of melt strategies on defect population in Ti-6Al-4V components manufactured by selective electron beam melting. *Materials Characterization*, 102, 47–61.
8. Fachinotti, V. D., Cardona, A., Baufeld, B., & Van Der Biest, O. (2012). Finite-element modelling of heat transfer in shaped metal deposition and experimental validation. *Acta Materialia*, 60(19), 6621–6630.
9. Nurhudan, A. I., Supriadi, S., Whulanza, Y., & Saragih, A. S. (2021). Additive manufacturing of metallic based on extrusion process: A review. *Journal of Manufacturing Processes*, 66, 228–237.
10. Mao, Y., Li, J., Li, W., Cai, D., & Wei, Q. (2021). Binder jetting additive manufacturing of 316L stainless-steel green parts with high strength and low binder content: Binder preparation and process optimization. *Journal of Materials Processing Technology*, 291, 117020.
11. Bai, Y., & Williams, C. B. (2018). Binder jetting additive manufacturing with a particle-free metal ink as a binder precursor. *Materials & Design*, 147, 146–156.
12. Do, T., et al. (2018). Additively manufactured full-density stainless steel 316L with binder jet printing. *In Volume 1: Additive Manufacturing; Bio and Sustainable Manufacturing*, College Station, TX.
13. Babu, S. S., et al. (2015). Additive manufacturing of materials: Opportunities and challenges. *MRS Bulletin*, 40(12), 1154–1161.
14. Mostafaei, A., et al. (2018). Binder jetting of a complex-shaped metal partial denture framework. *Additive Manufacturing*, 21, 63–68.
15. Javaid, M. H. A. (2018). Additive manufacturing applications in medical cases: A literature based review. *Alexandria Journal of Medicine*, 54(4), 411–422.
16. Díaz-Moreno, C. A., et al. (2019). Binder jetting additive manufacturing of aluminum nitride components. *Ceramics International*, 45(11), 13620–13627.
17. Nickels, L. (2015). AM and aerospace: An ideal combination. *Metal Powder Report*, 70(6), 300–303.
18. Manogharan, G., Kioko, M., & Linkous, C. (2015). Binder jetting: A novel solid oxide fuel-cell fabrication process and evaluation. *JOM*, 67(3), 660–667.
19. Seitz, H., et al. (2005). Three-dimensional printing of porous ceramic scaffolds for bone tissue engineering. *Journal of Biomedical Materials Research Part B: Applied Biomaterials*, 74B(2), 782–788.
20. Hodder, K. J., & Chalaturnyk, R. J. (2019). Bridging additive manufacturing and sand casting: Utilizing foundry sand. *Additive Manufacturing*, 28, 649–660.
21. Carneiro, V. H., Rawson, S. D., Puga, H., Meireles, J., & Withers, P. J. (2020). Additive manufacturing assisted investment casting: A low-cost method to fabricate periodic metallic cellular lattices. *Additive Manufacturing*, 33, 101085.
22. Bayat, M., Dong, W., Thorborg, J., To, A. C., & Hattel, J. H. (2021). A review of multi-scale and multi-physics simulations of metal additive manufacturing processes with focus on modeling strategies. *Additive Manufacturing*, 47(September), 102278.

23. Conner, B. P., Manogharan, G. P., Martof, A. N., Rodomsky, L. M., Rodomsky, C. M., Jordan, D. C., & Limperos, J. W. (2014). Making sense of 3-D printing: Creating a map of additive manufacturing products and services. *Additive Manufacturing*, 1, 64–76.
24. Gao, W., Zhang, Y., Ramanujan, D., Ramani, K., Chen, Y., Williams, C. B., Wang, C. C. L., Shin, Y. C., Zhang, S., & Zavattieri, P. D. (2015). The status, challenges, and future of additive manufacturing in engineering. *Computer-Aided Design*, 69, 65–89.

7 Additive Manufacturing of Lattice Structures

Xinwei Li and Wei Zhai
National University of Singapore

CONTENTS

7.1 INTRODUCTION

Additive manufacturing (AM), more commonly known as 3D printing, is one of the pillars of the fourth industrial revolution alongside the internet of things, artificial intelligence, etc. AM is essentially an advanced manufacturing technique that focuses on end user-customizable products. The fundamental concept of AM fabrication is associated with a layer-by-layer bottom-up building approach in accordance with a computer-aided design (CAD). AM is hence associated with extremely high product design freedoms, including that with highly complex geometries, such as interconnected features, pores and internal cavities. Thus far, numerous types of AM techniques based on different working principles have been developed, and almost all types of materials have also been manufactured via AM. With the advent of AM, there is hence a paradigm shift for materials design from that of being fully bulk to that being architectured. Such materials then constitute to a class of materials known as cellular solids, which are defined as porous solids or interpenetrating composites of solid and air. In particular,

DOI: 10.1201/9781003182146-7

lattice structures stand out as the most widely studied cellular solid. Various types of lattice structures exist, given the unlimited design possibilities in a three-dimensional space. That being said, with the unique solid and void architecture, the properties of lattice structures essentially derive from their physics rather than chemistry. Unique intrinsic features include being lightweight, having reduced mechanical rigidity, a large specific surface area (SSA) and unique fluid-flow properties. As such, lattice structures find diverse applications as structural materials, protective materials, sound absorbers, filters, electrochemical electrodes, etc., where all the above-mentioned attributes matter. With the high freedoms with materials selections and structural design, it is apparent to see that AM-ed lattice structures are becoming the state-of-the-art advanced functional materials of tomorrow. This chapter first aims to introduce the major types of lattice structures, their physical properties, design tools and methods used and various AM techniques and the materials used and finally an overview of four popular applications of cellular materials.

7.2 LATTICE STRUCTURES

In materials science, lattices are defined as any structures with repeating unit cells, and it is most commonly used to describe the atomic arrangements in crystalline solids. However, for the case of lattice structures as cellular materials, they are best agreed on by researchers to be defined as periodic and repeating 3D-structured unit cells. Other common terms found for lattice structures in literature also include microlattices, nanolattices, for the case of sub-milli- and sub-micro-printed lattices, or periodic foams, architectured metamaterials. As can be seen, in contrast to foams and honeycombs, lattices in turn allow immense geometrical design freedoms. Any combination of spatial arrangements of features constitutes a lattice structure. In principle, there are virtually unlimited possibilities for a lattice design. As such, lattices gain an advantage in being capable of highly tunable physical properties and hence functionalities. This section aims to give an overview on the main types of lattice structures, including the current state-of-the-art structural designs, followed by their mechanical properties and an illustration of their design tools.

7.2.1 Types of Lattice Structures

Numerous types of lattice structures exist in literature. Based on their physical appearance, herein, we classify them into five main classes: strut, triply periodic minimal surface (TPMS), variants and hybrids and structures based on meta-behaviors.

Strut lattices are lattices based on strut patterns with interconnected truss members joint at nodes. Among all classes of lattice structures, strut lattices have the highest design freedoms for the simplicity in design motifs. There can be virtually unlimited strut lattice designs given the boundless possibilities in three-dimensional spatial distributions. Strut lattices are characterized by strut

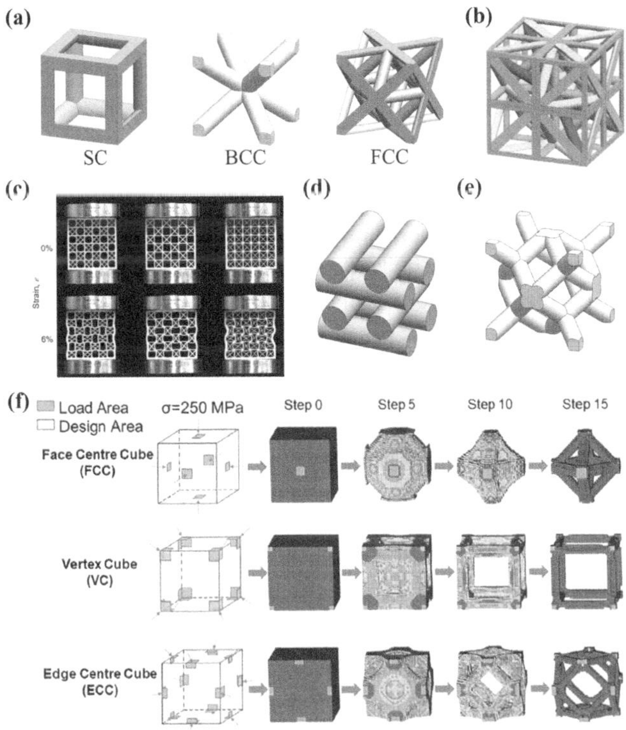

FIGURE 7.1 Some examples of strut lattices. (a) Illustration of fundamental crystal-structure mimicking lattices. (b) A proposed combination of SC+BCC+FCC lattice. Other types of strut lattices, including (c) bioinspired designs, (d) log-pile structures and (e) rhombic dodecahedron structures. (e) Topological optimized lattices using a commercial finite element software. (Adapted and reproduced with permission from Refs. [12], [16] and [24] for (b), (c) and (f) respectively.)

orientation, the number of struts and the number of nodes. Several fundamental and most studied designs are associated with crystal-structure mimicking, including the simple cubic (SC), body-centered cubic (BCC) and face-centered cubic (FCC) [1–4] (Figure 7.1a). Herein, nodes take up the lattice points of the crystal structure and struts, in turn, take the lattice constants. A wide range of variations exists from these fundamental designs. Prominent ones include designs with biased vertical struts (denoted as Z), notably BCCZ and FCCZ [5–11]. Researchers have also worked on numerous combinations or superpositions of SC, BCC and FCC lattices with interesting results including improved isotropy [12,13] and/or improved stiffness and strength [14] (Figure 7.1b). Other modifications include the stretching of an FCC cell to an orthorhombic symmetry [15].

Strut lattices based on other geometries deriving from bio- or physical inspirations also include the bioinspired [16], log-pile structures [17], polyhedrons [18–21] and covalent bonds [22,23] (Figure 7.1b–d). Apart from structural mimicking designs, strut lattices have also been generated from the finite element methods. For such structures, struts and nodes are derived from the highest stress distributions via topological optimization with pre-defined loads on bulk solid [24,25], most of the times using a commercial finite element software (Figure 7.1e).

Researchers have also worked on strut lattices with non-uniform struts. One popular method of strut heterogeneity is the tapering of struts with a larger diameter near nodes. The motivation behind strut tapering is for improved specific stiffness and strength for which regions of high-stress distributions, the nodes, are being reinforced. Tapered FSCC [14], BCC [26], rhombic dodecahedron lattices [18] and FCC [27] lattices have shown markedly improved stiffness and strength.

TPMS lattices are structures based on TPMS surfaces, a mathematical construct first devised by Hermann Schwarz back in 1865. Further contributions were made by Alan Schoen in 1970 on more structures based on crystallographic cells. TPMS surfaces are essentially surfaces with zero mean curvature and non-intersecting parts. They hence take on the form of smooth shells and are sometimes termed as shell or shellular lattices. Owing to this, TPMS structures hold unique properties, such as high passing fluid-flow rates [28] and the much-reduced stress localizations (as opposed to nodal and bending regions of struts), upon applied forces [29]. Numerous types of TPMS structures exist (Figure 7.2a), some of the most widely studied include the primitive, gyroid and I-WP structures and are parametrized with the following surface functions [30]:

$$\text{Primitive}: \quad \cos(x)+\cos(y)+\cos(z)=t$$

$$\text{Gyroid}: \quad \sin(x)\cos(y)+\sin(y)\cos(z)+\sin(z)\cos(x)=t$$

$$\text{Diamond}: \quad \sin(x)\sin(y)\sin(z)+\sin(x)\cos(y)\cos(z) +\cos(x)\sin(y)\cos(z)+\cos(x)\cos(y)\sin(z)=t$$

$$\text{I-WP}: \quad \cos(z)\cdot\sin(x+y)+\sin(z)\cdot\cos(x-y)=t$$

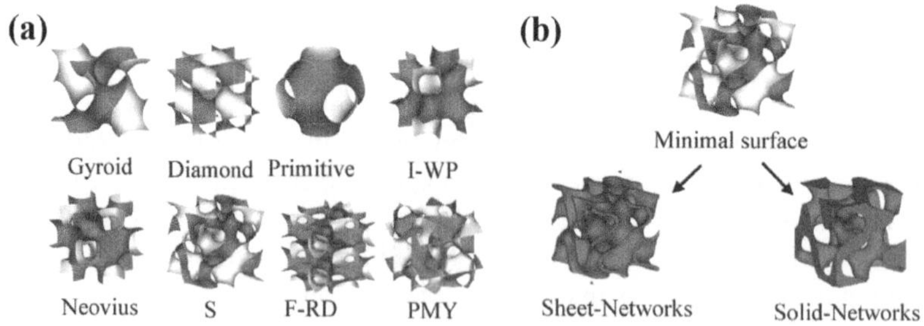

FIGURE 7.2 Some examples of TPMS lattices. (a) Various types of TPMS surfaces and (b) illustration of the sheet and skeletal structures possible from TPMS surfaces. (Adapted and reproduced with permission from Ref. [46].)

For Schwarz TPMS surfaces, the t parameter takes on the value of 0. Nonetheless, variations with respect to the equations and t have been studied too [31]. For their continuous surface morphology, TPMS structures can also be subdivided into a sheet or skeletal structure (Figure 7.2b). Sheet structures [28,32–35] take on the form of the parametric equation with thickened walls, while skeletal structures [36–43] take on (either) one part of the inverse of the sheet. Sheet structures generally have superior stiffness and strength as compared to their skeletal counterparts [44]. Strong TPMS structures, such as the sheet gyroid, diamond and I-WP, are generally stronger than most lattices too, although exceptions exist [44]. For their smooth bone-like biomorphic surfaces [45], TPMS lattices have also been widely studied for structural applications as bone implants [35,39–43].

Plate lattices are a relatively new concept brought out in recent years. Like strut lattices, fundamental plate lattices are also designed based on crystal structures. For crystal-structure mimicking plate lattices, plates are spatially added and joined as the close-pack planes of the respective crystal lattices (Figure 7.3a). The first plate lattice had been reported by Berger et al., for which a combination of SC+FCC plates showed to possess the Hashin-Shtrikman upper bounds on

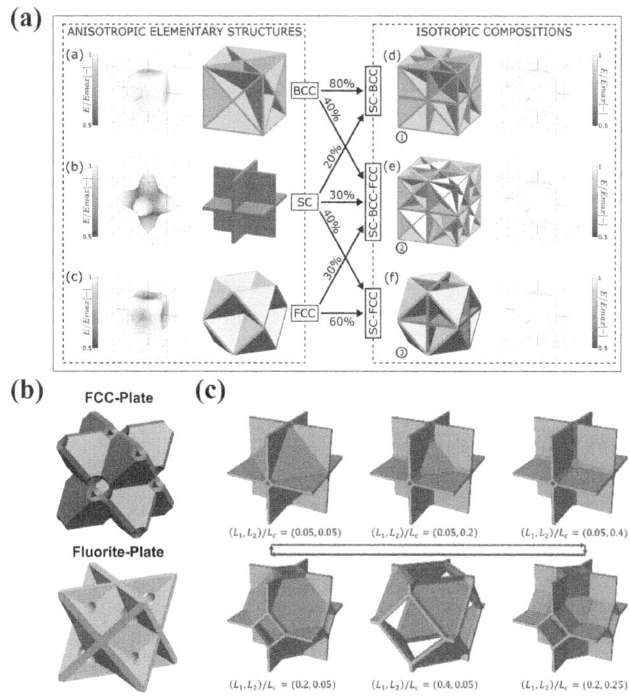

FIGURE 7.3 An overview of plate lattices. (a) Illustration of fundamental SC, BCC and FCC plate lattices and their permutations. (b) Close-cell FCC and fluorite plate lattices with holes added for manufacturability. (c) Novel open-cell plate lattice designs. (Adapted and reproduced with permission from Refs. [48], [51] and [52] for (a), (b) and (c), respectively.)

isotropic elasticity [47]. Tancogne-Dejean et al. worked on an entire series of SC, BCC and FCC plate lattices with different combinations. The theoretical limit of isotropic elasticity and strength is seemingly attainable by a combination of SC-BC-FCC [48]. Closed-cell plate lattices are however impossible to be built by the popular mainstream AM techniques, notably the vat-photopolymerization and powder-bed fusion processes. Hence, most closed-cell plate lattices are studied experimentally via the fused deposition modeling (FDM) printed [49,50]. Researchers have also worked on plates with holes introduced despite the tradeoff in weakened mechanical properties [51] (Figure 7.3b). Novel open-cell plate lattices have also been proposed (Figure 7.3c), with proper optimization, the theoretical maximum isotropic bulk modulus is also attainable [52].

Variants or combinations of concepts have been worked on for lattice design to achieve unique properties. Tube lattices, with struts made hollow internally, show designable isotropy based on the tune of wall thickness [53]. Hybrid lattices consisting of mixtures of struts, tubes, nodes and spheres have reduced localized stresses [54,55]. Variants of TPMS-like structures have also been reported. Researchers have worked on constant mean curvature TPMS surfaces [56], TPMS-like SC, BCC and FCC smooth shell structures based on minimization of curvature functions at nodes [57,58] and ultra-thin auxetic TPMS skeletal structures [59]. Apart from the unit cell level, variants and hybrids in designs have also been worked on the macro level. Hierarchical "lattice-in-lattice" fractal structures have been worked on, whereby the struts of lattices are composed of repeating lattices themselves [60–62] (Figure 7.4a). Researchers also leveraged the concept of functional grading, where features like the relative density or cell sizes vary either continuously or stepwise across unit cells (Figure 7.4b). Functional-graded lattices revealed unique and different mechanical response as compared to uniform ones, such as a reduced post-collapse valley with a sharply increasing plateau [63,64]. Functional-graded TPMS lattices have also been studied for artificial implants with varying local stiffness [65]. Recently, a new concept of multi-morphology macro-lattice has been brought up, including unique strengthening inclusions [66] and mixture of different unit cells [67] (Figure 7.4c), both showing superior mechanical properties as compared to their homogeneous counterparts.

Apart from the above-mentioned, there are also lattices designed with the meta-behavior in mind. For "meta" behaviors in lattice structures, researchers generally refer them to having physical properties not found in naturally occurring materials. Notable meta-behaviors include mechanical, consisting of negative Poisson's ratio, pentamode behavior, negative stiffness and acoustics, consisting of acoustic cloaking, lensing and absorbing. Lattices with a negative Poisson's ratio, also known as auxetics, are associated with an inward lateral shrinkage upon compression and outward expansion upon tension. Prominent auxetic cellular structures include the reentrant [68–70], chiral [71] and buckling [72–74] based ones (Figure 7.5a–c). Pentamode lattices are structures with an ideally zero shear modulus and hence an infinitely large bulk modulus with fluid-like behaviors. Most of them take on the infinitely small edge-joint strut design [75–78] (Figure 7.5d). Negative stiffness structures in turn take on several negative stress–strain gradients under deformation,

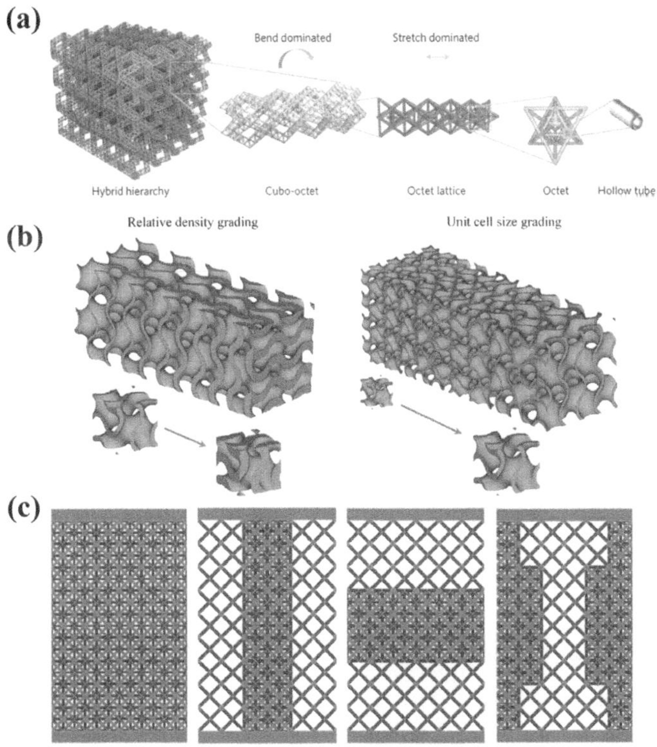

FIGURE 7.4 An overview of hybrid lattices. (b) Hierarchical octet-truss lattice. (b) Relative density and cell size graded sheet gyroid lattices. (c) Heterogeneous cell macrolattice of FCC and BCC unit cells. (Adapted and reproduced with permission from Refs. [62], [65] and [67], respectively, for (a), (b) and (c).)

such as structures designed for bifurcation buckling [79–81] (Figure 7.5e). Acoustic metamaterials are then designed to control sound wave propagations based on specific acoustic theories, such as resonance or equivalent fluid models. While most take on conventional lattice designs motifs, they are usually numerically designed with specific optimized geometries [82–84].

7.2.2 Physical and Mechanical Properties

There are several classifications associated to describe and characterize lattice structures. The most important among is their relative density (ρ^*). The relative density refers to the actual density of the lattice itself (ρ_c), to the density of its constituent solid material (ρ_s), and is given as:

$$\rho^* = \frac{\rho_c}{\rho_s}$$

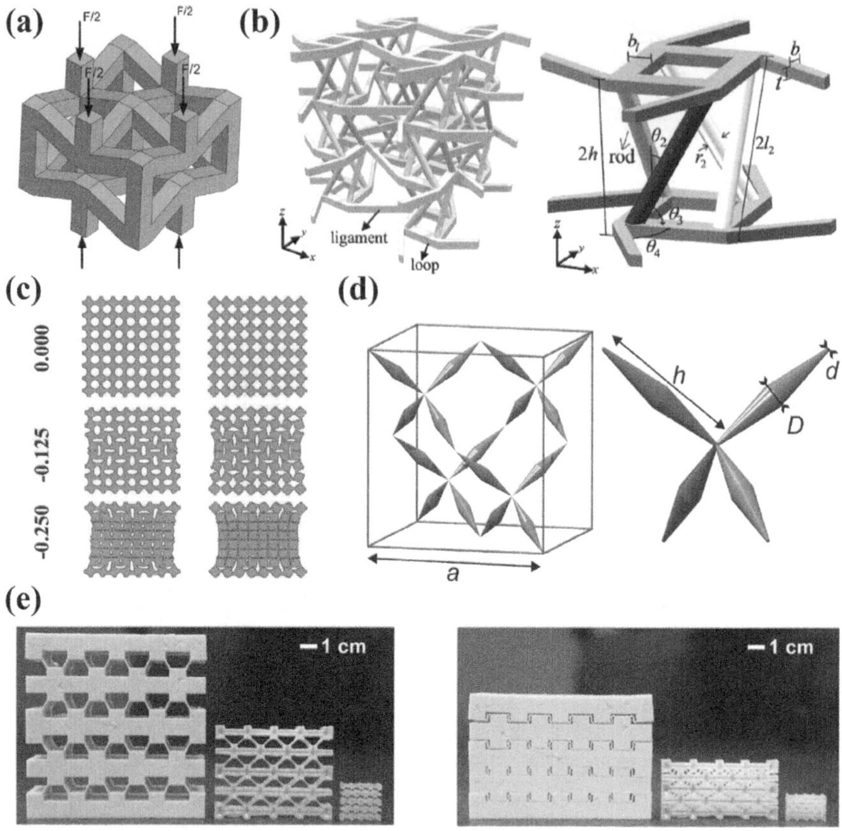

FIGURE 7.5 An overview of metamaterial lattices. Illustrations of auxetic structures, including (a) reentrant, (b) chiral and (c) buckling based. (d) Edge-joint pentamode lattice. (e) Negative stiffness structure, revealing its bi-stability. (Adapted and reproduced with permission from Refs. [70], [71], [74], [78] and [79] respectively for (a), (b), (c), (d) and (e), respectively.)

Conversely, the subtraction of relative density from 1 (or 100%) gives the porosity. Alternatively, ρ^* can also be derived by calculating the volume fraction of the lattice with respect to the volume of its surrounding bounding box. The cell size then refers to the length of a side of a smallest unit cell. The definition of cell size may then take on several values if a unit cell is not cubic, e.g., analogous to Bravais lattices. Feature size, depending on the type of lattice, then refers to either the strut diameter for strut lattices or wall thickness for TPMS and plate lattices. Lattices with unique or complex features, which are not easily characterized by strut or wall dimensions, such as heterogeneities, are then explicitly defined by researchers. However, it should be obvious that among these three parameters, only two are independent and the third must be dependent on the other two. It is most widely accepted in literature to compare lattices across the same relative densities.

Nonetheless, cell or feature sizes may be of interest sometimes due to the material properties' size dependency for some AM techniques.

One intrinsic property of lattice structure is its high SSA, with respect to a unit volume (and constitutively unit weight). The SSA may scale significantly with relative density depending on the lattice design. For instance, strut or plate lattices have significant node masses, which consume the free surface, while the TPMS lattices are associated with only thickening cell walls. In addition, it is worth noting that SSA is periodicity dependent. Another important intrinsic property of lattice structures is their reduced stiffness and strength. In fact, for most applications, the mechanical properties of lattice structures are also of utmost importance. Gibson and Ashby [85] devised an empirical rule, prominently known as the Gibson–Ashby model, for the scaling of stiffness (E) and strength (σ) with relative density for all cellular solids as:

$$\text{Stiffness}: \quad \frac{E_c}{E_s} = C_E\left(\rho^*\right)^{nE}$$

$$\text{Strength}: \quad \frac{\sigma_c}{\sigma_s} = C_E\left(\rho^*\right)^{n\sigma}$$

Researches nonetheless suggested the relationship to be largely true only in the low relative density region of below, a generally agreed on, 30%. Low-density lattices translate to thinner struts and walls where structural deformation is dominant as opposed to materials shearing in thicker ones. Despite this, it is worth noting that cellular materials of interest for various applications usually have low relative densities below 30%. From the Gibson–Ashby equations, it can also be shown that there are two constants, C and n. n refers to the scaling exponent that usually takes on values from 1 to 2. A value close to 1 suggests strong stretching dominated behavior while that larger than 1.5 is close to bending. The proportionality constant, C, ranges from 0 to 1, depending on the structural efficiency. For a cubic lattice with three-dimensional features, where only a proportion of the features may contribute to deformation resistance upon compression, C usually follows that proportion and takes on values between 0.1 to 0.5. For a directional structure like the honeycomb, where compression resist is directly proportional to the out-of-plane materials, C can be near to 1.

From the collective research efforts on cellular solids, it is conclusive that honeycombs are stretching while foams are bending dominated. Lattice structures can in turn be either stretching or bending or a mixture of both. As a rule of thumb, from the viewpoints of mechanics, bending dominated lattices deform via localized hinging or bending, while stretching dominated lattices are structurally very rigid, and lattice members deform via material compression or tension only. Common bending dominated lattices include the BCC, rhombic dodecahedron and skeletal TPMS structures. Those leaning toward stretching include the SC and FCC and many sheet TPMS structures. A common way to predict the deformation mechanism for strut lattices is the Maxwell's criterion. Given $M = b - 3j + 6$, where b and j refer to the number of complete struts and joints respectively, a calculated M larger than 0 constitutes to stretching while that below 0 to bending. For lattices of more complex morphologies, the behavior is hard to determine numerically and an investigation of the mechanical response across a range of relative densities is usually needed. However, whatever the deformation mode may tell us no information about the strength of the lattice (compared across the same relative density) of a cellular solid. As mentioned, strength in turn depends on C, a measure of the proportion of cellular members contributing to deformation resistance.

7.2.3 Computer-Aided Engineering Tools

The first step to 3D printing a lattice structure will be its CAD model. Various commercial software is fit-for-purpose: ranging from graphical user interface-based ones, such as Solidworks, AutoCAD and Rhino, to computing languages, such as MATLAB. Researchers have also developed free-to-use software for automatic lattice design generation, such as MSLattice [46] and FLatt Pack [86]. Following the CAD model, the relative density and SSA of the lattice structure are then also made known in the software. Conversely, it works vice-versa to design a lattice structure based on a desired relative density or SSA. After which, 3D object files are then exported to a dedicated 3D printing software to orientate and slice the CAD model before printing. As shown in Section 7.2.1, various lattice designs are based on mimic physical structures and their derivatives. This can also be classified as a manual way of designing lattice structures, i.e., using human intuition. In fact, the popularity of this method leads to it accounting for the majority of the published work on lattice structures [87]. Apart from manual design, computer-aided engineering designs based on either commercial finite element modeling (FEM) software or self-defined scripts have also been worked on. Optimization objectives have so far included a broad range of physical properties, e.g., mechanical, thermal and acoustics. Popular commercial FEM software widely used by researchers includes ABAQUS, ANSYS and COMSOL. The other purpose of FEM is to study mechanisms, such as stress distribution or failure under mechanical loads, thermal flux, wave propagation and dissipation.

7.3 ADDITIVE MANUFACTURING

7.3.1 Techniques

Mainstream AM techniques used for lattice structure fabrication can be classified into three classes: powder-bed fusion, vat-photopolymerization and extrusion based. For each of these classes, various subsets of techniques with varying material forming mechanisms and feedstock material exist.

The powder-bed fusion (PBF) technique is associated with the material forming via the fusion of powders in a powder bed. PBF techniques are mainly used for the AM of materials that can be made into powder forms and do not decompose

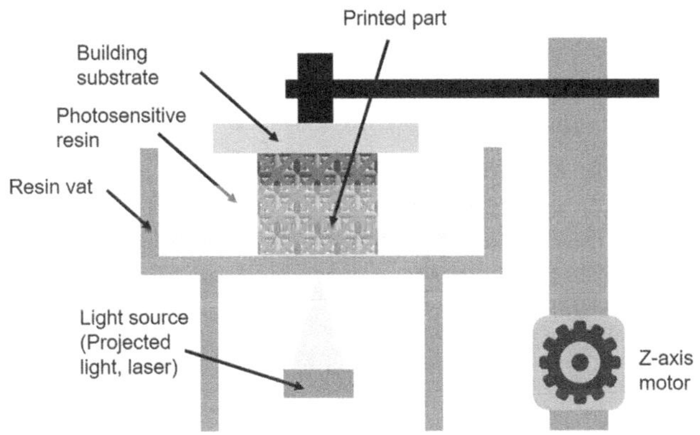

FIGURE 7.6 Schematic illustration of a powder-bed fusion-based AM process.

upon heating. A schematic representation of the general PBF process is shown in Figure 7.6. The main components of a PBF setup include the building substrate, energy source and powder feeding system. A layer of feedstock material powder is first paved onto the building platform during building. After which, an energy source, with a scan pattern determined by the slice of the CAD model, will then melt/sinter the powder and the target area of powder is then fused into a single solid entity. After one layer is complete, the platform gives way to its successive layer with a new spread of powder, and the process repeats itself until the desired 3D structure is obtained. In an ideal situation, melt scan tracks (resulting from scan patterns) should overlap with no gaps in between adjacent tracks and powders. As such, for each type of powder/material, parameter optimization is required by the operator to ensure that parts can be printed near net densities. Four main parameters for PBF process include the energy source power (P), scan speed (v), hatch distance (h) and layer thickness (t) (Figure 7.6). Generally, energy density, in terms of joules per cubic dimension, can be quantified with the following relationship: $E = \frac{P}{vht}$. Under simple terms, a low energy density results in insufficient bonding and porosity, while a too high energy density results in significant thermal stresses and possible *in situ* sublimation. An optimal set of parameters characterized by densification and mechanical properties are then required. The main difference between different PBF techniques lies with the energy source and fusion mechanisms. Three distinct types are (i) selective laser melting (SLM), (ii) electron beam melting (EBM) and (iii) selective laser sintering (SLS).

SLM is associated with high-intensity laser(s) as the energy source. The fusion mechanism is based on the melting upon absorbing the laser energy and the subsequent rapid cooling and solidification of the powder bed. EBM works similarly but with an electron source used in place of the high-intensity laser. Consequently, EBM functions only in a vacuum environment, whereas SLM does not need to. This also implies that different materials are better printed in one than another.

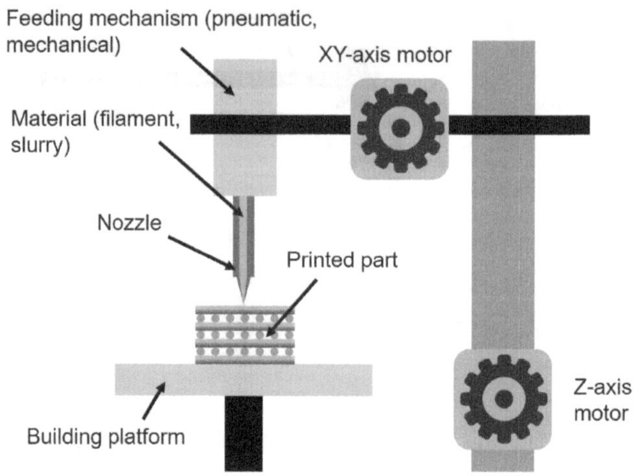

FIGURE 7.7 Schematic illustration of a vat-photopolymerization-based AM process.

Metals with a higher affinity to oxygen and nitrogen, such as Ti-6Al-4V alloy and high entropy alloys, are better printed with the EBM [88], while steels, aluminum and nickel alloys can be printed with the SLM [89]. Comparatively, SLM allows better resolutions for fine features and lower surface roughness as compared to EBM [90,91]. Operation costs are also lower without the need for vacuum. SLS in turns uses a low-intensity laser that only sinters particles together. SLS is hence useful for materials, which decompose upon melting or materials susceptible to extreme thermal gradients upon rapid solidification. Common SLS used materials hence include thermoplastics, such as polyamide, polycarbonate and polypropylene [92], and ceramics, such as silica and zirconia [93]. The above-mentioned are the most common or commercially used materials for PBF processes, it is worth noting that various novel and exotic materials are also being currently researched on for PBF possibilities [89].

Vat-photopolymerization techniques are in turn, as the name suggests, to be based on the polymerization of photosensitive resin in a resin container, also known as a vat. The working schematic representation of this class of techniques is shown in Figure 7.7. Differing from PBF, the most commonly adopted setup is based on a top-down building approach. Main components consist of a building substrate, light source and a resin vat coupled with a light transparent film. During operation, the building substrate will first be lowered into the resin vat, up till a height corresponding to the pre-defined layer thickness above the bottom film. The light source will then shine, with patterns in accordance with the sliced CAD layer, onto the sandwiched resin layer in between the film and substrate. The photosensitive resin will then solidify, and with an engineered preferential bonding to the substrate, detach from the bottom film and be pulled up by the substrate to form a layer of solidified print. Processes repeat until the desired 3D model is created. As such, to successfully cure a resin layer up till the substrate surface, proper build parameters need to be implemented. The main parameters include a matching light energy and input layer thickness. Essentially, in accordance with the Beer–Lambert law, increasing light energy increases the light penetration power and hence the photocuring depth in the resin. However, light energy should not be too high such that it decomposes the resin or potentially deteriorates the structural integrity of the as-build. As such, an optimal selection of light energy, which is usually based on a suitable combination of intensity and exposure time, and the substrate build layer thickness, will be needed for a successful build. Currently, three main types of vat-photopolymerization techniques exist, namely (i) digital light processing (DLP), (ii) stereolithography (SLA) and (iii) continuous liquid interface production (CLIP). All three techniques function essentially on the same basis with their only difference being the type of light source used.

As its name suggests, the light source in DLP is based on digital light, similar to the commonly used overhead projectors. Light is projected layer by layer according to the slices via a digital image created using pixels. A popular variant of DLP is liquid-crystal display (LCD) printing, in which instead of a projector, an LCD, similar to our daily computer, and smartphone displays are utilized. LCD printers are advantageous in terms of being low-cost, however, at the tradeoff with much weaker light sources. SLA is different from DLP, with a scanning laser being used in place of a projector. In terms of build speed, DLP and LCD are usually advantageous as compared to SLA. Regardless of the area and complexity of slice patterns, light is always exposed in a single projection for these techniques. However, considerably longer scanning times will be needed for a laser to scan large patterns. In terms of build quality, SLA has a better printing resolution as it is laser spot size and scanning pattern limited while that for DLP and LCD are pixel array limited. CLIP, also earlier known as continuous liquid interphase printing, is in turn an innovation from DLP. As opposed to a glass or film, here, an oxygen-permeable membrane lies at the bottom of a resin vat. This membrane creates a dead zone via oxygen inhibition where photopolymerization cannot take place and thus always allows a layer of liquid resin to exist. The existence of the resin layer ensures that the solidified part will never be in contact with the bottom film. As such, the building substrate can be continuously pulled up (slowly) to obtain a net 3D printed structure. CLIP is advantageous to DLP and SLA in terms of being less prone to failure (part stuck to bottom film) and a faster build rate. Despite the different types of light sources, light energies are all within the UV-visible spectrum with wavelengths ranging from 380 to 405 nm.

As mentioned, standard materials for vat-photopolymerization use are photosensitive resin. Such a type of resin consists of mainly a few standard components: oligomer/monomer, photoinitiator and crosslinkers. With the tune and varying combinations of oligomer/monomer and crosslinkers, photosensitive resins are capable of achieving a diverse range of mechanical properties, ranging from stretchable, tough to being rigid [94]. Apart from polymers, vat-photopolymerization techniques are also capable of printing ceramics. Ceramic printing is based on the printing of ceramic powder-doped resin and its subsequent debinding and sintering heat treatment to achieve a fully solid ceramic structure.

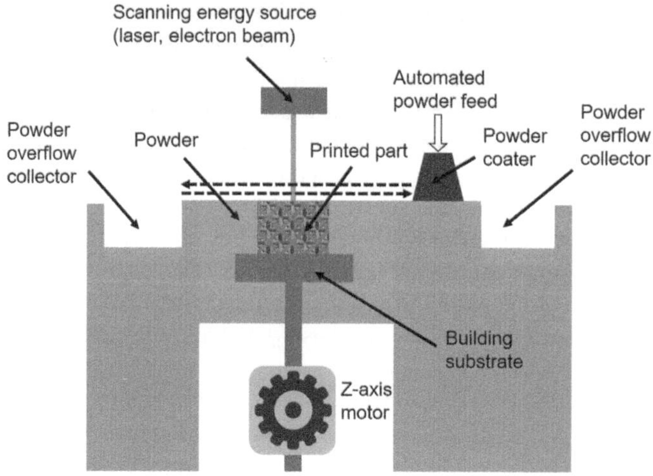

FIGURE 7.8 Schematic illustration of an extrusion-based AM process.

Apart from the mainstream, or commercial, other researches on materials also include biomaterials, hydrogels, advanced materials composites [95].

Like a pen that writes on a piece of paper, extrusion-based techniques utilize the same principle for extruding materials onto a substrate. A schematic representation of the extrusion-based technique is shown in Figure 7.8. The main features associated with this technique include a nozzle, for directing and extruding the material, a material feeding mechanism and a moving nozzle or substrate for controlling the deposited pattern. It can thus be seen that extrusion-based techniques are vastly different from PBF and vat-photopolymerization techniques which instead adopt a pool of feedstock materials. There are two main types of extrusion-based AM technique, namely direct ink writing (DIW) and FDM. The two techniques are different both in terms of their feed material and forming mechanism.

As its name suggests, DIW is based on the direct deposition of an ink. This also implies that a successful DIW print lies with the formulation of the ink. The fundamental requisite of a DIW ink is that it must exhibit shear thinning behavior, meaning under high shear rates, i.e., during extrusion, the ink is highly flowable, and when the shear forces are being lifted, i.e., being deposited and at rest, the ink is highly viscous. This ensures the ink to stay as an almost rigid solid upon being extruded out onto the substrate, such that it will not flow away, be able to bear its own weight as an overhanging, and to not collapse from subsequent build-up layers. Depending on the material choice, high loadings of ceramic or metal particles can be added into the ink formation to derive semi-solid green bodies, or for polymers, they can be directly extruded as it is. Post-processing, such as heat treatments, may then be done for ceramics or metals to sinter and make the prints fully dense. The ink feeding mechanism is usually based on pneumatic actuators or screw feed devices. For the less stringent requirements for an ink as compared to powder or resin, DIW has hence extremely high materials selection allowances. Almost all kinds of materials, ranging from polymer, hydrogel, ceramics and metals to advanced materials, such as metal-organic framework, MXenes and various carbonaceous materials, have been reported to be successfully printed with DIW [96]. Multimaterial printing with DIW is also better established than other AM techniques with the simplicity of a multi-nozzle implementation [97]. However, DIW generally trades off with poorer allowable design freedoms owing to its ink-based mechanism. As such, common cellular structures being printed with DIW remain at honeycomb or log-pile lattice structures, which are all geometrically simple. Hybrid DIW methods, which utilize multimaterial nozzles with sacrificial dissolvable supports, have been investigated to improve the printable structural design freedoms [98]. Differing from DIW, FDM in turn uses the principle of material melt. In contrary to the ink, a filament is used in FDM. The filament melts at a heating mechanism, passes through a nozzle, gets deposited onto a substrate and is then left to cool and solidify immediately. As such, common FDM materials mainly include thermoplastics, such as polylactic acid, acrylonitrile butadiene styrene and polyamides and composite filaments with metal, carbon fiber or wood dopants [99].

7.3.2 Materials

In principle, any material printable with AM can be used for the fabrication of lattice structures. As such, lattices of virtually all types of solid materials, ranging from elastomers, rigid polymers, ceramics, metals, composites and advanced materials have been studied and reported in literature [87,100]. The main motivation for the diverse range of materials being used lies with their functionalities.

Elastomeric materials mainly consist of photocurable elastomers and hydrogels. Such lattice structures hold properties, such as deformation recoverability and reusability, for being very soft, flexible and extremely ductile [101]. Elastomers are usually used as soft impact absorbers while hydrogels are for biomedical purposes. Rigid polymers in turn consist of various photocurable resins and thermoplastics, such as polylactic acid, acrylonitrile butadiene styrene, polycarbonate and polypropylene [102]. Tough polymers consist also of photocurable resins and thermoplastics, such as polyamide and polyurethane [103,104]. Polymers are usually investigated as energy absorbers or structural metamaterials. Ceramics consist of structural ceramics such as alumina, zirconia and silicon nitride; biocompatible ceramics such as hydroxyapatite and calcium phosphate; and functional ceramics like titania and yttrium barium copper oxide [105–109]. Like their host material, ceramic lattice structures are highly brittle too with very low toughness. Instead of structural applications, ceramic lattices are usually researched on as strong and lightweight substrates for membranes and catalysts through post-printing surface treatments. Common metals used for cellular materials include numerous grades of steel, aluminum alloys, Ti-6Al-4V and various superalloys [90]. With the inherent ductility of metals, most metal lattice structures are associated with high toughness. Metallic lattices are hence the ideal choice as lightweight structural, energy-absorbing materials and artificial bone implants. Advanced materials include numerous types of carbonaceous materials, such as graphene, carbon nanotubes, metal-organic framework and MXenes [110,111]. Such materials are usually studied as electrodes and catalysts given their capabilities for hierarchical porosity. Composite lattice structures are in turn associated with material heterogeneities at either the feedstock material or an inter-cell structural level [5,112,113]. In addition, composite lattices can be obtained via post-treatments, such as infill and coatings, with interesting properties, such as enhanced toughness [112] and recoverability [114,115] attainable. Composite lattices are hence usually studied for possible unique mechanical properties.

7.4 APPLICATIONS AND RESEARCH TRENDS

7.4.1 Energy Absorption

Intuitively, the introduction of pores into a solid material will both decrease its mechanical strength and change its deformation behavior. For instance, solid materials are expected to either shear, barrel or rupture during compression. With pores added, stress will be re-distributed such that the material fails with the collapse of pores first. Under compression, a cellular solid will collapse with a stress–strain curve consisting of three distinct regions (Figure 7.9a). Region I corresponds to the elastic region; region II to a characteristic plateau stress region from the collapse of the pores; and region III to the densification at the ε_d strain for which the solid materials are compacted together with the porosity (highly) eliminated. A large work done is then associated with the deformation mode in regions I and II. When the plateau stress is designed to be that of below the protection target's threshold, the energy absorber will fail in place of the protection target, absorbing all of the impact energy as work done for its own deformation without transmitting stresses above the target's threshold to it. Metrics to measure the performance of energy-absorbing materials are usually measured in terms of the specific energy absorption (SEA), with respect to a unit weight:

$$\text{SEA} = \frac{\int_0^{\varepsilon_d} \sigma \cdot d\varepsilon}{\rho_c}$$

Intuitively, the higher the value, the better the material is for energy absorption. Furthermore, since energy absorbers are characterized by their stress responses below that of a protection target's threshold, their effective working stresses can be described as a peak stress value in the plateau regime. Hence, an idealized energy absorber should always have a smooth and flat plateau stress to maximize the ratio of area under the stress–strain curve to a rectangular area described

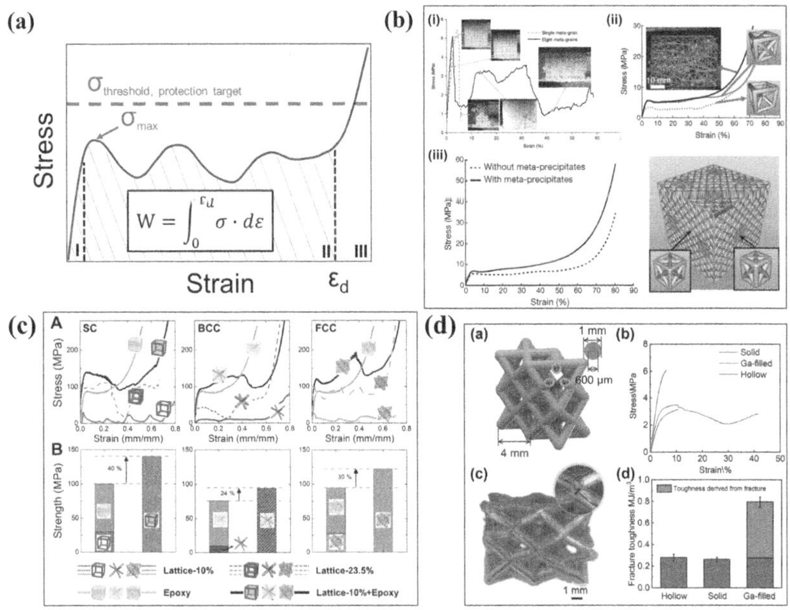

FIGURE 7.9 Illustration of examples of lattice energy absorbers. (a) A schematic illustration of the deformation of a typical cellular solid. (b) Strengthening of macro-lattice by mimicking metal mechanisms. b-i: grain boundary, b-ii: multiphase strengthening and b-iii: precipitation strengthening. (c) Metallic lattice reinforced with epoxy infiltration, revealing the improved compressive strengths. (d) Liquid metal infilled elastomeric lattice, revealing the improved toughness. (Adapted and reproduced with permission from Refs. [66], [4] and [115] for (b), (c) and (d), respectively.)

by the peak stress and densification strain. Such a ratio is termed as the energy absorption efficiency (EAE) and it refers to the effective energy absorbed as normalized to the maximum possible (Figure 7.9a).

$$\mathrm{EAE} = \frac{\int_0^{\varepsilon_d} \sigma \cdot d\varepsilon}{\sigma_{\max} \times \varepsilon_d}$$

From these, it is apparent that the stiffness, strength, plateau stress and hence the SEA and EAE are all directly dependent on the structural design and materials used. From the research point of view, the key interest lies with improving SEA and EAE.

The energy absorption performance of lattice structures lies with two key factors: its design and the base material used. As such, numerous works have focused on novelties with structural designs, materials and concepts. For their inherent combination of high strength and ductility, majority of the lattice energy-absorbing structures are based on metals and tough polymers. As such, the main AM techniques used here include SLM, EBM, SLA and DLP. Intuitively, various works have first studied the performance of fundamental designs such as crystal-structure mimicking and TPMS lattices. Bending dominated lattices, such as BCC and TPMS structures, usually present better EAE with a gradual and smooth deformation plateau stress as compared to stretch-dominated ones, such as SC and FCC. Despite this, stretch-dominated lattices still present superior EAE as compared to honeycombs owing to their three-dimensional connectivity. As compared, sheet TPMS lattice structures such as the gyroid, I-WP and diamond have usually better specific strength and SEA as compared to their skeletal counterparts or even in general, most of the strut lattices [44]. Apart from strut and TPMS lattices, the potentials of plate lattices have also been raised within these few years. Work in [51] worked on the energy absorption characteristics of plate lattices. Comparing across metallic lattices, FCC plate lattices present far superior SEA owing to the high membrane strength of the plates and the stabilized deformation.

Improvements to existing structures have been raised, with directions mainly focused on strut tapering [14,18,26], permutation and combination of designs [12] and inspirations [16,66,116]. Strut tapering refers to increasing the strut diameter toward nodes, reducing that in the middle, while keeping the relative density still the same. The motivation for the geometrical manipulation lies with strengthening the points taking the highest stress distributions under compression. A 47% increase in SEA has been reported for rhombic dodecahedron lattices of around 15% relative density [18]. Permutation and combination in turn refer to any combination of fundamental structures or concepts, such as inter-crystal-structure strut and plate lattices [12,48], inter-TPMS [117], struts and shells [54], struts and plates [51], etc. Analogous to composite materials, lattice derived from combinations usually possesses both properties from its constituents, e.g., both the dominant directional strength. Inspired by the strengthening mechanisms of metals, Pham et al. [66] adopted the concept of increased grain boundaries, precipitates and multiphase to achieve shear bands bowing and arrests in a macro-lattice (Figure 7.9b). Increased ductility and plateau strengths have both been achieved. Bioinspiration is also one popular route to cellular design. Leveraging on the skeletal structures of sea sponges, Fernandes et al. [16] created strut lattices with the highest theoretical buckling resistance. Such lattices should be expected to present stabilized large strain deformation for energy absorption. Bauer et al. [116] reported tensegrity-based lattices with delocalized deformation, which can mitigate overall material failure. Such structures presented a 25 times higher unfractured energy absorption as compared to a conventional FCC or Kelvin strut lattices.

Apart from structural design, efforts have also been put in on novelties in materials. For SLM metals, Li et al. [14] worked on a new AISI 4130 grade steel, which as compared to the current state-of-the-art AM metals like 316L stainless steel or Ti-6Al-4V alloys, possesses simultaneously both high specific strength and high ductility with a yielding strength of 1243 MPa and ductility of 15.5%. With an also synergistic new FCC+SC design, high specific strength and flat plateau constituted to a lattice with high SEA. Comparatively, 316L lattices are much weaker, while Ti-6Al-4V lattice have usually numerous points of fracture and hence low EAE. Traugutt et al. [118] proposed DLP-printable liquid-crystal elastomers (LCEs) as soft materials with an enhanced energy dissipation mechanism. Differing from elastomers, LCE consists of an anisotropic molecular order of liquid crystals with the entropic elasticity of a lightly cross-linked polymer network. LCEs are notable highly strain rate sensitive from an additional viscous dissipation mechanism in the rubbery regime via liquid-crystal domain rotations. Under high strain rates, the LCE lattices showed 5–27 times increase in SEA as compared to a same lattice made with a conventional elastomer.

Moving on from the improvements of the fundamental materials, various works have also focused on composite materials, such as infill [4,115,119,120] or coating. Infill into the porous phase of a cellular solid creates interpenetrating-phase composites (IPCs) where for such composites, both of its two phases are continuous, i.e., the cellular and the inverse of the cellular. Li et al. [4] worked on metallic lattice and epoxy IPC based on fundamental structures including the SC, BCC and FCC (Figure 7.9c). These lattices consist of struts which are inclined, constrained inclined, horizontal and vertical, corresponding to the different types of compressive deformation. The contributions from the epoxy reinforcements to each of these types of deformation are studied in detail in the work. Reinforcement was shown to be the most notable for the SC lattice for vertical struts, where instead of buckling, the strut instead takes on uniaxial compression, which is associated with much higher strain energies. The SEA of SC lattices (10% relative density) hence increased more than five-folds from 8.2 to 46 J/g. Infill can also be on the interior of the lattice. In particular, Zhang et al. [115] worked on tube lattices with infill into its hollow struts. A liquid metal, gallium, was used to infill into hollow FCC elastomer lattices (Figure 7.9d). At the solid state, gallium acts as a hard phase and this constituently improves the toughness of the composite lattice three-fold to 0.8 MJ/m^3. In addition, owing to the low melting point of gallium, the lattice can be recovered upon slightly heating the gallium for it to melt to take the shape of a pristine elastomer lattice again, making it also deformation recoverable.

7.4.2 Artificial Bone Implants

As mentioned in Section 7.2.2, reduced stiffness and strength are intrinsic of all cellular solids. Leveraging on this point, one particular application that specifically requires materials, which are less stiff and strong, is on artificial bone implants. This requirement derives from a particular phenomenon known as stress shielding. When an artificial bone implant is too stiff and strong as compared to the surrounding natural bone, the implant will help sustain loads in place of the natural bone, effectively "shielding" the bone from stresses. Without external stimulus, the natural bone will then degenerate and be resorbed back into the surrounding tissue. As such, stress shielding must be avoided at all costs to prevent the deterioration of the osteo-health. Stiffness, in terms of Young's modulus, and compressive strength of a human cortical bone range from around 15 to 35 Gpa and 188 to 222 Mpa, respectively, with variations according to health status and age. In contrast, for the soft and spongy cancellous bone, the respective properties range from around 0.02 to 1 Gpa and 1 to 11 Mpa [91,121]. As can be seen, even the hard cortical bone is notably much weaker than most metals and ceramics. Suitable biocompatible materials, such as Ti-6Al-4V, Co-Cr-Mo and stainless steel, have all high Young's modulus of more than 100 Gpa [121]. As such, several methods have been worked on for reduced stiffness of biocompatible materials. Several approaches include using bone cements to reduce the contact stiffness, researching on new materials with lower stiffness and using porous materials [122]. Nonetheless, adopting lattice structures as implants is still the state-of-the-art method owing to their porous nature, which allows nutrients exchanges, and that cells can seed and in-grow in the implants.

A diverse range of lattice structures with various biocompatible materials have been studied as possible implant materials. For such metallic structures, fabrications are mostly done through EBM owing to the oxygen sensitivity of the suitable metals. The biocompatibility of these materials has been affirmed through various work on the successful cell seeding and growth both *in vitro* and *in vivo* [123–125]. Intuitively, fundamental structures are first studied before a diverse variety of functional inspired designs. Early works have researched on honeycomb-like structures as possible implants [126–128]. Despite being capable of achieving bone-like mechanical properties with successful cell growth, however, such structures lack the superior structural design freedoms of lattices and were quickly outclassed. Lattice structures comparatively possess three-dimensional interconnectivities and pore morphologies, which are expected to better allow cell penetration and growth. Combined research efforts have also shown that there exist optimal pore sizes for different bio-functions. For reference, optimal sizes for cell seeding roughly range from 100 to 600 μm [129–132]. Large pores result in inefficient cell bridging with also low SSAs for high-density growth. On the other hand, nutrients and mass transports to support vascularization for the cell growth are in turn more optimal when the pores are larger [133–135]. Pores that are small, e.g., below 900 μm, are shown to lead to significant flow occlusions [130]. Given these limitations, it is apparent that lattice structures are the ideal type of cellular materials as an artificial bone implant.

A diverse range of strut-based structures, including the crystal-structure mimicking [136–140], rhombic dodecahedron [22,137], diamond [22], reentrant [141] and Voronoi pattern [142], have been studied with varying relative densities and bone-like mechanical properties. Novelties in structures also include heterogeneously porous [143] and optimally isotropic structures [144]. Kolken et al. presented a concept on an implant made of hybrid positive and negative Poisson's ratio (auxetic) cellular structures [141] (Figure 7.10a). The hybrid structure is able to be pressed onto the bone on both its medial and lateral sides under loading, unlike conventional structures which retracts. The implant and bone contacts will then be improved, which are expected to contribute to the longevity of the implant.

However, the current state-of-the-art structures as artificial bone implants are generally agreed on to be TPMS structures. TPMS stands out among cellular structures for their biomorphic design, smooth and continuous features with

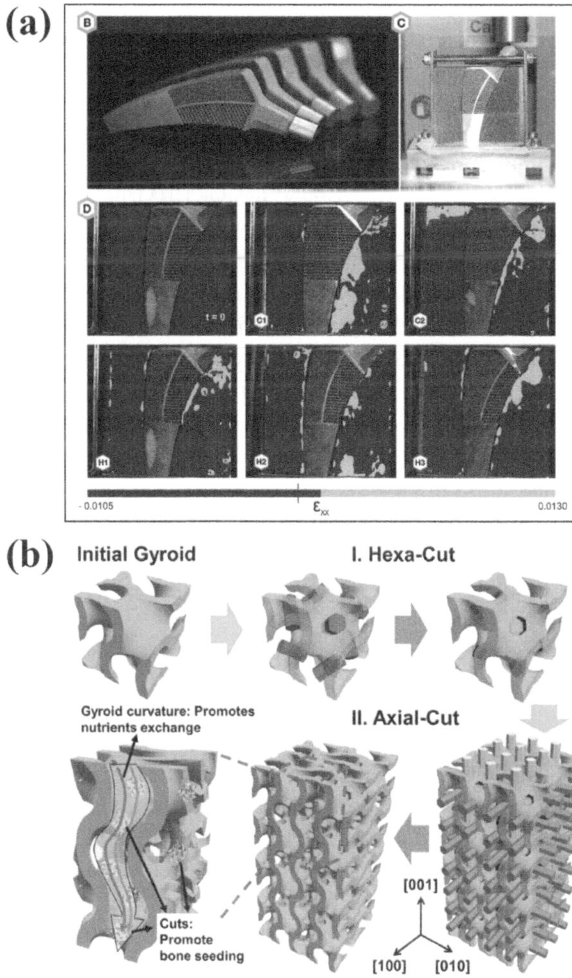

FIGURE 7.10 Illustration of examples of lattice artificial bone implants. (a) a-B: As-printed lattice-based implants, a-C, D: emulation mechanical tests, showing the strain distribution for different samples. (b) Illustration of a modified gyroid, with small pores introduced to the cell walls for cell growth. (Adapted and reproduced with permission from Refs. [141] and [145] for (a) and (b), respectively.)

high fluid flowability. They are hence also the most bone microstructure-like. Commonly studied TPMS includes the gyroid, neovius, diamond, primitive and I-WP structures [28,35,39–43]. In particular, the gyroid is the most widely studied for its combination of high specific strength, high flowability and continuous curvature. While keeping the pore sizes within the range for cell seeding, Alabort et al. [40] worked on skeletal gyroid, diamond and neovius structures with relative densities from 15% to 50% using the EBM. Possibilities with fabrication are affirmed and the investigated range of properties lies within that of the cancellous and cortical bones. Barba et al. [39] in turn studied the EBM processing and fabrication quality in detail and proposed a design guideline for these TPMS metallic implants. To better match the mechanical properties of the bone at different sections, the concept of functional grading is also introduced for various TPMS structures [65]. Numerous studies on the fatigue behavior of gyroid structures have been carried out [29,41]. In essence, proper EBM build parameter control and post-treatment to remove surface roughness are necessary to eliminate voids and defects which will accelerate fatigue failure. Experimental research has also been carried out for their fluid-flow performance [28]. The primitive structure is shown with the highest flowability at low densities while the gyroid with the highest at high densities. To achieve simultaneously optimal cell seeding and nutrients exchanges, the work of Wang et al. [145] introduced a new concept: heterogeneously porous sheet gyroid structures (Figure 7.10b). Differing from conventional sheet gyroids, micropores of around 600–800 μm are additionally added onto the cell walls for cell seeding. Gyroid wall spacings are in turn designed with a width ranging from 1.25 to 2 mm for effective mass transports. With proper relative density design, mechanical properties of the gyroids range from Young's modulus of 8–15 GPa and compressive strength of 150–250 MPa, all lying within the range of the cortical bone.

7.4.3 Acoustic Metamaterials

Acoustic metamaterials mainly refers to materials with one of the following capabilities: acoustic cloaking, lensing, abnormal transmission or sound absorption [82]. For the context of cellular materials, however, acoustic metamaterials are mainly focused on sound absorbers thus far. When an incoming sound wave propagation interacts with a material, the sound energy can be either reflected via impedance mismatch, be absorbed or transmitted. One major mechanism of sound absorption is the deceleration of sound wave propagation through flow resistance in porous materials. As such, cellular materials, with capabilities for a diverse range of complex porous architecture and hence designable flow resistivity, are of particular research interests as novel sound absorbers.

Although relatively new to the field of acoustic metamaterials, potentials of lattice structures have been identified. As compared, lattice sound absorbers hold advantages to traditional absorbers [146], such as foams, fabric, panels, with high fit-for-purpose design freedoms (tunable absorption, high rigidity, multi-functions). In addition, recent works have suggested the potentials of lattice structures to possess superior and more broadband sound absorption as compared with the traditional. Mechanisms for sound absorption mainly lie with viscous flow dissipation, and to a lesser extent, thermal flux and structural damping. Numerical models to design, or to analyze the performance of lattice absorbers, are mainly equivalent fluid models like the Johnson-Champoux and Allard [147], Delany-Bazley [148] and general viscothermal models [149,150], or resonance models like the theory of Maa [151]. As can be seen, the performance of cellular absorbers is then generally only geometry dependent and is hence (solid) materials and constituently fabrication route independent. As such, any AM technique can be used to manufacture cellular structures for sound absorption purposes; techniques, such as SLM, FDM, DLP and SLA, have all been reported in literature owing to their capabilities for complex geometries. Surface roughness and build defects are issues that should be addressed for the AM technique. An insight has nonetheless been given in the work of Boulvert et al. [152] where the FDM-built surface roughness and interlayer defects are actually shown to enhance absorption properties.

Conventional structures studied include the crystal-structure mimicking structures [51,147], log-pile [152], auxetic [150], Kelvin cell [149], TPMS [153] and pore-channel-cell structures [154,155]. Varying degrees of sound absorption performance have been shown. In general, for the high feature interconnectivities of lattices, many lattices consist of regions of heterogeneous pore geometries, e.g., narrow neck followed by a wide cavity [51,154,156,157]. As such, if the pore geometries sufficiently match the acoustic impedance of the incoming sound wave, lattices usually undergo some sort of resonance-based absorption mechanism at certain frequencies. Li et al. [51] worked on a series of plate/strut/hybrid lattices. Sound absorption coefficients for all have been accurately geometrically fit using the theory of Maa [151] as multi-layered resonators (Figure 7.11a). The lattices can hence also be regarded as a series of Helmholtz resonators. Of course, depending on the geometry, resonance may not always be present. For such cases, notably on the more strut-oriented lattices [147,152], absorption takes place with the dissipation from the high flow resistivities. Nonetheless, with the high complexity of the lattice structures, high flow resistivity with excellent sound absorption can also be achieved [147]. However, to achieve the maximum absorption coefficient of 1, resonance mechanisms are always necessary. Usually, the smaller the unit cell, or higher the relative density, the higher the sound absorption coefficients. Indeed, a highly "open-celled" auxetic lattice revealed very low sound absorption properties [150]. Comparing the highly open sheet primitive lattice structure with the highly torturous gyroid and diamond ones, the primitive also showed significantly poorer absorption properties [153]. In addition, sound absorption is thickness dependent in the direction of sound wave propagation. In general, higher thickness translates to higher absorption coefficients and a more broadband absorption curve. As such, unlike that for mechanical properties, sound absorbing lattices are also physical size dependent on top of the relative densities.

Another research direction is on heterogeneous, or graded, lattice structures for broadband sound absorption. For such lattices, different pore and cavity geometries are expected to work in tandem to provide a matching acoustic impedance for various frequency ranges to constitute broadband sound absorption. At the moment, attention has mainly been placed on the stepwise or continuous grading of struts for the log-pile [83] and cubic lattices [158,159] (Figure 7.11b). Numerical methods include transfer acoustic impedance using the equivalent fluid models or resonance models. For all these graded lattices, notable improvements with both

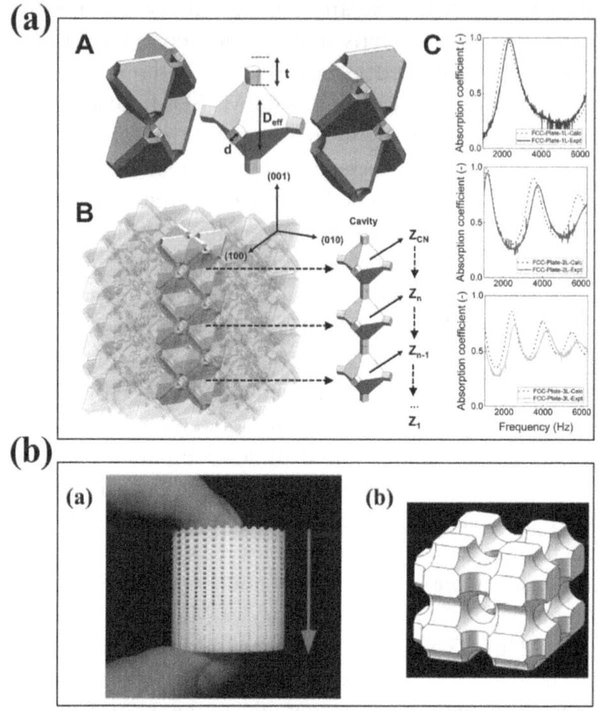

FIGURE 7.11 Illustration of examples of lattice acoustic metamaterials. (a) a-A: Illustration of a FCC plate lattices with dimensions dissected for analysis, a-B: transfer impedance function and a-C: correlation between the experimentally measured and numerical calculations. (b) Illustration of a continuous functional-graded lattice structure for effective sound absorption. (Reproduced from Refs. [51] and [159] for (a) and (b), respectively.)

higher localized peaks and broadband absorption have been observed. Inspired by microperforated panels, Li et al. [160] worked on a simultaneous in- and out-of-plane heterogeneously porous resonating lattice structure. The average absorption coefficient, across a broad frequency range from 1000 to 6300 Hz, increased from 0.55 for a homogeneously porous structure to 0.69 for an optimally heterogeneous one.

Although sound absorption is independent of material properties, however, a particular material choice may provide additional multifunctionalities. Li et al. worked on lattices with simultaneous energy absorption using a high strength steel [51]. The FCC plate structure, coupled with the high specific strength and ductility of the used material, presents high SEA of up to 50.3 J/g on top of excellent sound absorption coefficients with local peaks near 1. On using tough polymers, the same research team showed auxetic sound absorbing lattice with high deformation recovery of up to 30% strain for multifunctional use as recoverable impact absorbers [160].

7.4.4 Electrochemical Electrodes

Another intrinsic property of lattice structures is their significantly increased SSA, either with respect to a unit volume or weight, as compared to a bulk volume. In turn, electrochemical electrodes have a particularly high affinity for this property. The merit mainly attributes to the increased catalytic sites, host–medium interactions and improved materials exchanges. Herein, we define the scope for electrochemical applications to be those most widely researched on currently: water splitting, supercapacitors and batteries. Conventional methods to improve SSA are usually associated with using foams and fibrous electrodes; however, a bottleneck is often associated with those methods owing to their limited electrode design freedoms. As such, the potentials of lattice structures as electrodes have been gaining immense research interests with the advent of AM. Differing from the previous three mentioned applications, cellular materials for electrochemical applications are in turn materials motivated. That being said, the main purpose of making the material cellular is to increase its SSA. However, depending on the AM technique, numerous post-processing steps may be needed as compared to the three previous mentioned applications.

Various structures have been studied as electrochemical substrates; however, the complexity and type of structures adopted mainly depends on the AM technique used. The fundamental principle still lies with a high SSA. For processes that prints with the functional material itself, the AM technique adopted mainly lies with DIW which is highly versatile with feedstock materials selections. However, DIW is generally capable of only fabricating simple log-pile structures. In turn, for processes which adopt post-fabrication surface functionalization, AM techniques with superior design freedoms, such as SLM, SLA and DLP, can be used since the focus is on only the structure and not the material. Since the applications are electrochemical based, electrical conductivity is of also paramount importance. For processes such as SLA, DLP and DIW (of certain materials), the non-conductive polymers or ceramics produced need additional metallization steps to make them electrically conductive.

AM-fabricated porous electrodes have shown great potentials for water splitting. Early investigations included FDM-printed conductive graphene/PLA composites as electrodes [161]. However, performance is non-ideal given the simplicity in materials and design. Future works have since focused on surface functionalization and complex cellular structures. Su et al. [162] worked on an overall integrated water splitting full cell with FCC-structured electrodes (Figure 7.12a). As the electrodes were fabricated using photocurable polymers through SLA, nickel-phosphorous metallic alloy was first electroless-deposited to make the electrode conductive. After which, functional materials, such as NiFe-OOH, can then be electro-deposited to enhance the catalytic performances. As compared to a pure plate of the same functional material, the FCC electrode presents superior oxygen evolution reaction (OER) capabilities with the overpotential and Tafel slope being reduced from 273 and 52 to 197 and 51 mA/cm^2 and mV/dec, respectively. Performance owes to both the increased SSA for enhanced reactions and the porous architecture for enhanced kinetics. Such metal-coated polymer electrodes

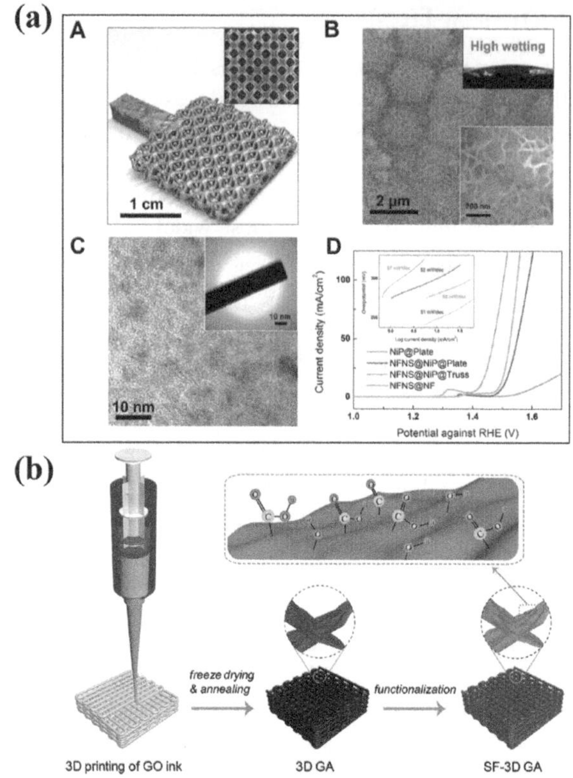

FIGURE 7.12 Illustration of examples of lattice electrochemical electrodes. (a) a-A: A metal-coated FCC strut lattice electrode, a-B, C: electron microscopy on the functionalized surfaces and a-D: illustration of the OER as compared with plate and non-functionalized electrodes. (b) Schematic illustration of DIW and post-treatment functionalization of a graphene aerogel cellular supercapacitor electrode. (Adapted from Refs. [162] and [168] for (a) and (b), respectively.)

hold advantages of being strong and lightweight. Chang et al. [163] worked directly on a metallic gyroid-structured electrode fabricated using SLM. With NiFe-OOH functionalization, the electrode shows excellent water splitting properties with a low overpotential of 261 mV at a current density of 1500 mA/cm^2. Other structures being worked on also include the honeycomb [164] and log-pile [165] electrodes.

Similar trends have also been achieved for electrochemical storage applications too. However, in contrast to water splitting, cellular electrodes for supercapacitors and batteries are usually printed with the intrinsically porous carbonaceous materials, and hence most of the time through DIW. As such, cellular electrodes for these applications are mainly based on log-pile structures. Thus far, materials successfully printed as supercapacitor electrodes include carbon aerogel [166], graphene aerogel [167,168] and MXene [169,170] (exemplified using Figure 7.12b). For batteries, the current state-of-the-art still lies with the lithium based. Base materials used for electrode fabrication include the metal-organic framework

[171], graphene [172] and carbon nanotubes [173]. Doping of lithium ions can be done either *in situ* as part of the printing ink [173] or to be post-printing deposited [171]. Apart from the cellular porosity, DIW-fabricated electrodes also possess microscopic porosity owing to the morphology of the template carbonaceous materials, making them effectively hierarchically porous. The derived very high SSA, and hence the electrochemically active surface area, hence allows them to possess both superior specific (per unit weight) energy and power density as compared to their solid bulk counterparts [174].

7.5 SUMMARY AND PERSPECTIVE

In summary, this chapter presented a comprehensive overview of the various types of lattice structures and AM techniques and highlighted several prominent applications associated with the AM of cellular materials. Till date, the potentials of AM-fabricated lattice structures have shown with capabilities to surpass the performance, or to possess multifunctionality, as compared to conventional bulk materials. These potentials are clearly illustrated for applications as energy absorbers, implants, acoustic metamaterials and electrochemical electrodes. With the continual research efforts being put into the developments of novel AM methods, materials and technological advances in topological designs, it should be optimistic that AM-ed lattice structures will be the prospective advanced functional materials of the future.

REFERENCES

1. T. Tancogne-Dejean, A.B. Spierings, D. Mohr, Additively-manufactured metallic micro-lattice materials for high specific energy absorption under static and dynamic loading, *Acta Materialia* 116 (2016) 14–28.
2. M. Leary, M. Mazur, J. Elambasseril, M. McMillan, T. Chirent, Y. Sun, M. Qian, M. Easton, M. Brandt, Selective laser melting (SLM) of AlSi12Mg lattice structures, *Materials & Design* 98 (2016) 344–357.
3. M. Leary, M. Mazur, H. Williams, E. Yang, A. Alghamdi, B. Lozanovski, X. Zhang, D. Shidid, L. Farahbod-Sternahl, G. Witt, Inconel 625 lattice structures manufactured by selective laser melting (SLM): Mechanical properties, deformation and failure modes, *Materials & Design* 157 (2018) 179–199.
4. X. Li, Y.H. Tan, P. Wang, X. Su, H.J. Willy, T.S. Herng, J. Ding, Metallic micro-lattice and epoxy interpenetrating phase composites: Experimental and simulation studies on superior mechanical properties and their mechanisms, *Composites Part A: Applied Science and Manufacturing* 135 (2020) 105934.
5. X. Li, H.J. Willy, S. Chang, W. Lu, T.S. Herng, J. Ding, Selective laser melting of stainless steel and alumina composite: Experimental and simulation studies on processing parameters, microstructure and mechanical properties, *Materials & Design* 145 (2018) 1–10.
6. M. Smith, W.J. Cantwell, Z. Guan, S. Tsopanos, M.D. Theobald, G.N. Nurick, G.S. Langdon, The quasi-static and blast response of steel lattice structures, *Journal of Sandwich Structures & Materials* 13(4) (2010) 479–501.
7. S. McKown, Y. Shen, W. Brookes, C. Sutcliffe, W. Cantwell, G. Langdon, G. Nurick, M. Theobald, The quasi-static and blast loading response of lattice structures, *International Journal of Impact Engineering* 35(8) (2008) 795–810.

8. V. Crupi, E. Kara, G. Epasto, E. Guglielmino, H. Aykul, Static behavior of lattice structures produced via direct metal laser sintering technology, *Materials & Design* 135 (2017) 246–256.
9. P. Köhnen, C. Haase, J. Bültmann, S. Ziegler, J.H. Schleifenbaum, W. Bleck, Mechanical properties and deformation behavior of additively manufactured lattice structures of stainless steel, *Materials & Design* 145 (2018) 205–217.
10. A. Großmann, J. Gosmann, C. Mittelstedt, Lightweight lattice structures in selective laser melting: Design, fabrication and mechanical properties, *Materials Science and Engineering: A* 766 (2019) 138356.
11. M. Mazur, M. Leary, S. Sun, M. Vcelka, D. Shidid, M. Brandt, Deformation and failure behaviour of Ti-6Al-4V lattice structures manufactured by selective laser melting (SLM), *The International Journal of Advanced Manufacturing Technology* 84(5–8) (2016) 1391–1411.
12. T. Tancogne-Dejean, D. Mohr, Elastically-isotropic truss lattice materials of reduced plastic anisotropy, *International Journal of Solids and Structures* 138 (2018) 24–39.
13. S. Xu, J. Shen, S. Zhou, X. Huang, Y.M. Xie, Design of lattice structures with controlled anisotropy, *Materials & Design* 93 (2016) 443–447.
14. X. Li, Y.H. Tan, H.J. Willy, P. Wang, W. Lu, M. Cagirici, C.Y.A. Ong, T.S. Herng, J. Wei, J. Ding, Heterogeneously tempered martensitic high strength steel by selective laser melting and its micro-lattice: Processing, microstructure, superior performance and mechanisms, *Materials & Design* 178 (2019) 107881.
15. M.M. Osman, M. Shazly, E.A. El-Danaf, P. Jamshidi, M.M. Attallah, Compressive behavior of stretched and composite microlattice metamaterial for energy absorption applications, *Composites Part B: Engineering* 184 (2020) 107715.
16. M.C. Fernandes, J. Aizenberg, J.C. Weaver, K. Bertoldi, Mechanically robust lattices inspired by deep-sea glass sponges, *Nature Materials* 20(2) (2021) 237–241.
17. Y. Liu, J.H. Campbell, O. Stein, L. Jiang, J. Hund, Y. Lu, Deformation behavior of foam laser targets fabricated by two-photon polymerization, *Nanomaterials* 8(7) (2018) 498.
18. X. Cao, S. Duan, J. Liang, W. Wen, D. Fang, Mechanical properties of an improved 3D-printed rhombic dodecahedron stainless steel lattice structure of variable cross section, *International Journal of Mechanical Sciences* 145 (2018) 53–63.
19. T.J. Horn, O.L.A. Harrysson, D.J. Marcellin-Little, H.A. West, B.D.X. Lascelles, R. Aman, Flexural properties of Ti6Al4V rhombic dodecahedron open cellular structures fabricated with electron beam melting, *Additive Manufacturing* 1–4 (2014) 2–11.
20. G. Epasto, G. Palomba, D. D'Andrea, E. Guglielmino, S. Di Bella, F. Traina, Ti-6Al-4V ELI microlattice structures manufactured by electron beam melting: Effect of unit cell dimensions and morphology on mechanical behaviour, *Materials Science and Engineering: A* 753 (2019) 31–41.
21. T. Zhong, K. He, H. Li, L. Yang, Mechanical properties of lightweight 316L stainless steel lattice structures fabricated by selective laser melting, *Materials & Design* 181 (2019) 108076.
22. S.A. Yavari, S. Ahmadi, R. Wauthle, B. Pouran, J. Schrooten, H. Weinans, A. Zadpoor, Relationship between unit cell type and porosity and the fatigue behavior of selective laser melted meta-biomaterials, *Journal of the Mechanical Behavior of Biomedical Materials* 43 (2015) 91–100.
23. K. Hazeli, B.B. Babamiri, J. Indeck, A. Minor, H. Askari, Microstructure-topology relationship effects on the quasi-static and dynamic behavior of additively manufactured lattice structures, *Materials & Design* 176 (2019) 107826.
24. Z. Xiao, Y. Yang, R. Xiao, Y. Bai, C. Song, D. Wang, Evaluation of topology-optimized lattice structures manufactured via selective laser melting, *Materials & Design* 143 (2018) 27–37.

25. Y. Xu, D. Zhang, S. Hu, R. Chen, Y. Gu, X. Kong, J. Tao, Y. Jiang, Mechanical properties tailoring of topology optimized and selective laser melting fabricated Ti_6Al_4V lattice structure, *Journal of the Mechanical Behavior of Biomedical Materials* 99 (2019) 225–239.
26. T. Tancogne-Dejean, D. Mohr, Stiffness and specific energy absorption of additively-manufactured metallic BCC metamaterials composed of tapered beams, *International Journal of Mechanical Sciences* 141 (2018) 101–116.
27. D. Qi, H. Yu, M. Liu, H. Huang, S. Xu, Y. Xia, G. Qian, W. Wu, Mechanical behaviors of SLM additive manufactured octet-truss and truncated-octahedron lattice structures with uniform and taper beams, *International Journal of Mechanical Sciences* 163 (2019) 105091.
28. F. Bobbert, K. Lietaert, A.A. Eftekhari, B. Pouran, S. Ahmadi, H. Weinans, A. Zadpoor, Additively manufactured metallic porous biomaterials based on minimal surfaces: A unique combination of topological, mechanical, and mass transport properties, *Acta Biomaterialia* 53 (2017) 572–584.
29. L. Yang, C. Yan, W. Cao, Z. Liu, B. Song, S. Wen, C. Zhang, Y. Shi, S. Yang, Compression–compression fatigue behaviour of gyroid-type triply periodic minimal surface porous structures fabricated by selective laser melting, *Acta Materialia* 181 (2019) 49–66.
30. E. Yang, M. Leary, B. Lozanovski, D. Downing, M. Mazur, A. Sarker, A. Khorasani, A. Jones, T. Maconachie, S. Bateman, Effect of geometry on the mechanical properties of Ti-6Al-4V Gyroid structures fabricated via SLM: A numerical study, *Materials & Design* 184 (2019) 108165.
31. X. Guo, J. Ding, X. Li, S. Qu, X. Song, J.Y.H. Fuh, W.F. Lu, W. Zhai, Enhancement in the mechanical behaviour of a Schwarz Primitive periodic minimal surface lattice structure design, *International Journal of Mechanical Sciences* 216 (2021) 106977.
32. Y. Wang, X. Ren, Z. Chen, Y. Jiang, X. Cao, S. Fang, T. Zhao, Y. Li, D. Fang, Numerical and experimental studies on compressive behavior of Gyroid lattice cylindrical shells, *Materials & Design* 186 (2020) 108340.
33. L. Zhang, S. Feih, S. Daynes, S. Chang, M.Y. Wang, J. Wei, W.F. Lu, Energy absorption characteristics of metallic triply periodic minimal surface sheet structures under compressive loading, *Additive Manufacturing* 23 (2018) 505–515.
34. D.W. Abueidda, M. Elhebeary, C.-S.A. Shiang, S. Pang, R.K.A. Al-Rub, I.M. Jasiuk, Mechanical properties of 3D printed polymeric Gyroid cellular structures: Experimental and finite element study, *Materials & Design* 165 (2019) 107597.
35. Z. Gorgin Karaji, M. Speirs, S. Dadbakhsh, J.-P. Kruth, H. Weinans, A. Zadpoor, S. Amin Yavari, Additively manufactured and surface biofunctionalized porous nitinol, *ACS Applied Materials & Interfaces* 9(2) (2017) 1293–1304.
36. L. Yang, R. Mertens, M. Ferrucci, C. Yan, Y. Shi, S. Yang, Continuous graded Gyroid cellular structures fabricated by selective laser melting: Design, manufacturing and mechanical properties, *Materials & Design* 162 (2019) 394–404.
37. Z. Ma, D.Z. Zhang, F. Liu, J. Jiang, M. Zhao, T. Zhang, Lattice structures of Cu-Cr-Zr copper alloy by selective laser melting: Microstructures, mechanical properties and energy absorption, *Materials & Design* 187 (2020) 108406.
38. J. Maszybrocka, B. Gapiński, M. Dworak, G. Skrabalak, A. Stwora, The manufacturability and compression properties of the Schwarz Diamond type Ti6Al4V cellular lattice fabricated by selective laser melting, *The International Journal of Advanced Manufacturing Technology* 105(7–8) (2019) 3411–3425.
39. D. Barba, E. Alabort, R. Reed, Synthetic bone: Design by additive manufacturing, *Acta Biomaterialia* 97 (2019) 637–656.

40. E. Alabort, D. Barba, R.C. Reed, Design of metallic bone by additive manufacturing, *Scripta Materialia* 164 (2019) 110–114.
41. A. Ataee, Y. Li, M. Brandt, C. Wen, Ultrahigh-strength titanium gyroid scaffolds manufactured by selective laser melting (SLM) for bone implant applications, *Acta Materialia* 158 (2018) 354–368.
42. A. Ataee, Y. Li, D. Fraser, G. Song, C. Wen, Anisotropic Ti-6Al-4V gyroid scaffolds manufactured by electron beam melting (EBM) for bone implant applications, *Materials & Design* 137 (2018) 345–354.
43. A. Yánez, A. Herrera, O. Martel, D. Monopoli, H. Afonso, Compressive behaviour of gyroid lattice structures for human cancellous bone implant applications, *Materials Science and Engineering: C* 68 (2016) 445–448.
44. O. Al-Ketan, R. Rowshan, R.K.A. Al-Rub, Topology-mechanical property relationship of 3D printed strut, skeletal, and sheet based periodic metallic cellular materials, *Additive Manufacturing* 19 (2018) 167–183.
45. S. Rajagopalan, R.A. Robb, Schwarz meets Schwann: Design and fabrication of biomorphic and durataxic tissue engineering scaffolds, *Medical Image Analysis* 10(5) (2006) 693–712.
46. O. Al-Ketan, R.K. Abu Al-Rub, MSLattice: A free software for generating uniform and graded lattices based on triply periodic minimal surfaces, *Material Design & Processing Communications* 21(10) (2019) e205.
47. J. Berger, H. Wadley, R. McMeeking, Mechanical metamaterials at the theoretical limit of isotropic elastic stiffness, *Nature* 543(7646) (2017) 533.
48. T. Tancogne-Dejean, M. Diamantopoulou, M.B. Gorji, C. Bonatti, D. Mohr, 3D plate-lattices: An emerging class of low-density metamaterial exhibiting optimal isotropic stiffness, *Advanced Materials* 30(45) (2018) 1803334.
49. H.Y. Sarvestani, A. Akbarzadeh, A. Mirbolghasemi, K. Hermenean, 3D printed meta-sandwich structures: Failure mechanism, energy absorption and multi-hit capability, *Materials & Design* 160 (2018) 179–193.
50. J.J. Andrew, P. Verma, S. Kumar, Impact behavior of nanoengineered, 3D printed plate-lattices, *Materials & Design* 202 (2021) 109516.
51. X. Li, X. Yu, J.W. Chua, H.P. Lee, J. Ding, W. Zhai, Microlattice metamaterials with simultaneous superior acoustic and mechanical energy absorption, *Small* 17(24) (2021) 2100336.
52. S. Duan, W. Wen, D. Fang, Additively-manufactured anisotropic and isotropic 3D plate-lattice materials for enhanced mechanical performance: Simulations & experiments, *Acta Materialia* 199 (2020) 397–412.
53. T. Tancogne-Dejean, D. Mohr, Elastically-isotropic elementary cubic lattices composed of tailored hollow beams, *Extreme Mechanics Letters* 22 (2018) 13–18.
54. C. Bonatti, D. Mohr, Large deformation response of additively-manufactured FCC metamaterials: From octet truss lattices towards continuous shell mesostructures, *International Journal of Plasticity* 92 (2017) 122–147.
55. Y. Liu, J. Zhang, X. Gu, Y. Zhou, Y. Yin, Q. Tan, M. Li, M.-X. Zhang, Mechanical performance of a node reinforced body-centred cubic lattice structure manufactured via selective laser melting, *Scripta Materialia* 189 (2020) 95–100.
56. B. Deng, G.J. Cheng, Soap film inspired mechanical metamaterials approaching theoretical bound of stiffness across full density range, *Materials Horizons* 8(3) (2021) 987–996.
57. C. Bonatti, D. Mohr, Mechanical performance of additively-manufactured anisotropic and isotropic smooth shell-lattice materials: Simulations & experiments, *Journal of the Mechanics and Physics of Solids* 122 (2019) 1–26.
58. C. Bonatti, D. Mohr, Smooth-shell metamaterials of cubic symmetry: Anisotropic elasticity, yield strength and specific energy absorption, *Acta Materialia* 164 (2019) 301–321.

59. C. Soyarslan, V. Blümer, S. Bargmann, Tunable auxeticity and elastomechanical symmetry in a class of very low density core-shell cubic crystals, *Acta Materialia* 177 (2019) 280–292.
60. L. Dong, Mechanical response of Ti–6Al–4V hierarchical architected metamaterials, *Acta Materialia* 175 (2019) 90–106.
61. L.R. Meza, A.J. Zelhofer, N. Clarke, A.J. Mateos, D.M. Kochmann, J.R. Greer, Resilient 3D hierarchical architected metamaterials, *Proceedings of the National Academy of Sciences* 112(37) (2015) 11502–11507.
62. X. Zheng, W. Smith, J. Jackson, B. Moran, H. Cui, D. Chen, J. Ye, N. Fang, N. Rodriguez, T. Weisgraber, Multiscale metallic metamaterials, *Nature Materials* 15(10) (2016) 1100.
63. D.S. Al-Saedi, S. Masood, M. Faizan-Ur-Rab, A. Alomarah, P. Ponnusamy, Mechanical properties and energy absorption capability of functionally graded F2BCC lattice fabricated by SLM, *Materials & Design* 144 (2018) 32–44.
64. I. Maskery, N. Aboulkhair, A. Aremu, C. Tuck, I. Ashcroft, R.D. Wildman, R.J. Hague, A mechanical property evaluation of graded density Al-Si10-Mg lattice structures manufactured by selective laser melting, *Materials Science and Engineering: A* 670 (2016) 264–274.
65. O. Al-Ketan, D.-W. Lee, R. Rowshan, R.K.A. Al-Rub, Functionally graded and multi-morphology sheet TPMS lattices: Design, manufacturing, and mechanical properties, *Journal of the Mechanical Behavior of Biomedical Materials* 102 (2020) 103520.
66. M.-S. Pham, C. Liu, I. Todd, J. Lertthanasarn, Damage-tolerant architected materials inspired by crystal microstructure, *Nature* 565(7739) (2019) 305.
67. R. Alberdi, R. Dingreville, J. Robbins, T. Walsh, B.C. White, B. Jared, B.L. Boyce, Multi-morphology lattices lead to improved plastic energy absorption, *Materials & Design* 194 (2020) 108883.
68. N. Novak, M. Vesenjak, L. Krstulović-Opara, Z. Ren, Mechanical characterisation of auxetic cellular structures built from inverted tetrapods, *Composite Structures* 196 (2018) 96–107.
69. F. Warmuth, F. Osmanlic, L. Adler, M.A. Lodes, C. Körner, Fabrication and characterisation of a fully auxetic 3D lattice structure via selective electron beam melting, *Smart Materials and Structures* 26(2) (2016) 025013.
70. L. Yang, O. Harrysson, H. West, D. Cormier, Compressive properties of Ti–6Al–4V auxetic mesh structures made by electron beam melting, *Acta Materialia* 60(8) (2012) 3370–3379.
71. M. Fu, F. Liu, L. Hu, A novel category of 3D chiral material with negative Poisson's ratio, *Composites Science and Technology* 160 (2018) 111–118.
72. S. Babaee, J. Shim, J.C. Weaver, E.R. Chen, N. Patel, K. Bertoldi, 3D Soft metamaterials with negative Poisson's ratio, *Advanced Materials* 25(36) (2013) 5044–5049.
73. X. Ren, J. Shen, P. Tran, T.D. Ngo, Y.M. Xie, Design and characterisation of a tuneable 3D buckling-induced auxetic metamaterial, *Materials & Design* 139 (2018) 336–342.
74. J.T.B. Overvelde, S. Shan, K. Bertoldi, Compaction through buckling in 2D periodic, soft and porous structures: Effect of pore shape, *Advanced Materials* 24(17) (2012) 2337–2342.
75. T. Bückmann, M. Thiel, M. Kadic, R. Schittny, M. Wegener, An elasto-mechanical unfeelability cloak made of pentamode metamaterials, *Nature Communications* 5 (2014) 4130.
76. R. Hedayati, A. Leeflang, A. Zadpoor, Additively manufactured metallic pentamode meta-materials, *Applied Physics Letters* 110(9) (2017) 091905.

77. M. Kadic, T. Bückmann, R. Schittny, P. Gumbsch, M. Wegener, Pentamode metamaterials with independently tailored bulk modulus and mass density, *Physical Review Applied* 2(5) (2014) 054007.
78. M. Kadic, T. Bückmann, N. Stenger, M. Thiel, M. Wegener, On the practicability of pentamode mechanical metamaterials, *Applied Physics Letters* 100(19) (2012) 191901.
79. S. Shan, S.H. Kang, J.R. Raney, P. Wang, L. Fang, F. Candido, J.A. Lewis, K. Bertoldi, Multistable architected materials for trapping elastic strain energy, *Advanced Materials* 27(29) (2015) 4296–4301.
80. T.A. Hewage, K.L. Alderson, A. Alderson, F. Scarpa, Double-negative mechanical metamaterials displaying simultaneous negative stiffness and negative Poisson's ratio properties, *Advanced Materials* 28(46) (2016) 10323–10332.
81. C.S. Ha, R.S. Lakes, M.E. Plesha, Cubic negative stiffness lattice structure for energy absorption: Numerical and experimental studies, *International Journal of Solids and Structures* 178 (2019) 127–135.
82. G. Liao, C. Luan, Z. Wang, J. Liu, X. Yao, J. Fu, Acoustic metamaterials: A review of theories, structures, fabrication approaches, and applications, *Advanced Materials Technologies* 6 (2021) 2000787.
83. J. Boulvert, T. Cavalieri, J. Costa-Baptista, L. Schwan, V. Romero-García, G. Gabard, E.R. Fotsing, A. Ross, J. Mardjono, J.-P. Groby, Optimally graded porous material for broadband perfect absorption of sound, *Journal of Applied Physics* 126(17) (2019) 175101.
84. X. Zhang, Z. Qu, H. Wang, Engineering acoustic metamaterials for sound absorption: From uniform to gradient structures, *Iscience* 23(5) (2020).
85. M. Ashby, The properties of foams and lattices, *Philosophical Transactions of the Royal Society of London A: Mathematical, Physical and Engineering Sciences* 364(1838) (2006) 15–30.
86. I. Maskery, L. Parry, D. Padrão, R. Hague, I. Ashcroft, FLatt Pack: A research-focussed lattice design program, *Additive Manufacturing* (2021) 102510.
87. M. Helou, S. Kara, Design, analysis and manufacturing of lattice structures: An overview, *International Journal of Computer Integrated Manufacturing* 31(3) (2018) 243–261.
88. S. Liu, Y.C. Shin, Additive manufacturing of Ti6Al4V alloy: A review, *Materials & Design* 164 (2019) 107552.
89. C.Y. Yap, C.K. Chua, Z.L. Dong, Z.H. Liu, D.Q. Zhang, L.E. Loh, S.L. Sing, Review of selective laser melting: Materials and applications, *Applied Physics Reviews* 2(4) (2015) 041101.
90. T. Maconachie, M. Leary, B. Lozanovski, X. Zhang, M. Qian, O. Faruque, M. Brandt, SLM lattice structures: Properties, performance, applications and challenges, *Materials & Design* 183 (2019) 108137.
91. X. Zhang, M. Leary, H. Tang, T. Song, M. Qian, Selective electron beam manufactured Ti-6Al-4V lattice structures for orthopedic implant applications: Current status and outstanding challenges, *Current Opinion in Solid State and Materials Science* 22(3) (2018) 75–99.
92. R. Brighenti, M.P. Cosma, L. Marsavina, A. Spagnoli, M. Terzano, Laser-based additively manufactured polymers: A review on processes and mechanical models, *Journal of Materials Science* 56 (2020) 961–998.
93. S.L. Sing, W.Y. Yeong, F.E. Wiria, B.Y. Tay, Z. Zhao, L. Zhao, Z. Tian, S. Yang, Direct selective laser sintering and melting of ceramics: A review, *Rapid Prototyping Journal* 23 (2017) 611–623.
94. M. Pagac, J. Hajnys, Q.-P. Ma, L. Jancar, J. Jansa, P. Stefek, J. Mesicek, A review of vat photopolymerization technology: Materials, applications, challenges, and future trends of 3D printing, *Polymers* 13(4) (2021) 598.

95. Z. Zhao, X. Tian, X. Song, Engineering materials with light: Recent progress in digital light processing based 3D printing, *Journal of Materials Chemistry C* 8 (2020) 13896–13917.
96. S. Tagliaferri, A. Panagiotopoulos, C. Mattevi, Direct ink writing of energy materials, *Materials Advances* 2 (2021) 540–563.
97. V.G. Rocha, E. Saiz, I.S. Tirichenko, E. García-Tuñón, Direct ink writing advances in multi-material structures for a sustainable future, *Journal of Materials Chemistry A* 8(31) (2020) 15646–15657.
98. C. Xu, B. Quinn, L.L. Lebel, D. Therriault, G. L'Espérance, Multi-material direct ink writing (DIW) for complex 3D metallic structures with removable supports, *ACS Applied Materials & Interfaces* 11(8) (2019) 8499–8506.
99. S. Vyavahare, S. Teraiya, D. Panghal, S. Kumar, Fused deposition modelling: A review, *Rapid Prototyping Journal* 26(1) (2020) 176–201.
100. J.U. Surjadi, L. Gao, H. Du, X. Li, X. Xiong, N.X. Fang, Y. Lu, Mechanical metamaterials and their engineering applications, *Advanced Engineering Materials* 21(3) (2019) 1800864.
101. E.C. Clough, T.A. Plaisted, Z.C. Eckel, K. Cante, J.M. Hundley, T.A. Schaedler, Elastomeric microlattice impact attenuators, *Matter* 1(6) (2019) 1519–1531.
102. C. Ling, A. Cernicchi, M.D. Gilchrist, P. Cardiff, Mechanical behaviour of additively-manufactured polymeric octet-truss lattice structures under quasi-static and dynamic compressive loading, *Materials & Design* 162 (2019) 106–118.
103. M. Mohsenizadeh, F. Gasbarri, M. Munther, A. Beheshti, K. Davami, Additively-manufactured lightweight Metamaterials for energy absorption, *Materials & Design* 139 (2018) 521–530.
104. F. Habib, P. Iovenitti, S. Masood, M. Nikzad, Fabrication of polymeric lattice structures for optimum energy absorption using Multi Jet Fusion technology, *Materials & Design* 155 (2018) 86–98.
105. H. Mei, R. Zhao, Y. Xia, J. Du, X. Wang, L. Cheng, Ultrahigh strength printed ceramic lattices, *Journal of Alloys and Compounds* 797 (2019) 786–796.
106. L. Zhang, S. Feih, S. Daynes, S. Chang, M.Y. Wang, J. Wei, W.F. Lu, Pseudo-ductile fracture of 3D printed alumina triply periodic minimal surface structures, *Journal of the European Ceramic Society* 40(2) (2020) 408–416.
107. J. Guo, Y. Zeng, P. Li, J. Chen, Fine lattice structural titanium dioxide ceramic produced by DLP 3D printing, *Ceramics International* 45(17) (2019) 23007–23012.
108. H. Jodati, B. Yılmaz, Z. Evis, A review of bioceramic porous scaffolds for hard tissue applications: Effects of structural features, *Ceramics International* 46(10) (2020) 15725–15739.
109. Q. Liu, W.F. Lu, W. Zhai, Toward stronger robocast calcium phosphate scaffolds for bone tissue engineering: A mini-review and meta-analysis, *Materials Science and Engineering: C* 134 (2021) 112578.
110. X. Zhou, C.J. Liu, Three-dimensional printing for catalytic applications: Current status and perspectives, *Advanced Functional Materials* 27(30) (2017) 1701134.
111. S. Lawson, X. Li, H. Thakkar, A.A. Rownaghi, F. Rezaei, Recent advances in 3D printing of structured materials for adsorption and catalysis applications, *Chemical Reviews* 121(10) (2021) 6246–6291.
112. J. Mueller, J.R. Raney, K. Shea, J.A. Lewis, Architected lattices with high stiffness and toughness via multicore–shell 3D printing, *Advanced Materials* 30(12) (2018) 1705001.
113. Z. Xu, C.S. Ha, R. Kadam, J. Lindahl, S. Kim, H.F. Wu, V. Kunc, X. Zheng, Additive manufacturing of two-phase lightweight, stiff and high damping carbon fiber reinforced polymer microlattices, *Additive Manufacturing* 32 (2020) 101106.

114. B. Deng, R. Xu, K. Zhao, Y. Lu, S. Ganguli, G.J. Cheng, Composite bending-dominated hollow nanolattices: A stiff, cyclable mechanical metamaterial, *Materials Today* 21(5) (2018) 467–474.
115. W. Zhang, J. Chen, X. Li, Y. Lu, Liquid metal-polymer microlattice metamaterials with high fracture toughness and damage recoverability, *Small* 16(46) (2020) 2004190.
116. J. Bauer, J.A. Kraus, C. Crook, J.J. Rimoli, L. Valdevit, Tensegrity metamaterials: Toward failure-resistant engineering systems through delocalized deformation, *Advanced Materials* 33(10) (2021) 2005647.
117. Z. Chen, Y.M. Xie, X. Wu, Z. Wang, Q. Li, S. Zhou, On hybrid cellular materials based on triply periodic minimal surfaces with extreme mechanical properties, *Materials & Design* 183 (2019) 108109.
118. N.A. Traugutt, D. Mistry, C. Luo, K. Yu, Q. Ge, C.M. Yakacki, Liquid-crystal-elastomer-based dissipative structures by digital light processing 3D printing, *Advanced Materials* 32(28) (2020) 2000797.
119. M. Zhang, Q. Yu, Z. Liu, J. Zhang, G. Tan, D. Jiao, W. Zhu, S. Li, Z. Zhang, R. Yang, 3D printed Mg-NiTi interpenetrating-phase composites with high strength, damping capacity, and energy absorption efficiency, *Science Advances* 6(19) (2020) eaba5581.
120. X. Li, K.M. Lim, W. Zhai, A novel class of bioinspired composite via ultrasound-assisted directed self-assembly digital light 3D printing, *Applied Materials Today* 26 (2022) 101388.
121. X. Wang, S. Xu, S. Zhou, W. Xu, M. Leary, P. Choong, M. Qian, M. Brandt, Y.M. Xie, Topological design and additive manufacturing of porous metals for bone scaffolds and orthopaedic implants: A review, *Biomaterials* 83 (2016) 127–141.
122. M. Prakasam, J. Locs, K. Salma-Ancane, D. Loca, A. Largeteau, L. Berzina-Cimdina, Biodegradable materials and metallic implants: A review, *Journal of Functional Biomaterials* 8(4) (2017) 44.
123. S. Ponader, E. Vairaktaris, P. Heinl, C.V. Wilmowsky, A. Rottmair, C. Körner, R.F. Singer, S. Holst, K.A. Schlegel, F.W. Neukam, Effects of topographical surface modifications of electron beam melted Ti-6Al-4V titanium on human fetal osteoblasts, *Journal of Biomedical Materials Research Part A: An Official Journal of the Society for Biomaterials, the Japanese Society for Biomaterials, and the Australian Society for Biomaterials and the Korean Society for Biomaterials* 84(4) (2008) 1111–1119.
124. S. Ponader, C. Von Wilmowsky, M. Widenmayer, R. Lutz, P. Heinl, C. Körner, R.F. Singer, E. Nkenke, F.W. Neukam, K.A. Schlegel, In vivo performance of selective electron beam-melted Ti-6Al-4V structures, *Journal of Biomedical Materials Research Part A: An Official Journal of the Society for Biomaterials, the Japanese Society for Biomaterials, and the Australian Society for Biomaterials and the Korean Society for Biomaterials* 92(1) (2010) 56–62.
125. X. Yan, Q. Li, S. Yin, Z. Chen, R. Jenkins, C. Chen, J. Wang, W. Ma, R. Bolot, R. Lupoi, Mechanical and in vitro study of an isotropic Ti6Al4V lattice structure fabricated using selective laser melting, *Journal of Alloys and Compounds* 782 (2019) 209–223.
126. X. Li, C. Wang, W. Zhang, Y. Li, Fabrication and compressive properties of Ti6Al4V implant with honeycomb-like structure for biomedical applications, *Rapid Prototyping Journal* 16(1) (2010) 44–49.
127. B.V. Krishna, S. Bose, A. Bandyopadhyay, Low stiffness porous Ti structures for load-bearing implants, *Acta Biomaterialia* 3(6) (2007) 997–1006.
128. A. Bandyopadhyay, F. Espana, V.K. Balla, S. Bose, Y. Ohgami, N.M. Davies, Influence of porosity on mechanical properties and in vivo response of Ti6Al4V implants, *Acta Biomaterialia* 6(4) (2010) 1640–1648.

129. N. Taniguchi, S. Fujibayashi, M. Takemoto, K. Sasaki, B. Otsuki, T. Nakamura, T. Matsushita, T. Kokubo, S. Matsuda, Effect of pore size on bone ingrowth into porous titanium implants fabricated by additive manufacturing: An in vivo experiment, *Materials Science and Engineering: C* 59 (2016) 690–701.
130. P.H. Warnke, T. Douglas, P. Wollny, E. Sherry, M. Steiner, S. Galonska, S.T. Becker, I.N. Springer, J. Wiltfang, S. Sivananthan, Rapid prototyping: porous titanium alloy scaffolds produced by selective laser melting for bone tissue engineering, *Tissue Engineering Part C: Methods* 15(2) (2008) 115–124.
131. A. Fukuda, M. Takemoto, T. Saito, S. Fujibayashi, M. Neo, D.K. Pattanayak, T. Matsushita, K. Sasaki, N. Nishida, T. Kokubo, Osteoinduction of porous Ti implants with a channel structure fabricated by selective laser melting, *Acta Biomaterialia* 7(5) (2011) 2327–2336.
132. Q. Ran, W. Yang, Y. Hu, X. Shen, Y. Yu, Y. Xiang, K. Cai, Osteogenesis of 3D printed porous Ti6Al4V implants with different pore sizes, *Journal of the Mechanical Behavior of Biomedical Materials* 84 (2018) 1–11.
133. X.-Y. Zhang, G. Fang, S. Leeflang, A.A. Zadpoor, J. Zhou, Topological design, permeability and mechanical behavior of additively manufactured functionally graded porous metallic biomaterials, *Acta Biomaterialia* 84 (2019) 437–452.
134. Y. Takahashi, Y. Tabata, Effect of the fiber diameter and porosity of non-woven PET fabrics on the osteogenic differentiation of mesenchymal stem cells, *Journal of Biomaterials Science, Polymer Edition* 15(1) (2004) 41–57.
135. S. Van Bael, Y.C. Chai, S. Truscello, M. Moesen, G. Kerckhofs, H. Van Oosterwyck, J.-P. Kruth, J. Schrooten, The effect of pore geometry on the in vitro biological behavior of human periosteum-derived cells seeded on selective laser-melted Ti6Al4V bone scaffolds, *Acta Biomaterialia* 8(7) (2012) 2824–2834.
136. H. Kolken, C. de Jonge, T. van der Sloten, A.F. Garcia, B. Pouran, K. Willemsen, H. Weinans, A. Zadpoor, Additively manufactured space-filling meta-implants, *Acta Biomaterialia* 125 (2021) 345–357.
137. S. Li, Q. Xu, Z. Wang, W. Hou, Y. Hao, R. Yang, L. Murr, Influence of cell shape on mechanical properties of Ti–6Al–4V meshes fabricated by electron beam melting method, *Acta Biomaterialia* 10(10) (2014) 4537–4547.
138. S. Arabnejad, R.B. Johnston, J.A. Pura, B. Singh, M. Tanzer, D. Pasini, High-strength porous biomaterials for bone replacement: A strategy to assess the interplay between cell morphology, mechanical properties, bone ingrowth and manufacturing constraints, *Acta Biomaterialia* 30 (2016) 345–356.
139. D. Melancon, Z. Bagheri, R. Johnston, L. Liu, M. Tanzer, D. Pasini, Mechanical characterization of structurally porous biomaterials built via additive manufacturing: Experiments, predictive models, and design maps for load-bearing bone replacement implants, *Acta Biomaterialia* 63 (2017) 350–368.
140. Y. Li, Y. Ding, K. Munir, J. Lin, M. Brandt, A. Atrens, Y. Xiao, J.R. Kanwar, C. Wen, Novel β-Ti35Zr28Nb alloy scaffolds manufactured using selective laser melting for bone implant applications, *Acta Biomaterialia* 87 (2019) 273–284.
141. H.M. Kolken, S. Janbaz, S.M. Leeflang, K. Lietaert, H.H. Weinans, A.A. Zadpoor, Rationally designed meta-implants: A combination of auxetic and conventional meta-biomaterials, *Materials Horizons* 5(1) (2018) 28–35.
142. H. Zhao, Y. Han, C. Pan, D. Yang, H. Wang, T. Wang, X. Zeng, P. Su, Design and mechanical properties verification of gradient Voronoi scaffold for bone tissue engineering, *Micromachines* 12(6) (2021) 664.
143. P. Wang, X. Li, Y. Jiang, M.L.S. Nai, J. Ding, J. Wei, Electron beam melted heterogeneously porous microlattices for metallic bone applications: Design and investigations of boundary and edge effects, *Additive Manufacturing* 36 (2020) 101566.

144. A. Takezawa, K. Yonekura, Y. Koizumi, X. Zhang, M. Kitamura, Isotropic Ti–6Al–4V lattice via topology optimization and electron-beam melting, *Additive Manufacturing* 22 (2018) 634–642.
145. P. Wang, X. Li, S. Luo, M.L.S. Nai, J. Ding, J. Wei, Additively manufactured heterogeneously porous metallic bone with biostructural functions and bone-like mechanical properties, *Journal of Materials Science & Technology* 62 (2021) 173–179.
146. M. Yang, P. Sheng, Sound absorption structures: From porous media to acoustic metamaterials, *Annual Review of Materials Research* 47 (2017) 83–114.
147. S. Deshmukh, H. Ronge, S. Ramamoorthy, Design of periodic foam structures for acoustic applications: Concept, parametric study and experimental validation, *Materials & Design* 175 (2019) 107830.
148. W. Zhai, X. Yu, X. Song, L.Y.L. Ang, F. Cui, H.P. Lee, T. Li, Microstructure-based experimental and numerical investigations on the sound absorption property of open-cell metallic foams manufactured by a template replication technique, *Materials & Design* 137 (2018) 108–116.
149. H. Rice, J. Kennedy, P. Göransson, L. Dowling, D. Trimble, Design of a Kelvin cell acoustic metamaterial, *Journal of Sound and Vibration* 472 (2020) 115167.
150. X. Sun, F. Jiang, J. Wang, Acoustic properties of 316L stainless steel lattice structures fabricated via selective laser melting, *Metals* 10(1) (2020) 111.
151. D.-Y. Maa, Potential of microperforated panel absorber, *The Journal of the Acoustical Society of America* 104(5) (1998) 2861–2866.
152. J. Boulvert, J. Costa-Baptista, T. Cavalieri, M. Perna, E.R. Fotsing, V. Romero-García, G. Gabard, A. Ross, J. Mardjono, J.-P. Groby, Acoustic modeling of micro-lattices obtained by additive manufacturing, *Applied Acoustics* 164 (2020) 107244.
153. W. Yang, J. An, C.K. Chua, K. Zhou, Acoustic absorptions of multifunctional polymeric cellular structures based on triply periodic minimal surfaces fabricated by stereolithography, *Virtual and Physical Prototyping* 15(2) (2020) 242–249.
154. T.G. Zieliński, K.C. Opiela, P. Pawłowski, N. Dauchez, T. Boutin, J. Kennedy, D. Trimble, H. Rice, B. Van Damme, G. Hannema, Reproducibility of sound-absorbing periodic porous materials using additive manufacturing technologies: Round robin study, *Additive Manufacturing* 36 (2020) 101564.
155. K.C. Opiela, T.G. Zieliński, Microstructural design, manufacturing and dual-scale modelling of an adaptable porous composite sound absorber, *Composites Part B: Engineering* 187 (2020) 107833.
156. L.-W. Cheng, C.-W. Cheng, K.-C. Chung, T.-Y. Kam, Sound absorption of metallic sound absorbers fabricated via the selective laser melting process, *Applied Physics A* 123(1) (2017) 37.
157. Z. Li, W. Zhai, X. Li, X. Yu, Z. Guo, Z. Wang, Additively manufactured dual-functional metamaterials with customisable mechanical and sound-absorbing properties, *Virtual and Physical Prototyping* 17(4) (2022) 864–880.
158. X. Zhang, Z. Qu, D. Tian, Y. Fang, Acoustic characteristics of continuously graded phononic crystals, *Applied Acoustics* 151 (2019) 22–29.
159. X. Zhang, Z. Qu, X. He, D. Lu, Experimental study on the sound absorption characteristics of continuously graded phononic crystals, *AIP Advances* 6(10) (2016) 105205.
160. X. Li, X. Yu, W. Zhai, additively manufactured deformation-recoverable and broadband sound-absorbing microlattice inspired by the concept of traditional perforated panels, *Advanced Materials* 33(44) (2021) 2104552.
161. C.W. Foster, M.P. Down, Y. Zhang, X. Ji, S.J. Rowley-Neale, G.C. Smith, P.J. Kelly, C.E. Banks, 3D printed graphene based energy storage devices, *Scientific Reports* 7(1) (2017) 1–11.

162. X. Su, X. Li, C.Y.A. Ong, T.S. Herng, Y. Wang, E. Peng, J. Ding, Metallization of 3D printed polymers and their application as a fully functional water-splitting system, *Advanced Science* 6(6) (2019) 1801670.
163. S. Chang, Y. Zhang, B. Zhang, X. Cao, L. Zhang, X. Huang, W. Lu, C.Y.A. Ong, S. Yuan, C. Li, Conductivity modulation of 3D-printed shellular electrodes through embedding nanocrystalline intermetallics into amorphous matrix for ultrahigh-current oxygen evolution, *Advanced Energy Materials* 11(28) (2021) 2100968
164. J.P. Hughes, P.L. dos Santos, M.P. Down, C.W. Foster, J.A. Bonacin, E.M. Keefe, S.J. Rowley-Neale, C.E. Banks, Single step additive manufacturing (3D printing) of electrocatalytic anodes and cathodes for efficient water splitting, *Sustainable Energy & Fuels* 4(1) (2020) 302–311.
165. X. Xu, C. Li, J.G. Lim, Y. Wang, A. Ong, X. Li, E. Peng, J. Ding, Hierarchical design of NiOOH@ amorphous Ni–P bilayer on a 3D mesh substrate for high-efficiency oxygen evolution reaction, *ACS Applied Materials & Interfaces* 10(36) (2018) 30273–30282.
166. S. Chandrasekaran, B. Yao, T. Liu, W. Xiao, Y. Song, F. Qian, C. Zhu, E.B. Duoss, C.M. Spadaccini, Y. Li, Direct ink writing of organic and carbon aerogels, *Materials Horizons* 5(6) (2018) 1166–1175.
167. C. Zhu, T. Liu, F. Qian, T.Y.-J. Han, E.B. Duoss, J.D. Kuntz, C.M. Spadaccini, M.A. Worsley, Y. Li, Supercapacitors based on three-dimensional hierarchical graphene aerogels with periodic macropores, *Nano Letters* 16(6) (2016) 3448–3456.
168. B. Yao, S. Chandrasekaran, H. Zhang, A. Ma, J. Kang, L. Zhang, X. Lu, F. Qian, C. Zhu, E.B. Duoss, 3D-printed structure boosts the kinetics and intrinsic capacitance of pseudocapacitive graphene aerogels, *Advanced Materials* 32(8) (2020) 1906652.
169. W. Yang, J. Yang, J.J. Byun, F.P. Moissinac, J. Xu, S.J. Haigh, M. Domingos, M.A. Bissett, R.A. Dryfe, S. Barg, 3D printing of freestanding MXene architectures for current-collector-free supercapacitors, *Advanced Materials* 31(37) (2019) 1902725.
170. L. Yu, W. Li, C. Wei, Q. Yang, Y. Shao, J. Sun, 3D printing of NiCoP/Ti 3C 2 MXene architectures for energy storage devices with high areal and volumetric energy density, *Nano-Micro Letters* 12(1) (2020) 1–13.
171. Z. Lyu, G.J. Lim, R. Guo, Z. Kou, T. Wang, C. Guan, J. Ding, W. Chen, J. Wang, 3D-printed MOF-derived hierarchically porous frameworks for practical high-energy density Li–O_2 batteries, *Advanced Functional Materials* 29(1) (2019) 1806658.
172. K. Shen, H. Mei, B. Li, J. Ding, S. Yang, 3D printing sulfur copolymer-graphene architectures for Li-S batteries, *Advanced Energy Materials* 8(4) (2018) 1701527.
173. L. Zhou, W. Ning, C. Wu, D. Zhang, W. Wei, J. Ma, C. Li, L. Chen, 3D-printed microelectrodes with a developed conductive network and hierarchical pores toward high areal capacity for microbatteries, *Advanced Materials Technologies* 4(2) (2019) 1800402.
174. J. Zhao, H. Lu, X. Zhao, O.I. Malyi, J. Peng, C. Lu, X. Li, Y. Zhang, Z. Zeng, G. Xing, Printable ink design towards customizable miniaturized energy storage devices, *ACS Materials Letters* 2(9) (2020) 1041–1056.

8 Antennas Manufactured from Conductive Fabric Materials

Terry Tao Ye, Mengxia Yu, and Yulong Liu
Southern University of Science and Technology

Anna Pui Yi Lau and Hai Tao Si
Laxcen Technology Inc.

CONTENTS

DOI: 10.1201/9781003182146-8

8.1 INTRODUCTION

Traditional smart-clothing applications mostly rely on solid-state devices that are embedded or sewn on apparels or garments, with metallic wires as interconnects and batteries or external energy sources as power supplies [1]; examples include the illuminated headband and the "VivoMetrics LifeShirt" [2]. These devices are cumbersome to use and difficult to wash, lacking esthetic appearance, textural comfort and flexibility of natural textiles. The introduction of conductive materials that are compatible with fabric and garment manufacturing processes, including embroidery, weaving and printing, has led to various e-textile devices for many applications. Compared with traditional wearable devices, e-textile-based sensors have obvious advantages in smart-clothing applications, such as vital signs monitoring, human activities tracking [3] and ambient environment sensing.

Among various e-textile devices, embroidered antennas are considered to be the ideal substitutes for conventional metallic antennas for wearable electronic devices. However, embroidered antennas exhibit some intrinsic limitations that should be taken into considerations as compared with conventional metallic antennas:

1. The conductive yarns have a higher resistivity than metallic conductors. Various types of conductive yarns are either made from carbon or nylon cores coated with silver plating, or from thin steel fibers twisted with other nonconductive fibers. The embroidered Radio Frequency Identification (RFID) antennas made with these conductive yarns have exhibited an order-of-magnitude higher resistivity than copper-based, aluminum-based or silver paste-printed antennas [4].
2. The geometries of embroidered antennas have lower resolution than those made from conventional antenna fabrication methods, such as plating, etching or printing. Typically, it is difficult to fabricate embroidered antennas with a resolution of less than one millimeter [5].
3. The embroidered antennas are easier to be deformed as compared to their metallic counterpart. This deformation effect, combined with the low resolution of the yarn geometries, makes fine geometry impractical and impossible to achieve [6].

The challenge of embroidery antenna design is to accommodate those limitations in the design exploration phase as well as the manufacturing phase while still ensuring that embroidered antennas possess satisfactory EM properties as compared to the metallic counterparts.

8.2 COMPARISON OF VARIOUS CONDUCTIVE TEXTILES

Wearable electronic devices manufactured from conductive textiles have attracted increasing interests in recent years. Conductive textile-based wearable devices can be divided into four categories: (i) mixed metal fabrics, (ii) conductive yarn-based fabrics, (iii) embroidered conductive fabrics and (iv) conductive inkjet fabrics. In general, in addition to good conductivity, the conductive textile material also demands good mechanical flexibility and robustness property to endure constantly repeated bending and deformation. Among these four categories, mixed metal fabrics, conductive yarn-based fabrics may develop metallized fracture, and inkjet-printed conductive textiles may exhibit poor Radio Frequency (RF) performance after repeated bending tests [4,7]. Studies [8,9] suggested that embroidered conductive fibers have the best combination of RF performance, mechanical strength and robustness.

Textile-based RFID antennas that can be embroidered directly onto fabric materials with conductive yarns using standard embroidery procedures have demonstrated considerable promise for wearable applications. Conductive fibers with inherently flexible and lightweight properties make them ideal for use in the construction of wearable electronics in smart garments. A previous study [10] showed that the electrical and mechanical characteristics of conductive yarns, such as the Q value and permittivity [11], are applicable for RF devices with high operation frequencies. Embroidered RFID antennas with appropriate geometric designs and electrical parameters still possess acceptable RF performance even after being mass-manufactured from the industrial-grade embroidery machines with high tension and assembly speeds [12].

Conductive yarns need to satisfy many requirements in order to be used for embroidered wearable antennas. They should be soft, lightweight, mechanically flexible and robust to compatible with the embroidery machine. Different types of conductive yarns have been designed and explored, such as metal filament bundles, metal filament composite yarn and metal-coated polymer filament plied yarn [13,14]. It was shown that nylon, polyester and cotton threads coated with silver nanowires (AgNws) have the best attributes of high conductivity and light weight, with a resistance of 0.8 Ω/cm. The threads were coated using the dip-coating technique. The coating procedure is depicted in Figure 8.1, and the density of the coating can affect the resistance.

Silverpan 250 is an AgNws-coated polymer yarn with a 0.08 mm thin diameter that can be used for RFID antenna embroidery [10]. Conductive yarns should be used on the upper bobbin to continually embroider antenna shapes. It is necessary for yarns with smaller diameters to glide smoothly across the needle during this procedure. The flat surface of Silverpan 250 decreases the interference between the yarns and machine. Antennas embroidered from this process need to have two layers, a loose layer covered by a thick layer, to reduce the antenna's resistance.

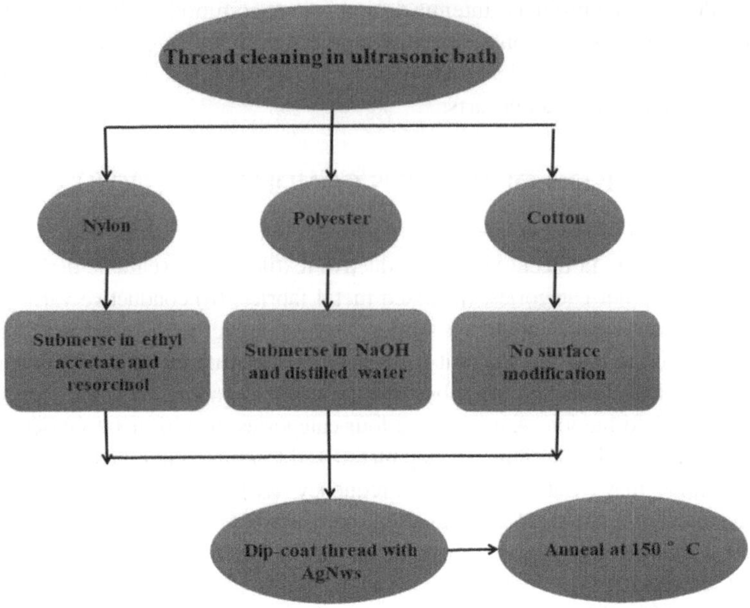

FIGURE 8.1 Conductive yarn coating procedure chart.

8.3 MANUFACTURING PROCEDURE OF EMBROIDERED DEVICES

Figure 8.2 illustrates the typical manufacture procedure for embroidered antennas using a computerized digital sewing machine (JCZA 0109-550 (700), ZSK Stickmaschinen GmbH). First, the design and optimization of the target antenna are performed in electromagnetic simulation software (such as HFSS or CST), followed by transforming the antenna and circuit design into the embroidery geometries and then being imported into the embroidered machine, as shown in Figure 8.2a. The computerized machine then performs programmed batch embroidery of antennas with fabric materials, including nonwoven cotton substrates, conductive yarns and top (bottom) yarns that maintain stable mechanical properties against tension and deformation (Figure 8.2b). Figure 8.2c presents the embroidered RFID antennas, with a magnified image (left) and antennas in mass production (right). The textile antenna prototypes show excellent mechanical flexibility, with comfortable touch-feeling and esthetic appearance.

To explore the feasibility of conductive fibers in building multilayer RF circuits, Wang et al. also developed a process for multilayer microstrip circuit structures using e-fiber fabricated pins [7]. They demonstrated that double layers required a specialized design to ensure maximum surface conductivity, minimum physical discontinuities and thread gaps. The gaps are always smaller than $\lambda/20$, where λ is the wavelength at the resonance frequency in free space.

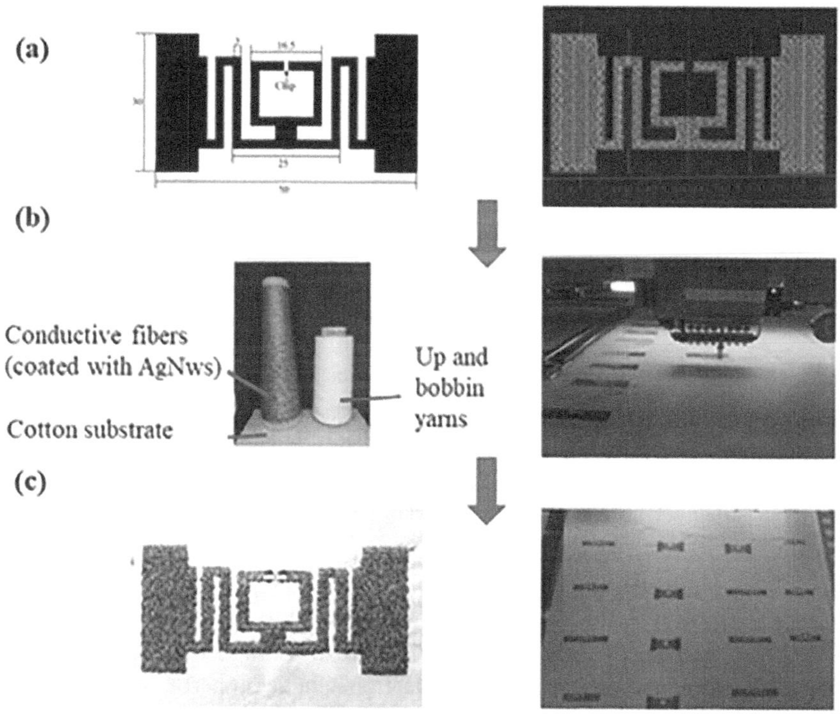

FIGURE 8.2 The fabrication process of embroidered textile antennas. (a) Computer model and digitization, (b) the yarns/fibers and machine used for embroidered antennas and (c) embroidered RFID antennas.

8.4 DESIGN CONSIDERATIONS FOR EMBROIDERED ANTENNAS

8.4.1 Parasitic Elements of Conductive Textile Threads

From the scanning electron microscopy (SEM) photograph (Figure 8.3a) of the conductive yarns, it can be observed that the conductive yarn consists of several intertwined wires instead of a single solid wire. Many small gaps are formed between the conductive wires. Through analysis and experiments [6], the embroidered RFID antenna geometry is composed of many yarns entangled with each other, contributing to the parasitic capacitance or inductance, and the equivalent circuit model is shown in Figure 8.3a.

With the parasitic capacitance in consideration, an equivalent circuit of an embroidered RFID transponder can be derived, as shown in Figure 8.3b, where R_a and L_a are the yarn's intrinsic resistance and intrinsic inductance, respectively, and C_a is parasitic capacitance induced from the gaps between the bundled wires. We will use this model to analyze RFID performance later.

In general, UHF RFID antenna designs are based on slotted patch and meander line dipole (MLD) geometries, while HF Near Field Communication (NFC)

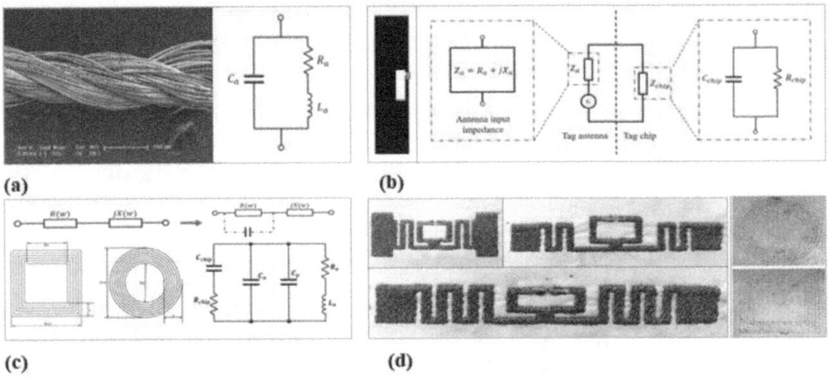

FIGURE 8.3 Equivalent circuit models and prototypes of embroidery yarns/antennas. (a) SEM photograph and equivalent circuit model of conductive yarns, (b) geometry design and circuit model of the RFID Tag, (c) two typical planar coils and their equivalent circuit model and (d) prototypes of embroidery antennas.

antennas typically use square or circular spiral conductive coils. The equivalent circuit of the coli is illustrated in Figure 8.3c. The embroidered prototype of RFID/NFC antennas is also shown in Figure 8.3d. Both of these UHF and HF embroidered geometries have different electromagnetic properties as well as different limitations in design and manufacturing.

8.4.1.1 Equivalent Circuit Model of a UHF RFID Antenna

The circuit model of a UHF RFID transponder tag consists of an antenna and a transponder chip, as illustrated in Figure 8.3b. The model is a matched source and load network connected by a transmission line, where Z_a is the input impedance of the tag antenna and Z_{chip} is the impedance of the chip. In addition, the frequency-dependent Z_a and Z_{chip} can be defined by the following equations:

$$Z_a(w) = R_a(w) + jX_a(w) \tag{8.1}$$

$$Z_{\text{chip}}(w) = \frac{1}{\dfrac{1}{R_{\text{chip}}} + jwC_{\text{chip}}} \tag{8.2}$$

$$\tau = \frac{4\,\text{Re}\{Z_a\}\text{Re}\{Z_{\text{chip}}\}}{\left|Z_a + Z_{\text{chip}}\right|^2} \tag{8.3}$$

$$Z_{\text{cal}} = \frac{1}{\dfrac{1}{R} + jwc_p} + jX = \frac{R}{1 - w^2R^2C_p^2} - j\frac{wR^2C_P}{1 - w^2R^2C_p^2} + jX \tag{8.4}$$

Equation (8.3) represents the power transfer coefficient τ between the chip and the antenna. $\tau=1$ means that the antenna is conjugate matched to the chip for maximum power transmission. The input impedance of the antenna is a key factor to consider when designing the antenna and performing electromagnetic simulations. The transponder chip impedance is normally provided from the datasheet by the chip manufacturer.

The impedance of an ideal UHF RFID antenna can be expressed in Equation (8.1), where both $R(\omega)$ and $jX(\omega)$ are frequency-dependent elements. However, the equivalent circuit of an embroidered UHF RFID antenna with parasitic capacitance caused by the yarn microstructure should be revised by applying parallel parasitic capacitance to the resistor, as shown in Figure 8.3b. With the parasitic elements, the antenna impedance can be expressed as Equation (8.4). Comparing Equations (8.1) and (8.4), the imaginary part becomes $X-\frac{wR^2C_P}{1-w^2R^2C_p^2}$. It is assumed that $wRC_p<1$, the value of C_p is very small and the antenna operates at relatively low frequencies (<10 GHz). Thus, the parasitic resistance in the conducting yarn will increase the real part of the impedance, while the parasitic capacitance will decrease the reactance part of the impedance.

8.4.1.2 Equivalent Circuit Model of a HF NFC Antenna

The equivalent circuit of a HF NFC antenna is depicted in Figure 8.3c, where the number of turns of the coil affects the inductance L_a, the cross section and length of the antenna wire affect the resistance R_a and the adjacent wire spacing and number of turns define the parasitic capacitance C_a of the antenna. The parasitic capacitance C_p generated by the conductive yarn also contributes to the spacing capacitance C_a between the coils, where the capacitances C_a and C_p are placed in parallel in the equivalent circuit model, as shown in Figure 8.3c.

The inductance of a coil is mainly determined by the geometric parameters defined by Wheeler's equation [14]. The geometry affects the coefficient factors in Wheeler's equation. Equations (8.5) and (8.6) formulate the inductance of a spiral coil and a square coil, respectively.

$$L_{\text{circle}}=31.33\quad \mu_0N^2\frac{a^2}{8a+11c}=31.33\quad \mu_0N^2\frac{a}{8+11\rho} \tag{8.5}$$

$$L_{\text{square}}=2.34\quad \mu_0N^2\frac{a^2}{a+1.875c}=2.34\quad \mu_0N^2\frac{a}{1+2.75\rho} \tag{8.6}$$

8.4.2 Bending Sensitivity of Embroidered Antennas with Different Intrinsic Resistance Values

Yu et al. employed two types of conductive yarns with different conductivities in the fabrication of embroidered slot patch antennas for bending sensor prototypes [15]. The parameters and performance of the two types of conductive yarns,

TABLE 8.1
Parameters of Antennas Constructed from Two Types of Conductive Yarns

	Tag Using HC 40	Tag Using HC 12
Linear density	290 dtex ± 6 dtex	610 dtex ± 15 dtex
Resistance	<300 Ω/m	<100 Ω/m
Performance	Lower sensitivity to bending	Higher sensitivity to bending

named HC40 and HC12, are shown in Table 8.1. Multiple sets of experimental results demonstrate that the bending sensitivity of slotted patch antennas varies with the inherent resistance of the embroidered yarn. The antenna embroidered with HC12 yarn (antenna with a smaller DC resistance) is more sensitive to bending and has better electromagnetic performance than the HC40-based antenna (antenna with a larger DC resistance). The different bending sensitivity of these two embroidered antennas can be exploited to construct a deformation sensor.

8.4.3 Deformation Sensitivity of Embroidered Antennas with Different Geometries

Sensitivity to deformation is an important characteristic of flexible antennas in wearable applications. Deformation mainly includes twisting, elongation (stretching) and bending, and the geometry of the embroidered RFID antenna affects its sensitivity to deformation. The MLD antenna is more sensitive to elongation, while the slotted patch antenna is more sensitive to bending [15]. The quantitative analysis of the effect of deformation on antenna's electromagnetic characteristics can be calculated by the following equations:

$$L_T = L_s + m * L \tag{8.7}$$

$$L_s = \frac{\mu_0}{2\pi} s\left(\log\frac{4s}{b} - 1\right) \tag{8.8}$$

$$L = \frac{\mu_0 h}{\pi}\log\frac{2\omega}{b}\left(1 + \frac{1}{3}(\beta h)^3\right) \tag{8.9}$$

where L_T is the total inductance, as the sum of the inductance L per turn caused by the MLD geometry and the self-inductance L_s of the equivalent conductive line of the dipole of length s. m is the number of turns of the meander line, and b is the thickness of the meander line geometry. h is the height of the turn, and w is the width and spacing of the turn. In addition, in Equation (8.9), β equals $2\pi/\lambda$, and μ_0 is the vacuum permeability. It can be deduced from these equations that the equivalent length of the embroidered MLD antenna becomes longer under elongation as the spacing and width of the meander geometry of each turn increase. This leads to a higher inductance, L_T, and consequently a lower resonant frequency of S11.

The above theoretical analysis shows that the MLD antenna is more sensitive to elongation. To investigate the curvature effect, the slender patch antenna geometry needs to be divided into many infinitesimal segments [16], each of which can be considered as a planar antenna. Equations (8.10) to (8.13) are the formulas to calculate the directivity, radiation, input resistance and maximum effective area of the curved antenna, where r_i is the distance from the pivot per segment to the center of the curvature circle, θ_i is the angle between r_i and each segment, W_{rad} is the total radiation power, I_0 is the total current, I_i is the current of each segment, k is the wavenumber, l is the length of the tag, and λ is the wavelength corresponding to the resonant frequency of the antenna.

$$D(\theta_i',\phi) = \frac{4\pi r_i^2 W_{\text{rad}}(r,\theta_i',\phi)}{\int_0^{2\pi}\int_0^{\pi} W_{\text{rad}}(r,\theta_i',\phi) r_i^2 \sin\theta_i' d\theta_i' d\phi} \tag{8.10}$$

$$R_r = \frac{2\int_0^{2\pi}\int_0^{\pi} W_{\text{rad}}(r,\theta_i',\phi) r_i^2 \sin\theta_i' d\theta_i' d\phi}{|I_0 - I_i|^2} \tag{8.11}$$

$$R_{\text{in}} = \frac{R_r}{\sin^2\left(\frac{kl}{2}\right)} \tag{8.12}$$

$$A_{\text{em}} = \frac{\lambda^2}{4\pi} D_{\max} \tag{8.13}$$

Based on these equations, it can be deduced that the geometric curvature changes the effective length of the antenna, as well as the input resistance and the radiation resistance. The effective length of the tag decreases with curvature, leading to a shift of S11 to higher frequencies. And the variation in radiation resistance due to bending will lead to impedance mismatch and resonant frequency mismatch between the antenna and the chip. In addition, the gain pattern of the bent antenna rotates from the X (or Y) direction to the Z direction [15]. Therefore, the electromagnetic characteristics of the antenna will be significantly affected by the antenna curvature.

The above theoretical analysis shows that the MLD antenna is more sensitive to elongation, while the slotted patch antenna is more sensitive to curvature. To further verify this assumption, we compare the deformation sensitivity of the two antennas through simulation and experiments (Figure 8.4a). An MLD antenna and a slotted patch antenna with different parameters are modeled in HFSS simulator and their performances are measured in experiments [15]. The simulation results show that the S11 resonance point is almost monotonically shifted to the

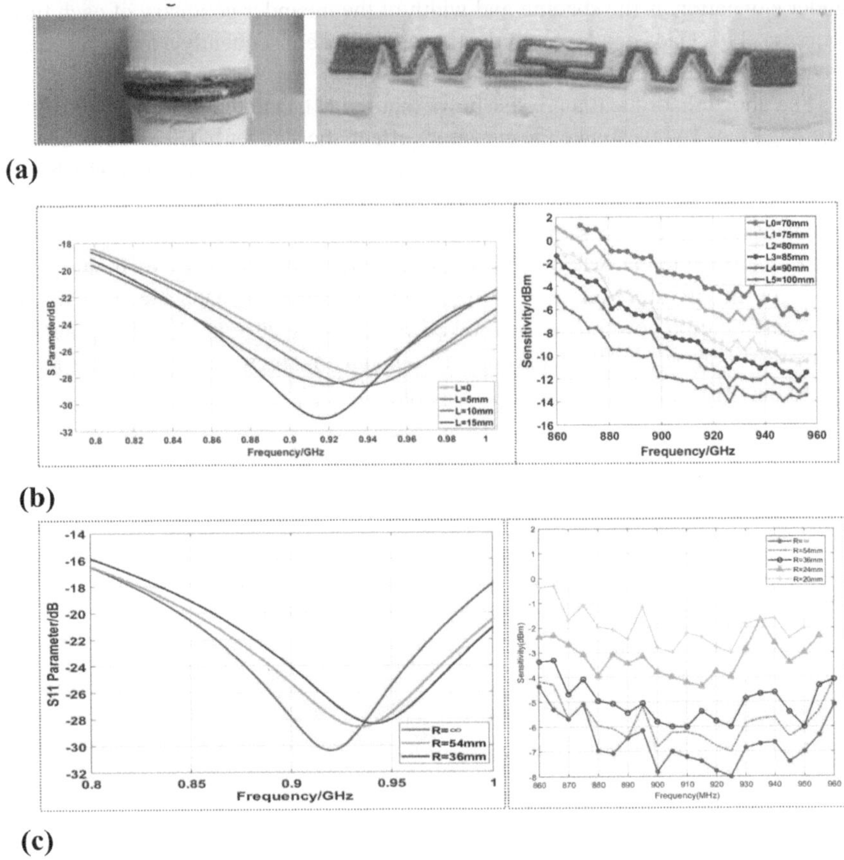

FIGURE 8.4 (a) Photos of two embroidered antennas under deformation, (b) S11 parameter and sensitivity variations caused by elongation, and (c) S11 parameter and sensitivity variations caused by bending.

left as the MLD antenna is elongated, while the S11 value and impedance of the patch antenna vary almost monotonically as the antenna bending curvature decreases, as shown in left plots of Figure 8.4b and c. In the experiment, all prototype tags were tested with different elongations and bending curvatures under the same setup. Right plots of Figure 8.4b and c illustrate that the sensitivity of the textile MLD antenna is improved with increasing elongation but almost remains constant with curvature variations, while the sensitivity of the embroidered patch antenna increases as the bending curvature increases, i.e., the performance of the antenna deteriorates. The sensitivity refers to the minimal input power required to activate the integrated chip; thus, a lower sensitivity level means more sensitivity and easier reading. As demonstrated in Ref. [15], different geometric designs have various sensitivities to deformation and should be taken into account when the antenna is designed as a deformation sensor.

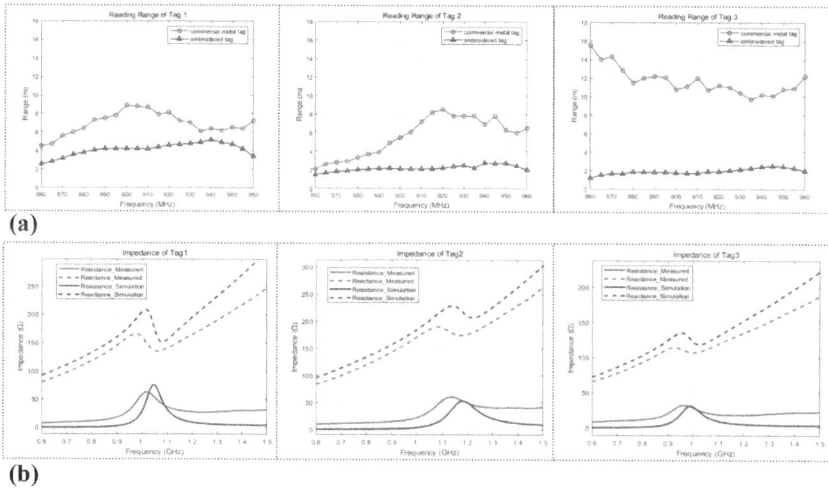

FIGURE 8.5 (a) Comparison between metallic and embroidered antennas and (b) comparison of the simulated and measured input impedance.

8.5 RF PERFORMANCE OF EMBROIDERED TEXTILE ANTENNAS

8.5.1 Comparison between Metallic and Embroidered Antennas

Figure 8.5a illustrates the comparison of the reading range between the embroidered MLD antennas and the metallic MLD antennas (made of etched aluminum) with the same geometry but different numbers of turns. It can be observed that the reading range attenuation between the embroidered antenna and the metallic antenna increases as the numbers of turns increases. This indicates that the increase in the end-to-end resistance of the antenna converts more energy into heat. In addition, the frequency of the maximum reading range shifts upward compared to the simulated resonating frequency (915 MHz), which can be attributed to the parasitic capacitance of the embroidered yarn. Figure 8.5b demonstrates that the measured input impedance of the braided antenna agrees well with the simulation results of the ideal conductor-based antenna.

The study in Ref. [15] shows that the impedance is conjugate matched at 915 MHz for the antenna without parasitic capacitance; however, the conjugate matching point moves to a higher frequency as the reactance decreases.

8.5.2 RF Performance of Embroidered Antennas

The study in Ref. [17] showed the RF and mechanical properties of the textile antenna. The antenna was fabricated by Wang et al. [17] on a polymer substrate and was measured on both planar and cylindrical surfaces. The resonant frequencies of the obtained planar textile patch antenna were measured in a pattern consistent with that of the simulated copper patch antenna. The textile patch antenna was mounted on an 80 mm diameter metal cylinder to measure the reflection coefficient and radiation pattern of the antenna, and it can be seen that the results are in good agreement with the simulated results of the equivalent copper patch antenna, but the gain is reduced. They show that the main reasons for the reduced gain are the curvature of the fabric surface and the higher resistance. The latter is due to the stretching of the e-fiber threads.

8.5.3 Performance Degradation on Human Bodies

Flexible, esthetic and comfortable embroidered antennas provide an ideal solution for smart clothing. However, similar to metallic UHF RFID tags, the performance of antennas, mainly including the antenna resonant frequency and radiation pattern [18], can be greatly affected by the loss and high dielectric constant of human biological tissues. The resonant frequency of the embroidered antenna is distorted if it is directly attached to or near the human body because the human body absorbs a large number of electromagnetic waves and prevents the RFID tag from having backscattered signals.

Previous studies [7–9] demonstrated that the human body would lead to antenna resonant frequency detuning and radiation shielding. Ouyang and Chappell [9] placed three layers of tissue substrate, including 1 mm thick skin, 3 mm thick fat and 40 mm thick muscle, under the antenna model for simulation. The simulation results all demonstrated the frequency dependence of the antenna [19]. In addition, Wang et al. [7] found that the frequency of the multiband antenna increases with decreasing air gap detuning and that the fabric antenna needs to be placed at least 10 mm away from the tissue to reduce body effects and maintain sufficient bandwidth. The experiments were performed by adding layers of the garment between the antenna and the tissue substrate in the simulation. The simulations and experiments of Wang et al. [17] also presented homo-polarized and cross-polarized radiation maps of the antenna and showed that the signal transmission of the omnidirectional antenna can be blocked by the human body. They concluded that the antenna placed on the shoulder position reduces the shielding effect of the human body on radiation.

8.6 APPLICATIONS OF EMBROIDERED TEXTILE DEVICES

8.6.1 Embroidered RFID Antenna as a Displacement Sensor for Human Activity Tracking

An embroidered, planar coupling feed (PCF) RFID tag antenna is designed to operate in the UHF band of 915 MHz as a wireless and passive displacement sensor [6]. Figure 8.6a illustrates the transmission mechanism of the antenna. The antenna structure consists of a dipole arm and a coupling loop connected with an RFID chip. The displacement between the loop and the dipole arm causes the variation of input impedance, thereby modulating the backscattered signal from the RFID chip to the reader, as shown in Figure 8.6a. Thus, the textile PCF RFID

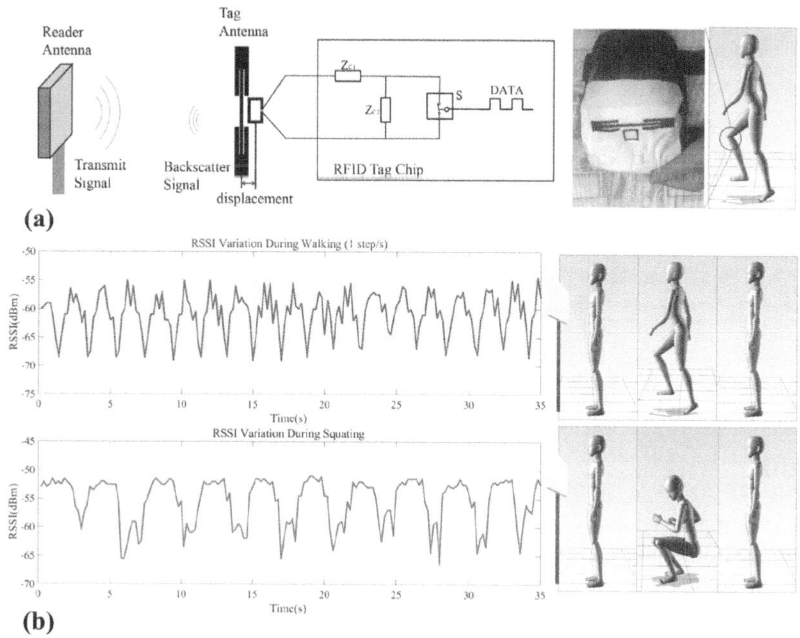

FIGURE 8.6 Embroidered RFID antennas for real-time body motion monitoring. (a) RFID data transmission mechanism and the photo of battery-less displacement sensor embedded inside a kneepad and (b) RSSI variation for slow walking and deep squatting.

tag antenna can be utilized as a passive, battery-free, wireless, flexible displacement sensor that does not require peripheral circuitry and batteries.

The embroidered antenna is highly sensitive to displacements caused by human motion, with a dynamic variation of 15 dB at a displacement of 1 cm, and can be easily distinguished from other environmental disturbances. The antenna is embedded in the knee area of the pants to monitor the knee motion, including walking on a treadmill and deep squatting, and the received signal strength indication (RSSI) from the antenna varies accordingly, as demonstrated in Figure 8.6a and b. In addition, Liu also showed that the displacement sensor can also be utilized as a wearable sensor for real-time monitoring of the respiratory rate when being embroidered on the human chest area.

8.6.2 Embroidered NFC Antenna as a Strain Sensor for Respiratory Rate Measurement

In recent years, there have been many studies on textile NFC coils to be used as energy harvesters. Researchers had proposed two embroidered coupling planar coils (CPCs) connected in series and stacked on top of each other to construct a strain sensor for respiratory rate measurement [20]. The relative position

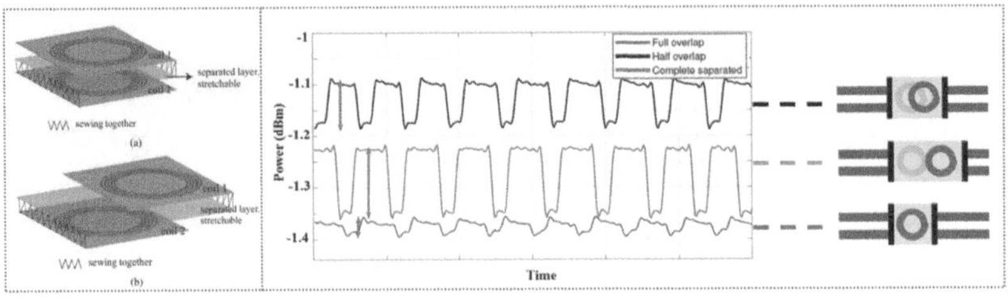

(a)

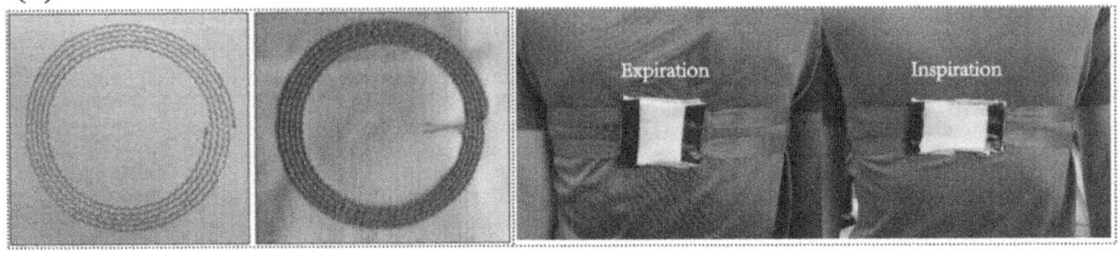

(b)

FIGURE 8.7 Embroidered NFC coils as the strain sensor for respiratory rate measurement. (a) Restorable structure and three displacement scenarios of CPC strain sensor and (b) CPC strain sensor on human chest for respiratory measurement.

and displacement between the pair of coupling coils determines the variation of mutual inductance with high sensitivity. The CPC coils can be used as a strain sensor for body movement and activity measurements. The displacement modulation of CPC coils can be detected by the changes of the amplitude of the back-coupling signals. Figure 8.7a presents the coupled signals of three displacement scenarios measured from a signal analyzer with peak-to-peak amplitudes of 0.125 dBm for the nonoverlapping condition, 0.09 dBm for the half-overlapping condition and 0.03 dBm for the fully overlapping condition. A passive displacement sensor can be constructed by using the CPC coils as the antenna of an NFC chip and extract sensing information from phase and amplitude changes. Experiments have shown that the wearable belt embedded with this sensor can reliably monitor human activities, such as normal breathing and deep breathing, as shown in Figure 8.7b.

8.7 ANTENNAS FOR SMART BOOKSHELF

8.7.1 Introduction

8.7.1.1 The Need for Smart Bookshelves and Its Role in a Modern Library

RFID technology has been proposed as a means to improve the ability to track inventory and locate objects. Developing a RFID system to locate a book buried in a pool of million books could make this fantasy materialized. Incidentally, the smart shelf (book-location system) is more than a time-saving scheme as it can identify and locate those books misplaced on wrong shelves, a tremendous property recovery. With the intelligent RFID system in the library, users can locate the desired materials with a user-friendly interface, it save time and human resources for book reservation. This is a new model of operation in the library, and it enhances the efficiency and improves the capability of all books in the library. Especially in Hong Kong, there is a need for smart bookshelves and its role in a modern library is essential.

8.7.1.2 Drawbacks of the Existing Systems and Antenna Solution

Actually, Hitachi High Technology Corporation (Tokyo, Japan) has developed a smart shelf consisting of an RFID reader with multiple antennas adapted to read Hitachi 13.56 MHz Coil-on-Chip chips [21]. As usual, loop antennas are used in HF-RFID smart shelf [22–23]. Leaking microstrip transmission line with extended ground plane has been proposed in the UHF RFID smart bookshelf [24]. However, the reading capability is not adequate and the RFID tag must be placed near the leaking transmission line to get the detection. It causes failure of detection if users misplace the book in the wrong direction. Conventional smart shelves with multiple RFID readers and antennas have also been installed and trialed in several retail stores in the United States and Europe, including the Metro Group Future Store in Rheinberg, Germany. Nevertheless, these shelf units are expensive, especially for handling large quantities of inventory in a retail facility. Especially in a library system, hundreds of bookshelves are needed. Reduced the number of readers and antennas are needed. Thus, antennas for smart bookshelves should have a wider beam width to have a wider coverage area in one shelf. However, the antenna gain should not be too high, say 4–6 dBi is good enough for the application. Else, the radiation may exist to the other shelves.

Therefore, to minimize the cost of the bookshelves, reader with multiple output ports and special antennas design is necessary. To conclude, specifications of the antennas for smart bookshelves should be wide beam width, say 80°–120°, low cost, ease of fabrications, small size and low profile.

In this chapter, two different types of antennas have been studied and compared. The first one is a small, low-cost and wide-beam-width CP patch antenna. The antenna is composed of four electromagnetically coupled patches sequentially fed with the phase angles of 0°, 90°, 180°, and 270°, respectively, and double short circuited to the ground. The proposed design is small in size and depicts a wide beam width of over 110°, besides preserving a wide axial ratio bandwidth of 11.6%.

To further provide a low-profile and low-cost solution to replace the conventional CP patch antenna, a printed CP patch antenna with electromagnetic band gap (EBG) structure is studied. EBG structures are studied first with an electric dipole antenna. It is found that the double-layer EBG structure can reduce the antenna height, enhance the bandwidth and increase the antenna gain effectively. Next, the double-layer EBG structure is applied on the CP patch antenna. Finally, an antenna with very low-profile and small-size CP patch printed on the low-cost material FR4 with a thickness of 4.8 mm (0.015 λ) is designed. This antenna achieves an 80% thickness reduction by comparing to the conventional patch antenna and yet maintains the antenna gain of about 5.5 dBi.

8.7.2 A Compact Sequential Rotated PIFA Array on the Back of the Shelf

8.7.2.1 Basic PIFA Antenna

The development of small integrated antennas plays a significant role in the progress of the rapidly expanding military and commercial communications applications. The Printed inverted F Antenna (PIFA) is an attractive antenna for wireless systems where the overall size of the antenna is quite limited. The basic PIFA is formed by a quarter-wave length monopole antenna folded coplanar to the ground plane and stretched to form a plate. Impedance matching of the antenna can be controlled by the feed line distance from the shorted edge of the PIFA [25]. Therefore, the conventional PIFA antenna is low profile and compact size. Also, wide bandwidth characteristic can be easily achieved. The radiation pattern of the PIFA antenna is slightly omnidirectional because of the monopole-like structure. To be an RFID reader antenna, radiation fields in the broadside direction are more appreciated. Therefore, a modified PIFA element is proposed.

The basic structure of the modified PIFA element is illustrated in Figure 8.8. This antenna is composed of an upper rectangular patch, which is shorted to the ground by a pair of shorting pins located at the corner of the patch. A small top-loaded rectangular patch is printed underneath of the upper patch, and coaxial probe feed is shorted to the ground plane.

The geometry of the proposed small wide-beam-width CP patch antenna is illustrated in Figure 8.9. The antenna comprises of four electromagnetically coupled shorted patches which are sequentially fed. The single element is a modified PIFA, composed of a small rectangular patch (L_2=35 mm or 0.1 λ_0, W_2=21 mm or 0.06 λ_0), and it is electromagnetically coupled to another rectangular patch (L_1=60 mm or 0.18 λ_0, W_1=45 mm or 0.14 λ_0). A pair of shorting pins (H_1 = 13 mm or 0.04 λ_0) is introduced to the upper patch, located on the edge of the patch shorted to the ground. The lower patch is excited by a vertical probe feed (H_2=12.2 mm or 0.037 λ_0). The upper and lower patches are printed on the same FR-4 material (ε_r = 4.4, thickness = 0.8 mm or 0.0024 λ_0). The four modified PIFAs are

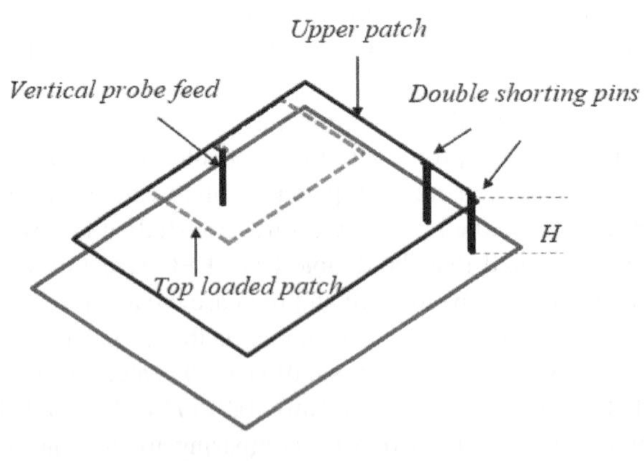

FIGURE 8.8 Geometry of the single modified PIFA.

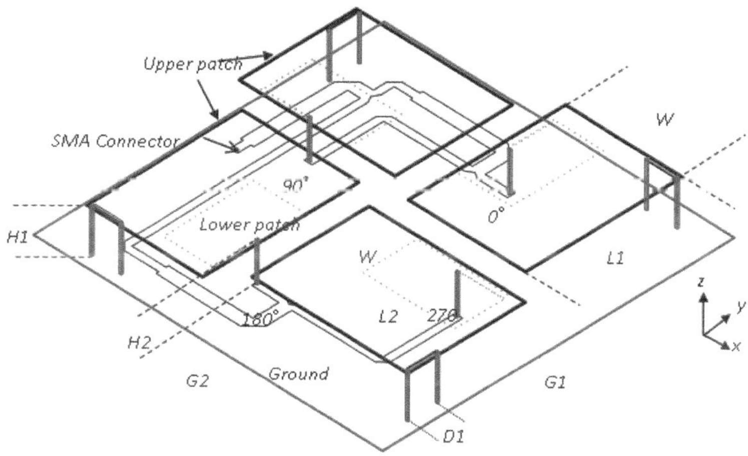

(a)

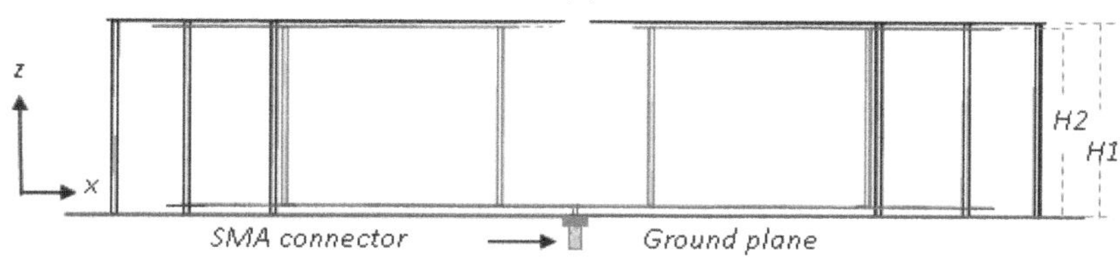

(b)

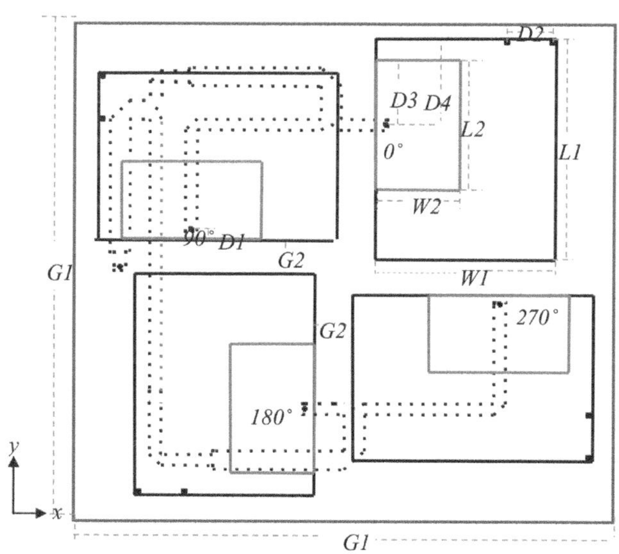

(c)

FIGURE 8.9 Antenna geometry of the proposed antenna: (a) 3D view, (b) side view and (c) top view.

TABLE 8.2
Parameters of the Proposed Wide-Beam-Width CP Patch Antenna

Parameters	L_1	L_2	W_1	W_2
Values (mm(λ))	60(0.18)	35(0.1)	45(0.13)	21(0.06)
Parameters	H_1	H_2	G_1	G_2
Values (mm(λ))	13(0.037)	12.2(0.035)	135(0.386)	9(0.026)
Parameters	D_1	D_2	D_3	D_4
Values (mm(λ))	2.5(0.007)	11.5(0.033)	17.5(0.05)	19(0.054)

sequentially rotated at phase angles of 0°, 90°, 180° and 270°, respectively, and are connected to the power divider. The gaps (G2) between the single elements are (9 mm or 0.027 λ_0). The square ground plane (G_1 = 135 mm or 0.4 λ_0) is printed on the same FR-4 with the power divider at the back. Details of the antenna parameters are presented in Table 8.2.

8.7.2.2 Principles of Operation

Sequential rotation technique is indeed one of the most useful ways to improve axial ratio bandwidth and polarization purity of circularly polarized antenna arrays [26–28]. It was first proposed by John Huang in 1986. For this kind of antenna array, the half power beam width is inherently narrow and the antenna array size is large. It is because they are composed of at least a 2×2 half-wave antenna array. For wide-angle beam scanning, the separation between each unit is almost half wavelength [26]. To reduce the antenna size, modified PIFAs are first introduced to the sequential rotating of the feed technique to achieve a circularly polarized antenna. Unlike the conventional sequentially rotating antenna, the proposed design is of wide beam width and small size.

The working mechanism is illustrated by scalar and vector current distributions as shown in Figure 8.10. The proposed array comprises four single antenna elements and consists of a top-loaded patch electromagnetically coupled to a short-circuited rectangular patch. The antenna size is therefore reduced by half for each basic unit. Also, the separation between each basic unit is only 9 mm (0.026 λ). This is also the key point to achieve a wide-beam-width characteristic. The shorting pins are designed in the outer part of each single unit.

From Figure 8.10, it can be observed that the induced fields are the strongest in the outer part of each basic unit and the ground plane. Therefore, a very wide radiation beam is obtained. As each patch has 90° phase difference, a good circularly polarized wave can be generated.

8.7.3 A Double-Layer EBG CP Patch Antenna on the Top of the Shelf

8.7.3.1 A Double-Layer EBG Structure

EBG is a kind of metamaterials, an artificial electromagnetic material that is normally realized by periodic dielectric materials and metallization patterns.

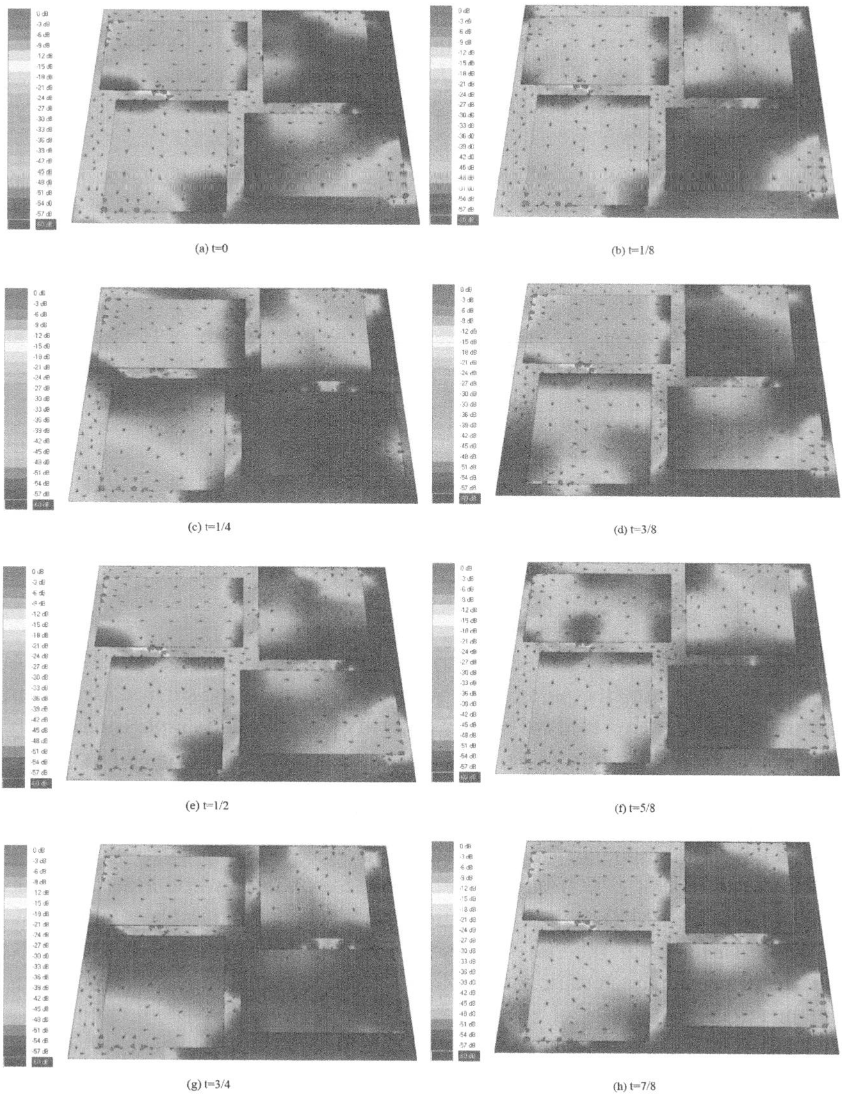

FIGURE 8.10 Instantaneous scalar and vector current distribution along the proposed antenna at f=915 MHz (t=1/f): (a) t=0, (b) t=1/8, (c) t=1/4, (d) t=3/8, (e) t=1/2, (f) t=5/8, (g) t=3/4 and (h) t=7/8.

The mushroom-like EBG structure was first proposed by Dan Sievenpiper in 1999 [29]. It is composed of a top layer of periodic patches, a dielectric substrate, a ground plane and a shorting vias from the patches to the ground. The proposed structure has a high-impedance surface on which no a/c current can be propagated for a stop-band for surface wave. The operational mechanism can be modeled as a parallel LC resonant circuit. The patches of the EBG form a cavity with the ground plane and provide the capacitance. The inductance, on the other hand, is generated by the current flowing through the shorting vias of the patches to the ground. For an EBG patch size *W*, thickness *h* and gap between patches *g*, values of capacitance and inductance can be calculated as follows [30]:

$$C = \frac{W\varepsilon_0(1+\varepsilon_r)}{\pi} co^{s-1}\left(\frac{2W+g}{g}\right) \tag{8.14}$$

$$L = \mu_0 h \tag{8.15}$$

where μ_0 is the permeability of free space, ε_0 is the permittivity of free space, and ε_r is the dielectric constant. And the impedance is:

$$Z = \frac{j\omega L}{1-\omega^2 LC} \tag{8.16}$$

The resonance frequency can be estimated as:

$$\omega_0 = \frac{1}{\sqrt{LC}} \tag{8.17}$$

By properly adjusting the parameters of the EBG patches, the surface wave from the antenna can be canceled out effectively. This surface wave suppression feature of EBG can be used to improve antenna performance in terms of back lope reduction, gain enhancement and bandwidth enhancement [31].

Another important feature of EBG is the reflection phase of the EBG surfaces. It is defined as the phase of the electric field reflected by the surface. A perfect dielectric conductor (PEC) has a reflection phase of 180°, while a perfect magnetic conductor (PMC) has a 0° reflection phase. The reflection phase of an EBG structure varies from 180° to –180°, which is more than that of a PEC or PMC surface. Such a characteristic makes EBG an ideal solution for a low-profile wire antenna [32].

Researchers have been applying EBG structure for a wide range of applications, such as performance enhancement in patch antennas and back lope suppression in low-profile spiral antennas. This technique can be applied to conceal a low-profile CP patch antenna for RFID application.

A printed electric dipole antenna backed with EBG structure is first studied. Mushroom-like EBG structure has many advantages, but it makes fabrication complicated. Therefore, in this chapter, a double-layer EBG structure is studied, instead of the mushroom-like structure. It is found that size reduction can be achieved with introduction of the double-layer EBG structure, as the surface wave suppression feature of EBG cancels out the effect of the mirror dipole. On the other hand, the extra capacitance and inductance generated by the EBG result in a wide impedance bandwidth (SWR<2) of 44.4%, and the antenna gain is stable across the entire operational frequency at 8.5 dBi.

Next, this double-layer EBG structure is applied to a printed truncated corners CP patch antenna. It is found that although the double-layer EBG structure cannot enhance the axial ratio of the antenna effectively, but it reduces the profile of the antenna by 80%. For RFID smart bookshelf application, a small-size and low-profile structure with good electric performance is more desirable. In the proposed design, the antenna is small size and the antenna gain is maintained at around 5.5 dBi in the operating band.

8.7.4 Antennas Arrangement on the Shelf

Figure 8.11 depicts a standard double-side five-layer bookshelf. Antenna installation is one of the most critical parts of an RFID smart shelf system. System implementation and cost are important considerations, as well. From a detection-accuracy point of view, the coupling zones of the antennas should be restricted to their own layers; else the extended coupling to others will give wrong information.

In this section, two antennas presented in Sections 8.7.2 and 8.7.3 are used to install into the smart shelf in different layers. These two antennas fulfill the smart shelf antenna requirements. A compact sequential rotated PIFAS array, which depicts a very wide beam width of 110°, is introduced on the back of the shelf.

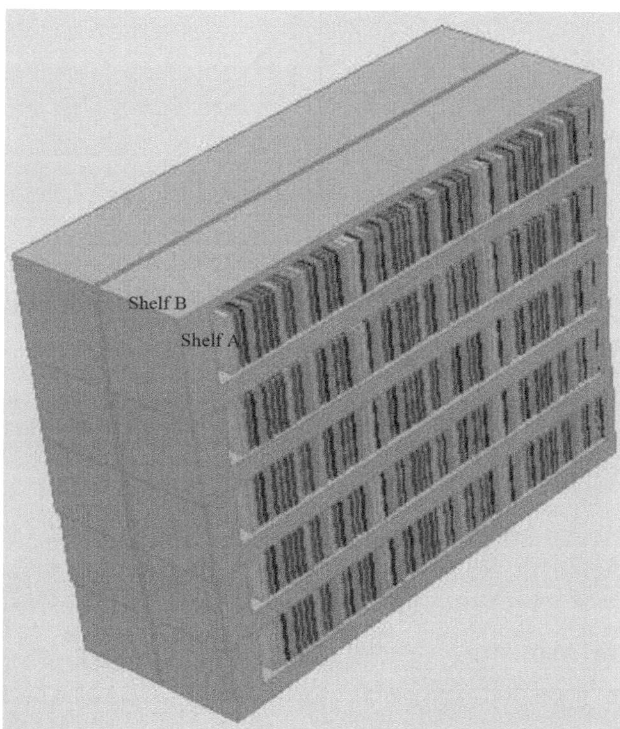

FIGURE 8.11 A double-side five-layer smart bookshelf in library.

Another one is a printed CP patch antenna backed with a double-layer EBG structure, which is designed to attach on the top of the shelf.

8.7.4.1 Antennas on the Back of the Shelf

The antenna presented in Section 8.7.2 is specially designed to be placing on the back of the smart bookshelves as shown in Figure 8.13. The dimension of the bookshelf is $2 \times 0.6 \times 0.4\,m^3$. The mathematically calculations are shown in Figure 8.12 and below.

The length of the smart shelf is 2 m, two antennas will be installed in one shelf, and therefore, DC ≥1 m and BC≥0.5 m. The antenna is located at the 1/3 and 2/3 of the shelf.

$$\text{Consider BC} = 0.5\,\text{m} \,\&\, \text{AB} = 0.4\,m \tan\theta = \text{BC/AB},$$

$$\theta \geq 51.3^\circ$$

$$\text{HPBW of the antenna} \geq 102.6^\circ$$

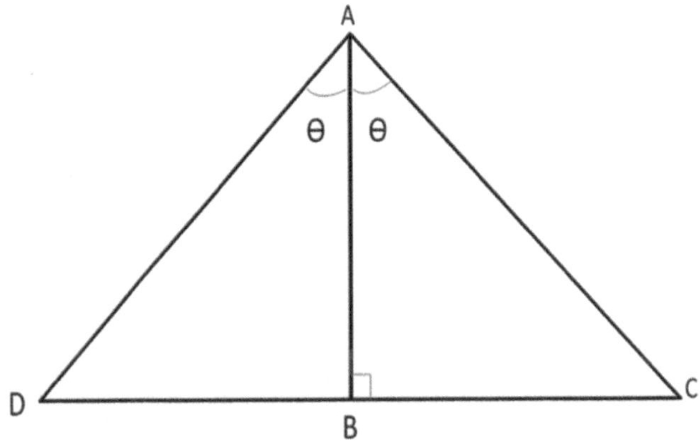

FIGURE 8.12 The diagram of mathematically calculations. A, antenna location; AB, width of the shelf=0.4 m; DC, distance of the radiation field; 2θ, half power beam width of the antenna: (a) front view and (b) side view.

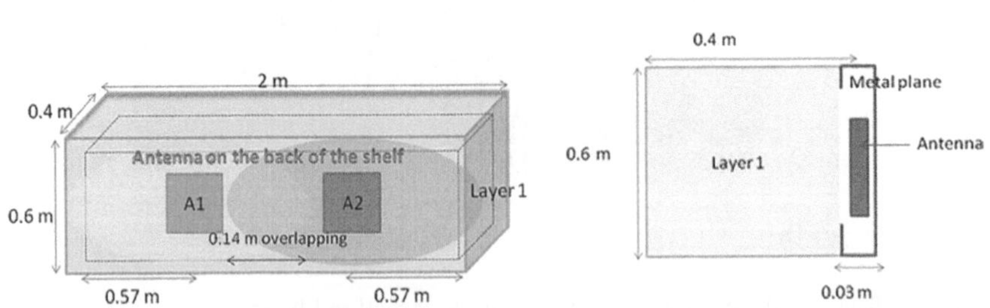

FIGURE 8.13 Antennas arrangement on the back of a layer in a smart bookshelf.

∴ A compact sequential rotated PIFAS array presented in Section 3.2 is the suitable candidate to install on the back of the shelf as its HPBW is about 110°.

$$\text{For the HPBW} = 110°,\ \text{BC} = \tan 55°\ \text{AB} = 0.57\,\text{m}$$

∴ The distance from the edge of the shelf to the antenna center is 0.57 m.

8.7.4.2 Antennas on the Top of the Shelf

The antenna presented in Section 8.7.3 is specially designed to be placing on the top of the smart bookshelves as shown in Figure 8.15. Also, it has a tilted radiation mechanically which is attached on the tiled metal grid by 30° on the top of the shelf. The purpose of the tilted radiation arrangement is to reduce the coupling between the shelf and the shelf on the adjacent side. The dimension of the bookshelf is $2\times0.6\times0.4\,\text{m}^3$. The mathematically calculations are shown in Figure 8.14 and below.

The length of the smart shelf is 2 m, two antennas will be installed in one shelf, and therefore, DC >1 m and BC ≥ 0.5 m. The antenna is located at the 1/3 and 2/3 of the shelf.

$$\text{Consider BC} = 0.5\,\text{m}\ \&\quad \text{AB} = 0.6\,\text{m}\ \tan\theta = \text{BC/AB}$$

$$\theta \geq 40°$$

$$\text{HPBW of the antenna} \geq 80°$$

∴ A printed CP patch antenna backed with a double-layer EBG structure presented in Section 3.3 is the suitable candidate to install on the back of the shelf as its HPBW is about 96°.

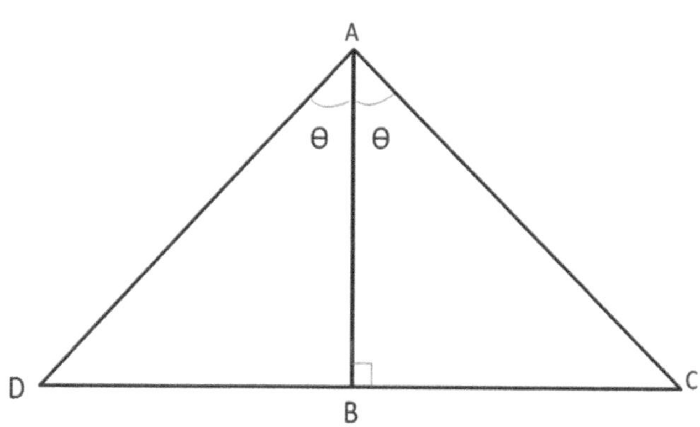

FIGURE 8.14 The diagram of mathematically calculations. A, antenna location; AB, height of the shelf=0.6 m; DC, distance of the radiation field; 2θ, half power beam width of the antenna.

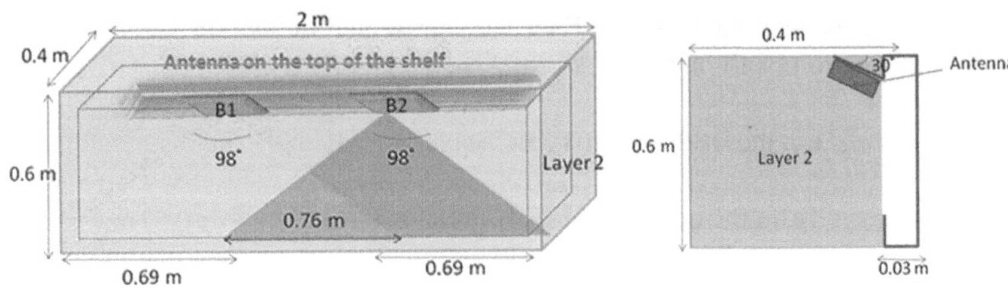

FIGURE 8.15 Antennas arrangement on the top of a layer in a smart bookshelf.

$$\text{For the HPBW} = 98°,$$

$$BC = \tan 49°\, AB$$

$$= 0.69\,\text{m}$$

∴ The distance from the edge of the shelf to the antenna center is 0.68 m.

8.7.4.3 System Arrangement of the Smart Shelves

The UHF RFID reader is developed from scratched by City University of Hong Kong. The low-cost RF switch L-Band High-Power SPDT (Single-Pole Double-Throw) Switch (UPG2009TB) is used to the reader. Therefore, one reader can provide more than eight output ports for antennas connection. This can reduce the system implementation, as well.

Time scanning methodology is introduced to smart bookshelf system. Figure 8.16a depicts the front view of a five-layer smart shelf. To enhance the detection accuracy, the antennas arrangement in each layer is different to the adjacent layers. Also, the scanning is always beginning from the bottom layer, say, from layer 5 to layer 1 and from left to right. It is because the radiation from the top of the shelf antenna is always going down, scanning from bottom layer can future prevent the radiation leakage to the under layers.

Figure 8.16b shows the side view of a double-sides smart shelf. As the smart shelf is double side, the scanning is always not at the same time for the shelves in the adjacent sides. For example, the scanning is always from Shelf A to Shelf B as shown in Figures 8.17 and 8.18.

8.8 SUMMARY AND DISCUSSION

This chapter studies RFID new technologies and their two main types of applications. In the first section of this chapter, we presented embroidered textile antennas, manufactured from conductive fibrous materials, for wireless identification and sensing applications. The properties and characteristics of various conductive textiles and antenna-embroidery methods based on conductive yarns and textile substrates have been compared and demonstrated. Design considerations for embroidered antennas, including parasitic effects, resistance of conductive yarns and deformation sensitivity of different geometries, are analyzed with examples.

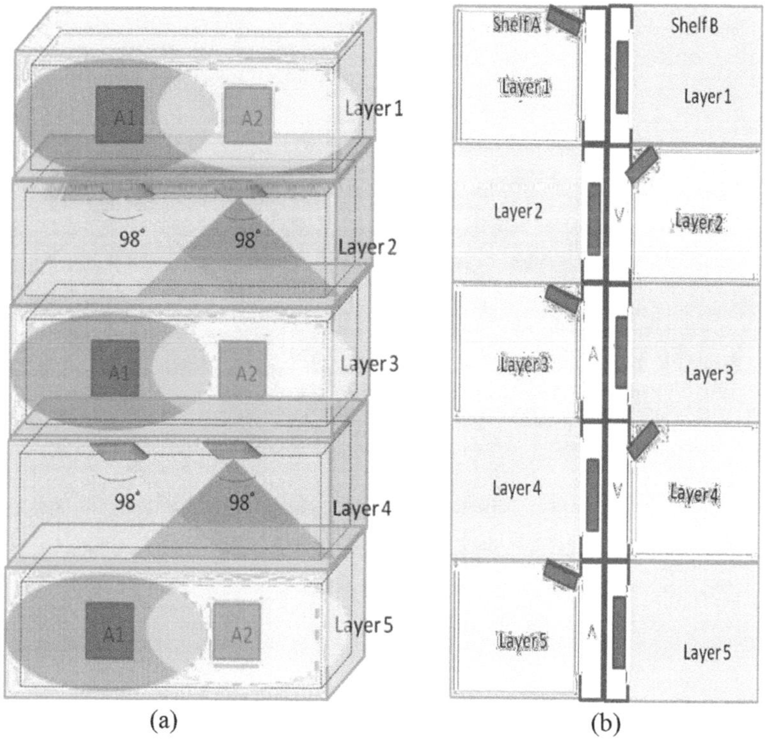

FIGURE 8.16 Antennas arrangement on the shelf of the smart bookshelf: (a) front view and (b) side view.

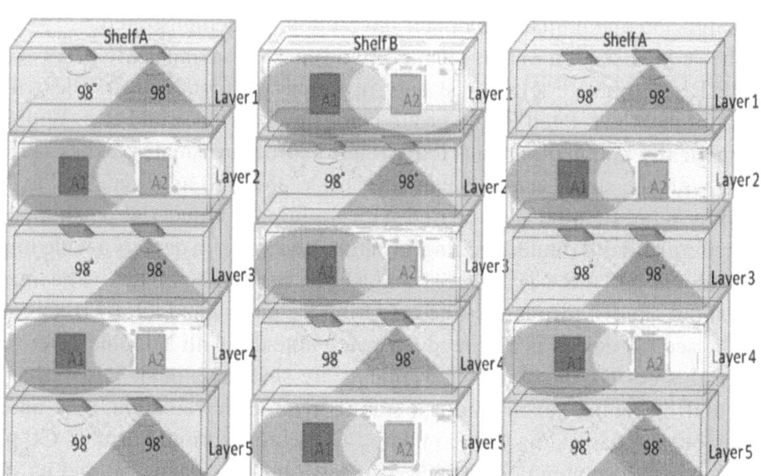

FIGURE 8.17 Antennas arrangement on the shelves of the smart bookshelf.

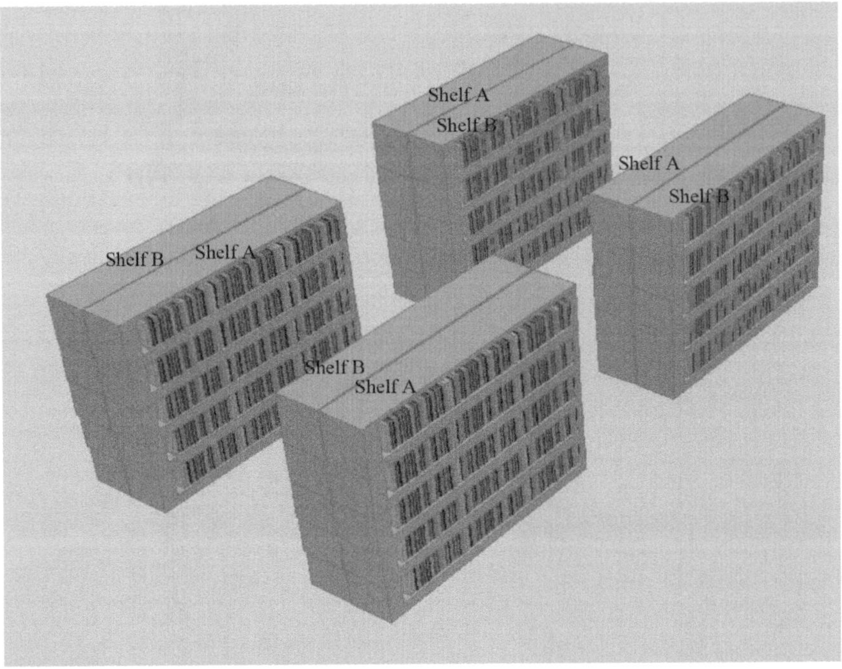

FIGURE 8.18 Smart shelves in library.

We have further measured the RF performance of embroidered textile antennas in various design cases. The experiments show that embroidered RFID/NFC antennas can also be used as sensing devices through the exploitation of antenna-sensing techniques (a.k.a. antenna impedance modulation). Several fabric-based wearable sensor applications including human-body movement tracking, vital sign monitoring and human–computer interaction are demonstrated.

In the second section of this chapter, two types of antennas are proposed for an UHF RFID smart bookshelf system in library system. The first design is a compact sequential rotated PIFAs printed on a low-cost material FR-4. With electromagnetically coupled and shorted-circuited structure small patches (PIFAs), together with the small gap between each basic unit, the antenna size has been reduced by more than 60% compared with traditional sequential patch antenna. With the sequentially rotated feeding structure, the antenna depicts a wide impedance band width (SWR<2) of 26.8% and an AR bandwidth of 12%. The antenna gain is about 4.5 dBi throughout the operating frequency and the HPBW is about 115°. The second design is a printed CP patch antenna with a double-layer structure. For this design, the main achievements are smallness of size as well as low profile and the ability to maintain a stable, symmetric radiation. The antenna gain has been enhanced as well compared to the conventional printed CP patch antenna. Also, this antenna is printed on a low-cost FR-4 material and its structure is simple from the perspective of fabrication. For smart bookshelf application, both antennas have been tested under the UHF RFID system developed by wireless communications research center, City University of Hong Kong. Both the antennas depicted similar performance under the UHF RFID system in bookshelf. The difference is the bandwidth of the antennas. The first design provides a wider axial ratio bandwidth than the second one. However, the size of the second antenna is smaller and gain is higher. In the implementation of the smart shelf system in the library, the two antennas specifically designed with different radiation beam widths are placed on the top and the back of the shelf, respectively. Therefore, the direction of the adjacent layers is always different and the coupling will be minimized. Together with the time scanning scheme, interference between the antennas in the adjacent layers will be greatly reduced. Besides being the reader antennas in UHF RFID smart bookshelf, these two antennas will find many applications in modern wireless communications, owned to their attractive electric characteristics.

In the future, we will concentrate on improving the geometric accuracy and electromagnetic properties of embroidery electronics without compromising the good electrical conductivity of the fabric for more applications in in healthcare, smart sports, consumer electronics, defense and other areas. In conclusion, this chapter can be utilized as guidelines for the design and fabrication of embroidered e-textile devices and smart city.

REFERENCES

1. Komolafe, A., Torah, R., Wei, Y., Nunes-Matos, H., Li, M., Hardy, D., ... Beeby, S. (2019). Integrating flexible filament circuits for e-textile applications. *Advanced Materials Technologies*, 4(7), 1900176.
2. Komolafe, A., Zaghari, B., Torah, R., Weddell, A. S., Khanbareh, H., Tsikriteas, Z. M., ... Beeby, S. (2021). E-textile technology review: From materials to application. *IEEE Access*, 9, 97152–97179. doi:10.1109/access.2021.3094303.
3. DicCbshfsE, J., Abd Rahim, M. K., Samsuri, N. A., Salim, H. A. M., & Ali, M. F. (2011). Embroidered fully textile wearable antenna for medical monitoring applications. *Progress in Electromagnetics Research*, 117, 321–337.
4. Atwa, Y., Maheshwari, N., & Goldthorpe, I. A. (2015). Silver nanowire coated threads for electrically conductive textiles. *Journal of Materials Chemistry C*, 3(16), 3908–3912.
5. Liu, Y., Yu, M., Xu, L., Li, Y., & Ye, T. T. (2020). Characterizations and optimization techniques of embroidered RFID antenna for wearable applications. *IEEE Journal of Radio Frequency Identification*, 4(1), 38–45. doi: 10.1109/jrfid.2019.2961189.
6. Liu, Y., Yu, M., Xia, B., Wang, S., Wang, M., Chen, M., ... Ye, T. T. (2021). E-textile battery-less displacement and strain sensor for human activities tracking. *IEEE Internet of Things Journal*, 8(22), 16486–16497. doi:10.1109/jiot.2021.3074746.
7. Wang, Z., Volakis, J. L., & Kiourti, A. (2015). Embroidered antennas for communication systems. In: Dias, T. (Ed.) *Electronic Textiles*, pp. 201–237. Woodhead Publishing: Sawston.
8. Kim, G., Lee, J., Lee, K. H., Chung, Y. C., Yeo, J., Moon, B. H., ... Kim, H. C. (2008). Design of a UHF RFID fiber tag antenna with electric-thread using a sewing machine. *Paper Presented at the 2008 Asia-Pacific Microwave Conference,* Hong Kong, China.
9. Ouyang, Y., & Chappell, W. J. (2008). High frequency properties of electro-textiles for wearable antenna applications. *IEEE Transactions on Antennas and Propagation*, 56(2), 381–389.

10. Xu, L., Liu, Y., Li, Y., Lau, P. Y., Si, H., & Ye, T. T. (2018). Design and fabrication of embroidered RFID antennas for wearable applications. *Paper Presented at the 2018 IEEE International Conference on Service Operations and Logistics, and Informatics (SOLI)*, Singapore.
11. Acti, T., Chauraya, A., Zhang, S., Whittow, W. G., Seager, R., Vardaxoglou, J., & Dias, T. (2014). Embroidered wire dipole antennas using novel copper yarns. *IEEE Antennas and Wireless Propagation Letters*, 14, 638–641.
12. Heo, E., Choi, K.-Y., Kim, J., Park, J.-H., & Lee, H. (2018). A wearable textile antenna for wireless power transfer by magnetic resonance. *Textile Research Journal*, 88(8), 913–921.
13. Roh, J.-S., Chi, Y.-S., & Kang, T. J. (2010). Wearable textile antennas. *International Journal of Fashion Design, Technology and Education*, 3(3), 135–153.
14. Roh, J.-S., Chi, Y.-S., Lee, J.-H., Nam, S., & Kang, T. J. (2010). Characterization of embroidered inductors. *Smart Materials and Structures*, 19(11), 115020.
15. Yu, M., Shang, X., Wang, M., Liu, Y., & Ye, T. T. (2020). Exploiting embroidered UHF RFID antennas as deformation sensors. *IEEE Journal of Radio Frequency Identification*, 4(4), 406–413. doi: 10.1109/jrfid.2020.3030790.
16. Rahman, M. M., & Sarkar, A. K. (2017). A method for calculating the resonant frequency of meander line dipole antenna by using antenna's geometrical parameters. *Paper Presented at the 2017 6th International Conference on Informatics, Electronics and Vision & 2017 7th International Symposium in Computational Medical and Health Technology (ICIEV-ISCMHT)*, Himeji, Hyogo, Japan.
17. Wang, Z., Zhang, L., Bayram, Y., & Volakis, J. L. (2012). Embroidered conductive fibers on polymer composite for conformal antennas. *IEEE Transactions on Antennas and Propagation*, 60(9), 4141–4147.
18. Lee, G.-Y., Psychoudakis, D., Chen, C.-C., & Volakis, J. L. (2010). Channel decomposition method for designing body-worn antenna diversity systems. *IEEE Transactions on Antennas and Propagation*, 59(1), 254–262.
19. Christ, A., Klingenbock, A., Samaras, T., Goiceanu, C., & Kuster, N. (2006). The dependence of electromagnetic far-field absorption on body tissue composition in the frequency range from 300 MHz to 6 GHz. *IEEE Transactions on Microwave Theory and Techniques*, 54(5), 2188–2195.
20. Liu, Y., Wang, M., Yu, M., Xia, B., & Ye, T. T. (2020). Embroidered inductive strain sensor for wearable applications. *Paper Presented at the 2020 IEEE International Conference on Pervasive Computing and Communications Workshops (PerCom Workshops)*, Austin, TX, USA.
21. Foster, P. R. & Burberry, R. A. (1999). Antenna problems in RFID systems. Institute of Electronics Engineers Colloquium on RFID Technology, pp. 31–35.
22. Qing, X. & Chen, Z. N. (2009). Characteristics of a metal-backed loop antenna and its application to a high-frequency RFID smart shelf. *IEEE Transactions on Antennas and Propagation*, 51(2), 26–38.
23. Qing, X. & Chen, Z. N. (2007). Proximity effects of metallic environments on high requency RFID reader antenna: Study and applications. *IEEE Transactions on Antennas and Propagation*, 55(11), 3105–3111.
24. Medeiros, C. R., Costa, J. R., & Fernandes, C. A. (2008). RFID smart shelf with confined detection volume at UHF. *IEEE Transactions on Antennas and Propagation*, 7, 773–776.
25. Virga, K. L. & Rahmat-Samii, Y. (1997). Low-profile enhanced-bandwidth PIFA antennas for wireless communications packaging. *IEEE Transactions on Microwave Theory and Techniques*, 45(10), 1879–1888.
26. Huang, J. (1986). A technique for and array to generate circular polarization with linearly polarized elements. *IEEE Transactions on Antennas and Propagation*, AP-34(9), 1113–1124.

27. Hall, P. S., Dahele, J. S., & James, J. R. (1989). Design principles of sequential feeding to wide bandwidth, circularly polarized microstrip antennas. *IEEE Proceedings*, 136(5), pp. 390–398.
28. Hall, P. S. (1989). Application of sequential feeding to wide bandwidth, circularly polarized microstrip patch arrays. *IEEE Proceedings*, 136(5), 390–398.
29. Sievenpiper, D., Zhang, L., Jimenez Broas, R. F., Alexopolous, N. G., Yablonovitch, E. (1999). High-impedance electromagnetic surfaces with a forbidden frequency band. *IEEE Transactions on Microwave Theory and Techniques*, 47(11), 2059–2074.
30. Yang, F., & Rahmat-Samii, Y. (2003). Microstrip antennas integrated with Electromagnetic Band-Gap (EBG) structures: A low mutual coupling design for array applications. *IEEE Transactions on Antennas and Propagation*, 51(10), 2936–2946.
31. Colburn, J. S., & Rahmat-Samii, Y. (1999). Patch antennas on externally perforated high dielectric constant substrates. *IEEE Transactions on Antennas and Propagation*, 47(12), 1785–1794.
32. Yang, F., & Rahmat-Samii, Y. (2003). Reflection phase characterizations of the EBG ground plane for low profile wire antenna applications. *IEEE Transactions on Antennas and Propagation*, 51(10), 2691–2703.

Index